The Era of Artificial Intelligence, Machine Learning, and Data Science in the Pharmaceutical Industry

The Era of Artificial Intelligence, Machine Learning, and Data Science in the Pharmaceutical Industry

Edited by

Stephanie Kay Ashenden

Data Sciences and Quantitative Biology, Discovery Sciences,
R&D, AstraZeneca, Cambridge, United Kingdom

ELSEVIER

ACADEMIC PRESS
An imprint of Elsevier

Academic Press is an imprint of Elsevier
125 London Wall, London EC2Y 5AS, United Kingdom
525 B Street, Suite 1650, San Diego, CA 92101, United States
50 Hampshire Street, 5th Floor, Cambridge, MA 02139, United States
The Boulevard, Langford Lane, Kidlington, Oxford OX5 1GB, United Kingdom

Notices
Knowledge and best practice in this field are constantly changing. As new research and experience broaden
our understanding, changes in research methods, professional practices, or medical treatment may become
necessary.

Practitioners and researchers must always rely on their own experience and knowledge in evaluating and
using any information, methods, compounds, or experiments described herein. In using such information or
methods they should be mindful of their own safety and the safety of others, including parties for whom they
have a professional responsibility.

To the fullest extent of the law, neither the Publisher nor the authors, contributors, or editors, assume any
liability for any injury and/or damage to persons or property as a matter of products liability, negligence or
otherwise, or from any use or operation of any methods, products, instructions, or ideas contained in the
material herein.

Library of Congress Cataloging-in-Publication Data
A catalog record for this book is available from the Library of Congress

British Library Cataloguing-in-Publication Data
A catalogue record for this book is available from the British Library

ISBN: 978-0-12-820045-2

For information on all Academic Press publications
visit our website at https://www.elsevier.com/books-and-journals

Publisher: Andre Gerhard Wolff
Acquisitions Editor: Erin Hill-Parks
Editorial Project Manager: Samantha Allard
Production Project Manager: Punithavathy Govindaradjane
Cover Designer: Greg Harris

Typeset by SPi Global, India

Contents

v

CHAPTER 4   Target identification and validation................................**61**
Stephanie Kay Ashenden, Natalie Kurbatova, and Aleksandra Bartosik

CHAPTER 5   Hit discovery...**81**
Hannes Whittingham and Stephanie Kay Ashenden

CHAPTER 10 Clinical trials, real-world evidence, and digital medicine 191

Jim Weatherall, Faisal M. Khan, Mishal Patel, Richard Dearden, Khader Shameer, Glynn Dennis, Gabriela Feldberg, Thomas White, and Sajan Khosla

CHAPTER 11 Beyond the patient: Advanced techniques to help predict the fate and effects of pharmaceuticals in the environment..........................217

Stewart F. Owen and Jason R. Snape

Contributors

Paul-Michael Agapow
Oncology R&D Real World Evidence, AstraZeneca, Cambridge, United Kingdom

Stephanie Kay Ashenden
Data Sciences and Quantitative Biology, Discovery Sciences, R&D, AstraZeneca, Cambridge, United Kingdom

Aleksandra Bartosik
Clinical Data and Insights, Biopharmaceuticals R&D, AstraZeneca, Warsaw, Poland

Kaustav Bera
Center for Computational Imaging and Personalized Diagnostics, Department of Biomedical Engineering, Case Western Reserve University, Cleveland, OH; Maimonides Medical Center, Brooklyn, NY, United States

Krishna C. Bulusu
Bioinformatics and Data Science, Translational Medicine, Oncology R&D, AstraZeneca, Cambridge, United Kingdom

Fabiola Cecchi
Translational Medicine, Research and Early Development, Oncology R&D, AstraZeneca, Gaithersburg, MD, United States

Adam M. Corrigan
Data Sciences and Quantitative Biology, Discovery Sciences, R&D, AstraZeneca, Cambridge, United Kingdom

Richard Dearden
Digital Health, Oncology R&D, AstraZeneca UK Ltd, Cambridge, United Kingdom

Glynn Dennis
AI and Analytics, Data Science and Artificial Intelligence, Biopharma R&D, AstraZeneca, Gaithersburg, MD, United States

Sumit Deswal
Genome Engineering, Discovery Sciences, BioPharmaceuticals R&D, AstraZeneca, Gothenburg, Sweden

Laura A.L. Dillon
Translational Medicine, Research and Early Development, Oncology R&D, AstraZeneca, Gaithersburg, MD, United States

Gabriela Feldberg
Digital Health, Oncology R&D, AstraZeneca UK Ltd, Durham, NC, United States

Jason Hipp
Translational Medicine, Research and Early Development, Oncology R&D, AstraZeneca, Gaithersburg, MD, United States

Faisal M. Khan
AI and Analytics, Data Science and Artificial Intelligence, Biopharma R&D, AstraZeneca, Gaithersburg, MD, United States

Sajan Khosla
Oncology Data Science, AstraZeneca, Gaithersburg, MD, United States

Natalie Kurbatova
Data Infrastructure and Tools, Data Science and Artificial Intelligence, R&D, AstraZeneca, Cambridge, United Kingdom

Anant Madabhushi
Center for Computational Imaging and Personalized Diagnostics, Department of Biomedical Engineering, Case Western Reserve University; Louis Stokes Cleveland Veterans Administration Medical Center, Cleveland, OH, United States

Armin Meier
Translational Medicine, Research and Early Development, Oncology R&D, AstraZeneca, Munich, Germany

Stewart F. Owen
Global Sustainability, AstraZeneca, Cambridge, United Kingdom

Mishal Patel
Imaging and Data Analytics, Clinical Pharmacology & Safety Sciences, R&D, AstraZeneca, Cambridge, United Kingdom

Günter Schmidt
Translational Medicine, Research and Early Development, Oncology R&D, AstraZeneca, Munich, Germany

Elizaveta Semenova
Data Sciences and Quantitative Biology, Discovery Sciences, R&D, AstraZeneca, Cambridge, United Kingdom

Khader Shameer
AI and Analytics, Data Science and Artificial Intelligence, Biopharma R&D, AstraZeneca, Gaithersburg, MD, United States

Jason R. Snape
Global Sustainability, AstraZeneca, Cambridge, United Kingdom

Daniel Sutton
Imaging and Data Analytics, Clinical Pharmacology & Safety Sciences, R&D, AstraZeneca, Cambridge, United Kingdom

Jim Weatherall
Data Science and Artificial Intelligence, R&D, AstraZeneca UK Ltd, Macclesfield, United Kingdom

Thomas White
AI and Analytics, Data Science & Artificial Intelligence, Biopharma R&D, AstraZeneca, Cambridge, United Kingdom

Hannes Whittingham
Data Sciences and Quantitative Biology, Discovery Sciences, R&D, AstraZeneca, Cambridge, United Kingdom

Johannes Zimmermann
Translational Medicine, Research and Early Development, Oncology R&D, AstraZeneca, Munich, Germany

Preface

This book aims to pull together the full pharmaceutical research and development process, from concept to patient. It explores how the data science field, particularly areas such as artificial intelligence and machine learning, aids this process, helping us to make better decisions faster.

In addition, this book is a platform for those wanting to understand the who, why, where, when, and what of artificial intelligence, machine learning, and data science throughout the drug discovery process.

Acknowledgments and conflicts of interest

Stephanie Ashenden would like to thank Ian Barrett, Stan Lazic, Yinhai Wang, Delyan Ivanov, Afshan Ahmed, Ola Engkvist, Claus Bendtsen, and Aurelie Bornot for their tremendous support, feedback, and guidance. She would also like to thank Rory and her parents Paul and Tania. Finally, Stephanie would like to thank all the authors and Erin Hill-Parks, Samantha Allard, and Punithavathy Govindaradjane at Elsevier.

Paul Agapow would like to thank Samit Kundu, Mansoor Saqi, and Paul Metcalfe for the comments and suggestions.

Aleksandra Bartosik would like to thank Paweł Gunerka, Patryk Hes, Zbyszek Pietras and Tomasz Zawadzki for reviews as well as insightful discussions. Aleksandra is also grateful to Andreas Bender for computational toxicology inspiration.

Stewart F. Owen and Jason R. Snape are employed by AstraZeneca, a global innovation-based biopharmaceutical company that discovers, develops, and markets pharmaceuticals. Necessarily, AstraZeneca conducts environmental risk assessments of its products and submits these as part of the registration process as well as making the summary data available on their website. S.F. Owen and J.R. Snape thank all their past and present partners from academic, NGO and industry scientists, and regulators who collaborated on a wide range of projects. In particular, the authors thank Dr. Tom Miller and Dr. Leon Barron at Kings College London for introducing them to this field of ML and AI. They thank Jack Owen for use of his photograph. J.R. Snape and S.F. Owen thank Simomics for their partnership on the Innovate UK, "National Centre for the Replacement, Refinement & Reduction of Animals in Research" (NC3Rs) funded project number 102519 "Virtual Fish EcoToxicology Laboratory." S.F. Owen and J.R. Snape also acknowledge the support of colleagues on the European Union Innovative Medicines Initiative 2 Joint Undertaking "Prioritisation and Risk Evaluation of Medicines in the EnviRonment (PREMIER)," project number 875508. This Joint Undertaking receives the support from the European Union's Horizon 2020 research and innovation program and EFPIA. Their time preparing this chapter represents a contribution toward this project.

Introduction to drug discovery

Stephanie Kay Ashenden

Data Sciences and Quantitative Biology, Discovery Sciences, R&D, AstraZeneca, Cambridge, United Kingdom

The drug discovery process

The drug discovery and development process (here meaning the pharmaceutical research and development pipeline from concept to beyond the patient) is a long, expensive, and complex process[1]; only a small proportion of molecules that are identified as a candidate drug are approved as new drugs each year.[2] It has been estimated that it costs approximately US$2.6 billion to develop a new treatment.[3] In addition to creating a finished product costing over $1 billion, it can take up to 15 years.[1] The cost of research and development continues to increase.[4] Considering the long timelines, the increasing cost and complexity of the drug discovery process, efforts to aid in reducing these concerns are of interest. Having said this, new drug modalities have been explored beyond small molecules, which can result in new methods of treatment, patient stratification has potential to speed discovery as well as more focused development. In combination with experimental methodologies, artificial intelligence is hoped to improve the drug discovery process.[5] Artificial intelligence can help beyond areas of research and develop, such as in finance but these areas are beyond the scope of this book.

A general overview of a typical drug discovery process (Fig. 1) is split up into several different stages,[1] namely: target identification and target validation (focusing on de-risking experiments), lead discovery, lead optimization, preclinical testing, and clinical testing.[6]

While there is no one definite way to arrive at a novel drug, depending on whether a specific target of interest in present, both target-led (target is known) or phenotypic (target is not known) screening methods[7] can be used. Today, drug discovery is being led by techniques such as high-throughput screening and empirical screening which involves screening libraries containing chemicals against targets in a physical way. However virtual screening, which screens libraries computationally for compound chemicals that target known structures and having them tested experimentally, has become a leading method to predict new compound structures.[8] Experimental testing confirms that interactions between the known target and the desired compound is therefore optimized to achieve desirable properties[1] including biological activity, while reducing or eliminating negative properties (such as toxicity).[6]

Target identification

Target identification is involved in the process of identifying targets that are hypothesized to be linked to a disease and will also be suitably druggable. Ultimately the purpose of assessing drugabillity is to

The Era of Artificial Intelligence, Machine Learning, and Data Science in the Pharmaceutical Industry. https://doi.org/10.1016/B978-0-12-820045-2.00002-7

FIG. 1

A general overview of a drug discovery process from target identification to the clinical testing.

identify hits (compounds that have activity against a desired target) and could be developed into lead compounds (a compound that has potential to be developed into a drug due to desirable properties and observed activity).

To achieve target identification, often a combination of both experimental and computational methods are used.[9] Target identification uses methods such as genetic associations such as connecting genes with a disease and data mining methods for searching through literature and databases.[1] A target may be identified for further investigation, from both clinical and academic research as well as from the commercial setting as well.[1] However, it has been shown that generally, industry follows the research conducted by academia.[10] Computational approaches and tools such as gene prioritization tools have been produced to try to guide research in the right direction, especially because so few drugs progress to an approved status.[11]

Of course, it is possible to have a therapeutic compound before knowing the target and in those situations target deconvolutional strategies are used.[12] These strategies involve the investigation of phenotypic changes within a biological system having been exposed to a compound.[12]

Even if you can identify a druggable target, review articles have been published discussing what actually makes a good drug target.[13] Approval of a drug requires proof that it is safe and that it is suitably efficacious.[14] Therefore, understanding the target is beneficial for guiding development and reducing risk.[14]

Target validation

In this phase of the drug discovery process, we are concerned with whether the identified target, is worth further investigation and development of therapeutics.[15] Target validation is concluded once an effective drug is in the clinic with a verified suspected target.

Methods for validating targets include genetic methods such as RNA interference that are applied to potential targets.[1] ThermoFisher have published an overview of RNA interference (RNAi) and explains that it works by knocking down the expression of a target of interest allowing for evaluation of its response to a compound.[16] Other methods to aid in developing understanding include biophysical technique of which there are a wide variety of them. Moustaqil and coauthors discussed the techniques available for target validation specifically related to transcription-targeted therapy and explain that new therapeutic candidates have different requirements to understanding activity than classic inhibitors so new methods are required.[17]

One of the most well-known recent processes in target validation is that of CRISPR. CRISPR stands for Clustered Regularly Interspaced Short Palindromic Repeats. This technology is used for the precising editing of DNA in the genome.[18] This allows for the application gene editing such as knockout methods. The purpose of a gene knockout in the study of disease is to allow for observations of the effect of a particular gene. By removing the gene, phenotypic changes can be observed. However, genetic methods are not the only way to modulate a target. It is important to try to modulate the target in the way that we want to drug it. Therefore compounds and a variety of experimental tools, including omics technologies are available that can help to assess the modulation of dose response. Genetic evidence can be used to predict the mechanisms of drugs and can aid in increasing success in development.[19]

Ultimately, target validation is concerned with risk management by understanding the target a compound is interacting with and thus understanding potential downstream effects. However, risk management continues throughout the drug discovery process which contributes to decision-making on whether to continue with a project or halt it.

Hit identification and lead discovery

Hit identification and lead discovery is based on the identification of a hit (a compound that has desired activity[1]) and the identification of a lead compound that is taken forward for further optimization and analysis.

In relation to identifying compound hits, typically, assays are performed to identify whether a compound has an effect and can be cell based or biochemical based assays.[1] Michelini and coauthors[20] wrote a paper describing cell-based assays and explaining that these assays can be used for a variety of observations including, among others toxicity, morphological changes, and activation of signaling pathways.[20] The authors explain that many of these assays are based on reporter gene technology, although other technologies have been developed to try to deal with their limitations, particularly related to the issue that these technologies can be slow.[20] Biochemical assays consider the test compound to a target protein and the follow-up assays can be less complicated than cell-based screens.[21] Both, types of assays have been summarized and compared previously with the authors concluding that both are useful in lead discovery[22] (identifying the hit compound that will be taken forward to be improved for further exploration).

In terms of computational approaches used in hit identification, DNA-encoded chemical libraries also known as DELs are of interest. Madsen and coauthors describe DELs as collections of small molecules that are covalently attached to DNA tags.[23] The authors also say that these DNA tags are amplifiable and contain information about the structure[23] and using combinatorial approaches, the libraries are constructed iteratively of which allows the compounds synthetic history to be tracked by DNA sequencing.[23] There are various other strategies for identifying hits as listed by Sygnature[24] such as virtual screening, high-throughput screening, fragment screening and knowledge based design. However, there are other drug modalities beyond compounds.

There are a variety of technologies and modalities applied in drug discovery for the identification of a target and a lead compound. These modalities will help to move into areas that were previously considered undruggable. Dang and co-authors[25] discuss drugging cancer targets that were considered undruggable. The authors explain that rather than a target being called undruggable, meaning it cannot be targeted, it should be referred to as "difficult to drug" and that there have been targets previously thought to be undruggable that have since reached market.[25]

Blanco and Gardinier highlighted the different chemical modalities available.[26] Such discussed examples by the authors include targeted protein degradation and RNA-based drugs.[26] Such methods include the use of PROTACs. PROTAC stands for PROteolysis TArgeting Chimera. Pei and coauthors explain that the method works by inducing protein degradation in a targeted manner by using ubiquitin systems.[27] It is thought that the use of this technology may help researchers to move into what may have been considered the undruggable proteome, however, challenges are present including the small number of E3 ubiquitin ligases that have been currently utilized, off-target effects, and size.[27] Our understanding of what is considered the undruggable genome is changing thanks to new drug modalities.[25]

Another modality is that of antisense oligonucleotides which, as discussed by Rinaldi and Wood, can alter RNA and change or modify protein expression.[28] Therefore they can focus on the source of pathogenesis directly.[28] The authors explain that antisense oligonucleotides have been modified to overcome previous issues such as lack of activity or off-target toxicity and are now of great interest with approvals for use in Duchenne muscular dystrophy and spinal muscular atrophy.[28]

Antibody drug discovery focuses on the search for therapeutic antibodies of which new ones are approved each year.[29] Antibodies selectively bind and can also be engineered which have made them key in the drug discovery pipeline.[30] Antibody drug conjugates are composed of three parts, the antibody, the small molecule (payload), and then a linker.[31] Sadekar et al. discussed antibody drug conjugates explaining that the small molecule can be potent and released once internalized at the desired location.[31] Because of new improved technology such as large-scale data availability bioinformatics is required to aid in the discovery process.[29]

Genome editing is not only limited to CRISPR but also involves methods such as TALENs (transcription activator-like effector nucleases) which allow for the alteration of a targeted DNA sequence.[32] In a paper describing TALENS, the authors explain that they, contain a DNA-binding domain that is bound to a nonspecific DNA-cleaving nuclease.[32] Zinc-finger nucleases (ZFNs) are another technology that is used and allow editing in live cells.[33]

Cell therapy is a technology where cells are injected or transplanted and used for a therapeutic purpose.[34] Stem cell therapy has shown potential thanks to their differentiating capabilities, as well as their prolific nature.[35] A key challenge in using stem cells relates to ethical considerations.[36] However, the concern of uncontrolled proliferation is another area requiring further assessment.[37]

Virtual screening

Virtual screening analyses databases computationally with the aim to identify potential hit candidates. It is primarily split into two types, namely, ligand-based virtual screening and structure-based virtual screening.[38]

The method chosen depends on how much information is available, for example, if little information is known about the disease target, ligand-based virtual screening would be applied.[38] Although if you know the structure of your target then structure-based methods such as protein ligand docking is the way forward.[38] Knowing the structure of an active compound can allow for the search of similar compounds based on their structure to be conducted.

Compound libraries

Libraries of compounds are sometimes used in the process of identifying candidates. These libraries can vary in their size and content by being either focused in nature (such as assessing all the compounds within the library against a particular target or target class) or broad in their exploration by not being focused on a particular target family.[39] Focused libraries are advantageous for needing fewer compounds for screening and often there are higher hit rates compared with screening diverse sets of compounds.[40] However, they do require knowledge of the compound target or target family.

Diversity is one of the key focuses of designing a screening library and there have been efforts to move into more knowledge-based screening library design.[39] Cheminformatic techniques can be applied to compounds to assess how chemically similar they are in terms of their structure. Approaches to analyze diversity in a screening library have been undertaken, such as by Akella and DeCaprio[41] who wrote about cheminformatic approaches to do just this. The authors note that various visualization techniques can be applied to aid in the identification of unexplored areas of chemical space.[41] Furthermore, metrics to quantify properties of the compounds in a collection such as how similar to a natural product a compound is can be applied. The authors explain that applying these different approaches can help to increase diversity.[41]

When creating compound libraries, there are several key considerations that need to be made. Dandapani and co-authors discussed creating compound libraries and noted that practical considerations like cost and assay objective need to be taken into consideration.[42] The authors discussed several steps that should be performed to create a screening library and is summarized here.[42] Compounds that have inappropriate qualities can removed. PAINS (Pan Assay INterference compoundS)[43] are an example of this of which have an increased likelihood of being identified as a hit in an assay known as frequent hitters.[44] Furthermore, it is also possible to predict the properties of compounds and remove those that are unlikely to be suitable as drugs.[42] Following this the authors suggest to perform complexity analysis and consider how complex the compounds should be and follow this by assessing how diverse the collection is and then considering how unique the structures are.[39,42]

High-throughput screening

High-throughput screening (HTS) refers to the process of screening and assaying compounds against targets on a large scale.[45] Following the development of an assay, high throughput screening is one of the most commonly applicable methods that allow for the identification of a lead compound. HTS utilizes robotics and automatized technologies that allow for rapid tests, such as pharmacological tests, to be conducted.[6]

Before the birth of this technique, the approach was done manually and only allowed for between 20 and 50 compounds to be analyzed each week.[46] However, with new and improved techniques for identifying potential targets began emerging, it became clear that this methodology could not be sustainable and more efficient technologies and methodologies that were cost effective would need to be introduced. The fact that large numbers of compounds can be screened in small assays against biological targets at the same time has made HTS a powerful tool in the process of discovering new medicines.[6]

The future of HTS was discussed recently by Mayr and Fuerst.[47] This paper notes that over time, particularly in the past 20 years, HTS has adapted to the needs and requirements of lead discovery such as improved quality (fewer false-positives and false-negatives for example), whereas previously the focus had been on quantity by implementing miniaturization techniques. Yet, in recent years there has been some disagreement between achieving "quantity" and consideration of the relevance of the data. Mayr and coauthors argue that with the implementation of plates with larger numbers of wells being used, such as 384-well plates to conduct the assays, focus will move away from miniaturization and toward increasing the relevance of each hit-finding strategy.[47] An essential ingredient to the successful improvements for this technique will be the curation of adequate chemical libraries that contain good diversity and drug-like properties.[6,47,48]

Structure-based drug discovery

Structure-based drug discovery utilizes the structure of the therapeutic target to aid in drug discovery. The 3D structure of the biological target[49] and a variety of computational methods (docking, molecular dynamic simulations, etc.) are utilized in structure-based drug discovery. Batool and coauthors[50] described the structure-based drug discovery process and is summarized here. The authors summarize the process by explaining that target proteins 3D structure is determined after the target has been extracted and purified from genome sequencing.[50] Following this the authors explain that the active compounds are compiled in a database and the target and its binding sites are identified.[50] Docking and structure-based virtual screening methods will be applied to screen the active compounds against the target binding site and then the top hits are synthesized and is evaluated to determine the lead compound.[50]

Fragment-based drug discovery

Fragment-based drug discovery is used to identify ligands that are below approximately 150 Da that could be biologically important.[51] The identification of these small fragments could be good starting points for lead compounds and are advantageous due to their low complexity.[52] Congreve and coauthors explained that, it is more efficient to explore chemical space (all possible compounds) using small fragments instead of larger molecules.[53] Kirsch and coauthors described fragment-based drug discovery and explain that it starts with screening small compounds against a particular target to observe activity.[52] The authors explain that these structures may bind to multiple targets due to their small size and so this must be taken into consideration when building a library.[52] A study published in 2015 highlighted that fragment-based drug discovery has led to marketed drugs including Vemurafenib.[54]

Phenotypic drug discovery

Phenotypic drug discovery methods do not need knowledge of the drug target.[55] Zheng and coauthors[56] explained that a specific disease characteristic is exploited, and a cell-based assay is developed. Active compounds are then identified that improve the observed disease characteristic.[56]

This can be very beneficial in progressing a compound where no target is known but therefore, does have challenges in hit validation and target deconvolution.[55] Such challenges also are that it can be more expensive and more complex compared with target-based screening methods For example, to support SAR analysis, phenotypic screening may need additional assays.[56] It has been shown that phenotypic screening methods have had success in the identification of first in class drugs.[57,58]

Natural products

A natural product in drug discovery is a compound that naturally occurs and is produced by living organisms. These are sources of inspiration and represent over a third of all FDA-approved new molecular entities.[59] A new molecular entity can be described as a compound that is a product of the drug discovery process and are not derived of any previously known substance. Natural products have offered the greatest number of origins for leads that are taken into further development.[60]

Despite this, interest from pharmaceutical companies has declined due to technical difficulties with screening them in high-throughput assays.[61]

Lead optimization

In this phase of the drug discovery process, attempts are made to improve a compounds property in terms of its absorption, distribution, metabolism, excretion, and toxicity (referred to as ADMET) but

Table 1 Description of the key focus areas for lead optimization.

Property	Description
Absorption (pharmacology)	The ability of a compound to move from the target site and into the bloodstream[62]
Distribution	Concerned with the movement of a drug[62]
Metabolism	The metabolic breakdown of a drug within the body[63]
Excretion	The removal of the drug[64]
Activity (biological)	The ability of a compound to have an effect on a biological process[65]
Toxicity	Complications that arise from administration of a drug[66]

also its potency and activity (Table 1). Changes to a compound to understand and improve its ADMET properties, potency and desired biological activity profile, are analyzed in what is known as the DMTA cycle (design, make, test, and analyze cycle). The DMTA cycle helps us to drive structure-activity relationship studies (SAR) can be conducted to develop understanding between a compounds structure and its activity. Again, biophysical methodologies are used to aid in our understanding.

Modeling in lead optimization

Preclinical models can be either human or nonhuman and the reasons for using either depend on the question and requirements needed.[67] Several types of models exist including in vivo and in vitro models. Ibarrola-Villava and coauthors explained that these models are used to predict the efficacy of a treatment.[68] There is also ex vivo models and in silico models. In vivo models represent experiments performed within a living model, such as a mouse. Ex vivo means outside the living model, for example, removing a particular tissue or organ from the living model. In vitro differs from ex vivo in that we explore the cells (such as cell culture systems). In silico means computational models. Fröhlich and Salar-Behzadi[69] compared in vivo, ex vivo, in vitro, and in silico studies to assess inhaled nanoparticles toxicology. The authors explain there have been concerns that cell culture systems cannot be representative of a multicellular organism, however, in vitro methods have many advantages including ethics and economic compared in in vivo studies.[69] The authors noted that ex vivo is often not used in nanoparticular studies because isolated lungs are difficult to prepare and maintain, hence why in vitro and in vivo methods are preferred.[69]

Various assays can be performed to assess the biological activity (ability to cause an effect in the biological process[65]) and other ADMET properties but also these properties can also be modeled and predicted. These prediction methods can involve the use of the chemical structure as a whole or even include methods such as matched molecular pairs. Matched molecular pairs are two compounds that are identical with exception of a single molecular entity difference in the same location and can reveal how properties can be altered by adding or removing parts of a chemical's structure.

Toxicity prediction is of great interest to prevent serious adverse events occurring that could result in patient harm and then market withdrawal. Preclinical safety evaluation aims to define toxicity endpoints before the drug entering human trials. There is difficult due to the vast number of mechanisms that toxicity can occur, and some responses may be dose-related and do not affect everyone. Additionally, the type of toxicity must be considered, for example, whether the toxicity is target mediated or compound mediated. This considers whether the toxicity is originating from the target itself, or whether the toxicity is off-target. Therefore, understanding whether the aim is to understand the type of toxicity or toxicity overall is an important concept. Computational methods may make unprecedented changes to the drug discovery process, providing solutions that are cheaper, reliable, and fast.[3]

Pharmacokinetic and pharmacodynamic (PK/PD) modeling is another area of consideration. Pharmacodynamics is how the body processes a drug in terms of ADME[70] and is specifically concerned with the time course of a drug concentration after a drug has been administered into the body,[71] whereas pharmacodynamics is the observed effects from the drug.[71] PK/PD modeling can be used to understand both the efficacy of a drug as well as its safety and is advantageous as it can be used throughout the drug discovery process, even in clinical phases of the development process.[72]

Precision medicine

Precision medicine aims to guide selection of treatments for patients based on the genetic, molecular or clinical-pathological understanding of the patient's disease. Recent technological advances have improved the ability to sequence genetic material and identify genetic changes such as mutations within an individual; however, different drugs may work better for different genetic changes. Oncology is a well-known field where precision medicine is applied,[73] however, there are also advances in fields such as nephrology[74] and cardiovascular medicine.[75] Patient stratification is an important task where patients are stratified by risk or response by means of a companion diagnostic test. It is important to mention that there are also prognostic tests where we can assess how does the disease progress naturally and theragnostic tests which help to predict response to therapy.[76] There may be differences in the underlying pathology and disease segmentation can identify different endotypes of disease. Various pieces of information are used such as omic data (genomic, proteomic, etc.) and information such as the patient's demographic to capture a more personal picture of the patient's condition. To give an example, proteomics has been discussed for personalized medicine in cases such as kidney disease,[77] neurodegenerative disease and neurotrauma[78] as well as oncology.[79]

Clinical testing and beyond

Clinical testing is involved with the development stage. In this phase of the drug discovery process, the testing is moved into human subjects. It is split into different phases.

The earlier phases are focused on assessing the safety of the compound, followed by efficacy studies and determination that this new medicine is an improvement on any other similar licensed medicine.[80] Phase 4 is typically conducted after the drugs licensing and is used to further understand the drugs side effects and general effects in various populations. Safety and efficacy are a key theme throughout the entire drug discovery pipeline.[80]

Beyond clinical testing, a therapeutics impact on the environment once disseminated into the general population is carefully monitored. The aquatic environment is a place where pharmaceuticals are observed in low concentrations and understanding any hazards and risks this may pose is a complex question.

There are many key considerations during the clinical trials stages including site selection and recruitment modeling for participants for the trials. Predicting participant recruitment is important for the design of clinical trials as well as planning and monitoring ongoing trials to ensure that important data are captured. This also helps to adapt trials that deviate from the original plan. The recruitment start date is another key consideration because not all centers recruit at the same time on trials. This modeling needs to consider the start-up time for each center as well as recruitment rate once the centers have started. Recruitment modeling also captures the clinical supply chain which needs to have a good idea of the patient numbers involved. Clinical event adjudication is another area of consideration in clinical trials. Clinical event classification is important for high quality health care, involved in diagnostic perspectives and risk assessments and with use of digital transformation approach can aid in event adjudication prediction that is faster and more accurate than manual processes. However, careful consideration needs to be applied due to risk of biases. The adoption of new medicines into the marketplace is increasingly becoming reliant on evidence-based criteria. Electronic health records contain details of an individual's health including demographic information, any treatments they are on and testing results and such data can be used to predict clinical scenarios as shown by Rajkomar and coauthors.[81] The authors used deep learning to achieve high accuracy when predicting things such as in-hospital mortality and whether there would be a prolonged length of stay.[81] Advances in technologies have allowed for real-world data, data generated in clinical practice, to be collected more readily and can be used to derive real-world evidence.

Another consideration is that of digital medicine which includes the use of digital apps and digital devices such as wearable technology.[82] Neurodegenerative diseases are an area where there is interest in using wearable technology for a variety of reasons such as widespread use and sensitivity on sensors.[82]

Because of the technological advances, patient monitoring has evolved from measurements being done as time snapshots—when a patient visits a hospital—to continuous monitoring. Wearables,[83] mobile devices,[84] and embedded biosensors allow to capture multiple parameters and understand how everyday behavior and health interact. Such observation methods benefit patients as they have the freedom to be mobile and monitored in their usual environment. Glucose monitors, blood pressure monitors, pulse oximeters, ECG monitors, and accelerometers can measure and allow to collect continuous data. Gait speed,[85] cardiovascular monitoring, arrhythmia[86] are a few parameters being measured both in and out of the hospital.

References

1. Hughes JP, Rees SS, Kalindjian SB, Philpott KL. Principles of early drug discovery. *Br J Pharmacol* 2011;**162**:1239–49. https://doi.org/10.1111/j.1476-5381.2010.01127.x.
2. Fishman MC, Porter JA. Pharmaceuticals: a new grammar for drug discovery. *Nature* 2005;**437**:491–3.
3. Fleming N. How artificial intelligence is changing drug discovery. *Nature* 2018. https://doi.org/10.1038/d41586-018-05267-x.
4. Shaw DL. Is open science the future of drug development? *Yale J Biol Med* 2017;**90**:147–51.
5. Chan HCS, Shan H, Dahoun T, Vogel H, Yuan S. Advancing drug discovery via artificial intelligence. *Trends Pharmacol Sci* 2019;**40**:592–604.
6. Ashenden S. *On the dissemination of novel chemistry and the process of optimising compounds in drug discovery projects.* University of Cambridge; 2019. https://doi.org/10.17863/CAM.38232.
7. Palmer M. Phenotypic screening. In: Werngard C, Hamley P, editors. *Small molecule medicinal chemistry: strategies and technologies.* Wiley; 2015. p. 281–304. https://doi.org/10.1002/9781118771723.ch10.
8. Shoichet BK. Virtual screening of chemical libraries. *Nature* 2004;**432**:862–5.
9. Schenone M, Dančík V, Wagner BK, Clemons PA. Target identification and mechanism of action in chemical biology and drug discovery. *Nat Chem Biol* 2013;**9**:232–40.
10. Ashenden SK, Kogej T, Engkvist O, Bender A. Innovation in small-molecule-druggable chemical space: where are the initial modulators of new targets published? *J Chem Inf Model* 2017;**57**:2741–53.
11. Paliwal S, de Giorgio A, Neil D, Michel JB, Lacoste AM. Preclinical validation of therapeutic targets predicted by tensor factorization on heterogeneous graphs. *Sci Rep* 2020;**10**:18250. https://doi.org/10.1038/s41598-020-74922-z.

12. Terstappen GC, Schlüpen C, Raggiaschi R, Gaviraghi G. Target deconvolution strategies in drug discovery. *Nat Rev Drug Discov* 2007;**6**:891–903.

13. Gashaw I, Ellinghaus P, Sommer A, Asadullah K. What makes a good drug target? *Drug Discov Today* 2011;**16**:1037–43.

14. Mechanism matters. *Nat Med* 2010;**16**:347. https://doi.org/10.1038/nm0410-347.

15. Blake RA. Target validation in drug discovery. *Methods Mol Biol* 2007;**356**:367–77.

16. Interference, RNA & Silencing, RNA. *RNA interference overview*. Available at: https://www.thermofisher.com/uk/en/home/life-science/rnai/rna-interference-overview.html. [Accessed 30 August 2019].

17. Moustaqil M, Gambin Y, Sierecki E. Biophysical techniques for target validation and drug discovery in transcription-targeted therapy. *Int J Mol Sci* 2020;**21**:2301.

18. *Everything you need to know about CRISPR-Cas9*. Available at: https://www.synthego.com/learn/crispr?utm_term=crispr&utm_campaign=Q12020+Crispr+General&utm_source=adwords&utm_medium=ppc&hsa_tgt=kwd-298744624982&hsa_grp=102616848434&hsa_src=g&hsa_net=adwords&hsa_mt=e&hsa_ver=3&hsa_ad=428280457673&hsa_acc=6964378581&hsa_kw=crispr&hsa_cam=9730955713&gclid=EAIaIQobChMIh5qZzpLF6gIV34BQBh1w-whaEAAYASAAEgJ51PD_BwE. [Accessed 11 July 2020].

19. Nelson MR, et al. The support of human genetic evidence for approved drug indications. *Nat Genet* 2015;**47**:856–60.

20. Michelini E, Cevenini L, Mezzanotte L, Coppa A, Roda A. Cell-based assays: fuelling drug discovery. *Anal Bioanal Chem* 2010;**398**:227–38.

21. Westby M, Nakayama GR, Butler SL, Blair WS. Cell-based and biochemical screening approaches for the discovery of novel HIV-1 inhibitors. *Antivir Res* 2005;**67**:121–40.

22. Moore K, Rees S. Cell-based versus isolated target screening: how lucky do you feel? *J Biomol Screen* 2001;**6**:69–74.

23. Madsen D, Azevedo C, Micco I, Petersen LK, Hansen NJV. An overview of DNA-encoded libraries: a versatile tool for drug discovery. *Prog Med Chem* 2020;**59**:181–249.

24. Sygnature Discovery. *Hit identification in drug discovery*. https://www.sygnaturediscovery.com/drug-discovery/integrated-drug-discovery/hit-identification/. [Accessed 11 July 2020].

25. Dang CV, Reddy EP, Shokat KM, Soucek L. Drugging the 'undruggable' cancer targets. *Nat Rev Cancer* 2017;**17**:502–8.

26. Blanco MJ, Gardinier KM. New chemical modalities and strategic thinking in early drug discovery. *ACS Med Chem Lett* 2020;**11**:228–31.

27. Pei H, Peng Y, Zhao Q, Chen Y. Small molecule PROTACs: an emerging technology for targeted therapy in drug discovery. *RSC Adv* 2019;**9**:16967–76.

28. Rinaldi C, Wood MJA. Antisense oligonucleotides: the next frontier for treatment of neurological disorders. *Nat Rev Neurol* 2018;**14**:9–21.

29. Shirai H, et al. Antibody informatics for drug discovery. *Biochim Biophys Acta, Proteins Proteomics* 2014;**1844**:2002–15.

30. Marsden CJ, et al. The use of antibodies in small-molecule drug discovery. *J Biomol Screen* 2014;**19**:829–38.

31. Sadekar S, Figueroa I, Tabrizi M. Antibody drug conjugates: application of quantitative pharmacology in modality design and target selection. *AAPS J* 2015;**17**:828–36.

32. Joung JK, Sander JD. TALENs: a widely applicable technology for targeted genome editing. *Nat Rev Mol Cell Biol* 2013;**14**:49–55.

33. Gabriel R, et al. An unbiased genome-wide analysis of zinc-finger nuclease specificity. *Nat Biotechnol* 2011;**29**:816–23.

34. Arrighi N. Stem cells at the core of cell therapy. In: *Stem cells*. Elsevier; 2018. p. 73–100. https://doi.org/10.1016/b978-1-78,548-254-0.50003-3.

35. Fleifel D, et al. Recent advances in stem cells therapy: a focus on cancer, Parkinson's and Alzheimer's. *J Genet Eng Biotechnol* 2018;**16**:427–32.

36. Zakrzewski W, Dobrzyński M, Szymonowicz M, Rybak Z. Stem cells: past, present, and future. *Stem Cell Res Ther* 2019;**10**:68.

37. George L, Eumorphia R. Proliferation versus regeneration: the good, the bad and the ugly. *Front Physiol* 2014;**5**:10. https://doi.org/10.3389/fphys.2014.00010 ISSN=1664-042X. https://www.frontiersin.org/article/10.3389/fphys.2014.00010DOI=10.3389/fphys.2014.00010ISSN=1664-042X.

38. Gillet V. *Ligand-based and structure-based virtual screening*. University of Sheffield. Available at: https://www.ebi.ac.uk/sites/ebi.ac.uk/files/content.ebi.ac.uk/materials/2013/131209DrugDiscovery/1_-_val_gillet_-_ligand-based_and_structure-based_virtual_screening.pdf. [Accessed 18 October 2019].

39. Ashenden SK. Screening library design. *Methods Enzymol* 2018;**610**:73–96. https://doi.org/10.1016/bs.mie.2018.09.016.

40. John Harris C, Hill RD, Sheppard DW, Slater MJ, Stouten PFW. The design and application of target-focused compound libraries. *Comb Chem High Throughput Screen* 2011;**14**:521–31.

41. Akella LB, DeCaprio D. Cheminformatics approaches to analyze diversity in compound screening libraries. *Curr Opin Chem Biol* 2010;**14**:325–30.

42. Dandapani S, Rosse G, Southall N, Salvino JM, Thomas CJ. Selecting, acquiring, and using small molecule libraries for high-throughput screening. *Curr Protoc Chem Biol* 2012;**4**:177–91.

43. Baell JB, Holloway GA. New substructure filters for removal of pan assay interference compounds (PAINS) from screening libraries and for their exclusion in bioassays. *J Med Chem* 2010;**53**:2719–40.

44. Ertl P. An algorithm to identify functional groups in organic molecules. *J Cheminform* 2017;**9**:36.

45. Szymański P, Markowicz M, Mikiciuk-Olasik E. Adaptation of high-throughput screening in drug discovery-toxicological screening tests. *Int J Mol Sci* 2012;**13**:427–52.

46. Pereira DA, Williams JA. Origin and evolution of high throughput screening. *Br J Pharmacol* 2007;**152**:53–61.

47. Mayr LM, Fuerst P. The future of high-throughput screening. *J Biomol Screen* 2008;**13**:443–8.

48. Schreiber SL, Nicolaou KC, Davies K. Diversity-oriented organic synthesis and proteomics: new frontiers for chemistry & biology. *Chem Biol* 2002;**9**:1–2.

49. Lionta E, Spyrou G, Vassilatis D, Cournia Z. Structure-based virtual screening for drug discovery: principles, applications and recent advances. *Curr Top Med Chem* 2014;**14**:1923–38.

50. Batool M, Ahmad B, Choi S. A structure-based drug discovery paradigm. *Int J Mol Sci* 2019;**20**:2783.

51. Murray CW, Rees DC. The rise of fragment-based drug discovery. *Nat Chem* 2009;**1**:187–92.

52. Kirsch P, Hartman AM, Hirsch AKH, Empting M. Concepts and core principles of fragment-based drug design. *Molecules* 2019;**24**:4309.

53. Congreve M, Chessari G, Tisi D, Woodhead AJ. Recent developments in fragment-based drug discovery. *J Med Chem* 2008;**51**:3661–80.

54. Renaud JP, Neumann T, Van Hijfte L. Fragment-based drug discovery. In: Czechtizky W, Hamley P, editors. *Small molecule medicinal chemistry: strategies and technologies*; 2015. https://onlinelibrary.wiley.com/doi/abs/10.1002/9781118771723.ch8.

55. Moffat JG, Vincent F, Lee JA, Eder J, Prunotto M. Opportunities and challenges in phenotypic drug discovery: an industry perspective. *Nat Rev Drug Discov* 2017;**16**:531–43.

56. Zheng W, Thorne N, McKew JC. Phenotypic screens as a renewed approach for drug discovery. *Drug Discov Today* 2013;**18**:1067–73.

57. Owens J. *Phenotypic versus target-based screening for drug discovery*. Technology Networks; 2018. Available at: https://www.technologynetworks.com/drug-discovery/articles/phenotypic-versus-target-based-screening-for-drug-discovery-300037. [Accessed 22 July 2020].

58. Swinney DC. Phenotypic *vs.* target-based drug discovery for first-in-class medicines. *Clin Pharmacol Ther* 2013;**93**:299–301.

59. Patridge E, Gareiss P, Kinch MS, Hoyer D. An analysis of FDA-approved drugs: natural products and their derivatives. *Drug Discov Today* 2016;**21**:204–7.

60. Harvey AL. Natural products in drug discovery. *Drug Discov Today* 2008;**13**:894–901.

61. Harvey AL, Edrada-Ebel R, Quinn RJ. The re-emergence of natural products for drug discovery in the genomics era. *Nat Rev Drug Discov* 2015;**14**:111–29.
62. Jennifer L. Drug Absorption; 2020. Available at: https://www.msdmanuals.com/en-gb/home/drugs/administration-and-kinetics-of-drugs/drug-absorption. [Accessed July 2020].
63. Wilkinson G, Drug R. Metabolism and variability among patients in drug response. *N Engl J Med* 2005;**352**:2211–21.
64. Lu J-D, Xue J. Poisoning: kinetics to therapeutics. In: Ronco C, Bellomo R, Kellum JA, Ricci Z, editors. *Critical care nephrology*. 3rd ed. Elsevier; 2019. p. 600–629.e7. https://doi.org/10.1016/B978-0-323-44942-7.00101-1.
65. Jackson CM, Esnouf MP, Winzor DJ, Duewer DL. Defining and measuring biological activity: applying the principles of metrology. *Accred Qual Assur* 2007;**12**:283–94.
66. Stolerman I.P. (eds). Drug toxicity. In: Riley AL, Kohut S, Stolerman IP, editors. *Encyclopedia of psychopharmacology*. Berlin, Heidelberg: Springer; 2010.
67. Geyer MA, Markou A. The role of preclinical models in the development of psychotropic drugs. In: Davis KL, Charney D, Coyle JT, Nemeroff C, editors. *Neuropsychopharmacology: The Fifth Generation of Progress*. American College of Neuropsychopharmacology; 2002. p. 445–55.
68. Ibarrola-Villava M, Cervantes A, Bardelli A. Preclinical models for precision oncology. *Biochim Biophys Acta Rev Cancer* 2018;**1870**:239–46.
69. Fröhlich E, Salar-Behzadi S. Toxicological assessment of inhaled nanoparticles: role of in vivo, ex vivo, *in vitro*, and in silico studies. *Int J Mol Sci* 2014;**15**:4795–822.
70. Shafer SL. Principles of pharmacokinetics. In: *Wylie and Churchill-Davidsons: a practice of anesthesia. 7th ed.* BC Decker; 2003. p. 29–43. https://doi.org/10.1201/9781351072472-1.
71. Meibohm B, Dorendorf H. Basic concepts of pharmacokinetic/pharmacodynamic (PK/PD) modelling. *Int J Clin Pharmacol Ther* 1997;**35**:401–13.
72. Rajman I. PK/PD modelling and simulations: utility in drug development. *Drug Discov Today* 2008;**13**:341–6.
73. Moscow JA, Fojo T, Schilsky RL. The evidence framework for precision cancer medicine. *Nat Rev Clin Oncol* 2018;**15**:183–92.
74. Wyatt CM, Schlondorff D. Precision medicine comes of age in nephrology: identification of novel biomarkers and therapeutic targets for chronic kidney disease. *Kidney Int* 2016;**89**:734–7.
75. Leopold JA, Loscalzo J. Emerging role of precision medicine in cardiovascular disease. *Circ Res* 2018;**122**:1302–15.
76. McGeough CM, Bjourson A. Diagnostic, prognostic and theranostic biomarkers for rheumatoid arthritis. *J Clin Cell Immunol* 2013. https://doi.org/10.4172/2155-9899.s6-002.
77. Siwy J, Mischak H, Zürbig P. Proteomics and personalized medicine: a focus on kidney disease. *Expert Rev Proteomics* 2019;**16**:773–82.
78. Alaaeddine R, Fayad M, Nehme E, Bahmad HF, Kobeissy F. The emerging role of proteomics in precision medicine: applications in neurodegenerative diseases and neurotrauma. *Adv Exp Med Biol* 2017;59–70. https://doi.org/10.1007/978-3-319-60,733-7_4.
79. Doll S, Gnad F, Mann M. The case for proteomics and phospho-proteomics in personalized cancer medicine. *Proteomics Clin Appl* 2019;**13**:e1800113.
80. Fernández-Avilés F, et al. Phases I-III clinical trials using adult stem cells. *Stem Cells Int* 2010;**579**:142.
81. Rajkomar A, Oren E, Chen K, et al. Scalable and accurate deep learning with electronic health records. *npj Digit Med* 2018;**1**:18. https://doi.org/10.1038/s41746-018-0029-1.
82. Kourtis LC, Regele OB, Wright JM, Jones GB. Digital biomarkers for Alzheimer's disease: the mobile/wearable devices opportunity. *npj Digit Med* 2019;**2**:9. https://doi.org/10.1038/s41746-019-0084-2.
83. Banaee H, Ahmed MU, Loutfi A. Data mining for wearable sensors in health monitoring systems: a review of recent trends and challenges. *Sensors (Switzerland)* 2013;**13**:17472–500.

84. Boursalie O, Samavi R, Doyle TE. M4CVD: mobile machine learning model for monitoring cardiovascular disease. *Procedia Comput Sci* 2015;**63**:384–91.
85. Mueller A, et al. Continuous digital monitoring of walking speed in frail elderly patients: noninterventional validation study and longitudinal clinical trial. *J Med Internet Res* 2019;**7**:e15191.
86. Devadharshini MS, Heena Firdaus AS, Sree Ranjani R, Devarajan N. Real time arrhythmia monitoring with machine learning classification and IoT. In: *2019 international conference on data science and engineering, ICDSE 2019 1–4*. Institute of Electrical and Electronics Engineers; 2019. https://doi.org/10.1109/ICDSE47409.2019.8971792.

Introduction to artificial intelligence and machine learning

2

Stephanie Kay Ashenden[a], **Aleksandra Bartosik**[b], **Paul-Michael Agapow**[c], **and Elizaveta Semenova**[a]

[a]Data Sciences and Quantitative Biology, Discovery Sciences, R&D, AstraZeneca, Cambridge, United Kingdom,
[b]Clinical Data and Insights, Biopharmaceuticals R&D, AstraZeneca, Warsaw, Poland,
[c]Oncology R&D Real World Evidence, AstraZeneca, Cambridge, United Kingdom

Artificial intelligence is the term used to describe the automation of intellectual tasks. Tasks that are commonly performed by a human are called as intellectual tasks. Machine learning (ML) is a set of techniques that describe methods used by computers to learn how to do the tasks. Deep learning is a specific subset of ML focused on layered techniques. This chapter reviews some typical areas, whereas some others will be discussed within their own context later in the book.

Artificial intelligence lends its beginning to symbolic AI that is comprised of explicit rules that are used for manipulating knowledge. However, symbolic AI will not capture the complexities of many situations, particularly in medicine where it is not always a simple yes or no answer. Symbolic AI takes data and rules as input to give answers, whereas, ML takes data and the answers as an input to outputs rules. This difference highlights that ML is trained by learning from many different examples and finding patterns within these examples. Finding and learning these patterns allows for the development of rules that allow for automation of a given task.

ML projects tend to follow a path from understanding the problem to presenting the results.[1] Fig. 1 shows a guide to complete an ML project, which involves asking questions, exploring the data, building models and understanding the outcome and often iterating this. Although data are the key to ML algorithms, before looking at the data, it is important to ensure that there is clarity in the problem statement and to frame the question in such a way that we work toward a meaningful answer. What you want to do with the data will depend on what questions you are trying to answer.

For example, in attempting to understand a tumor we may be able to predict whether the tumor is cancerous or not. However, the question may relate to the tumor growth rate instead and this would require a different form of analysis. Hence, it is important to ensure that we understand the end purpose of the analysis.

Another example is the need to analyze the age of different patients. Is this because you want to predict the onset of a disease for a patient with a given risk factor? Or is there an attempt to identify the cause of a disease by analyzing genetic and phenotypic variables but it is important to take into consideration any affect that the individual's age may have on these variables.

Asking these questions will be essential to provide the client with the correct model to answer their questions. Different models serve different questions better depending on the task at hand. In addition, building multiple models to compare against each other is crucial for creating the best predictions

The Era of Artificial Intelligence, Machine Learning, and Data Science in the Pharmaceutical Industry. https://doi.org/10.1016/B978-0-12-820045-2.00003-9

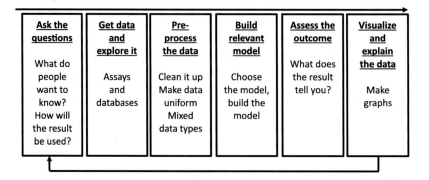

FIG. 1

Guide to complete a machine learning project.

possible. Once the results are available, assessing them for the accuracy and making any general conclusions in a way that can be easily interpreted is the next key step. Sometimes linking this into the original exploratory analysis can help build a full complete picture. Often this process is iterative.

Once you have framed the question, getting hold of the data, exploring it, and processing it is the next important steps. Quite often, ML algorithms require data of an acceptable quality because this is what the algorithm learns from. This can be hard to find, and often is the case that the data need to be preprocessed, which may involve removing or imputing missing data and ensuring that values are in a suitable format (such as numeric). For example, if you want to use the measured activity (in nanomolars) of a compound against a target, you may have a numeric value such as 75. However, sometimes, it can be recorded as "≤75" as the value and the computer will read the value "≤75" as a character value rather than a numeric value. You may need to decide whether to separate the relation (≤) from the activity value (75) so that you can read the activity value into an algorithm as a numeric value. At the same time, you may decide to just remove this particular example as saying that compounds activity on a particular target is ≤75 (nM) is not very informative, as it is not clear what the actual value recorded was. In addition, you may have a column indicating the units of your activity value, if they are not same you may want to convert them so that they are the same.

Exploratory analysis is where you explore the data to observe missing data and any obvious patterns in the analysis. You may produce plots to see how the data are distributed, for example, if you are looking at the measured activities of a compound-target association, what is the distribution of these activities. Are there some outliers that need to be taken into consideration? Therefore, exploratory analysis cannot only help you understand what your data look like, but it can also help detect oddities in the data. For example, you may need to decide how to deal with missing data, do you just remove all rows that have missing data in one of their columns, or do you try to impute it given other values. Understanding our data is important as it is what the algorithm will learn from.

Five core ingredients of ML are as follows:

- question/task
- input data
- model/algorithms
- measurement of whether the algorithm is working
- examples of the expected output

Finally, how do we assess whether the algorithm is actually doing what it should be. How do we know that the algorithm is not randomly guessing whether the new compound is active? To do this we use different metrics which will be explained in more detail later. However, as the input data that we give to the algorithm include the data and the answers (labels), which tell us whether the compound-target association is active, we generally can do a direct comparison. We can compare the actual answer to the predicted answer. The more times these match or are correlated (in the case of predicting values known as regression modeling) the more we are confident that the model is likely doing a good job. In addition, we can also try predicting on a test set, a set of data that the algorithm has never seen and determine whether the accuracy is still high as trying it out on the set that the data were trained on (training set). The training set is the data in which the algorithm learns from and is the data that we use to fit the model. The test set is a set of data that has not been shown to the algorithm and is used to evaluate the fit of the final model. Another set, known as the validation set, is used to evaluate the fit of a model and helps to tune the hyperparameters of a model. The test set would only be used after the model is completely trained which may use both the training set and the validation set.

Before presenting the results in a final manner, attempts may be made to improve the predictions by tuning the parameters or trying different models. Producing visualizations can help to explain the outcomes in a clear and constructive manner.

Supervised learning

In supervised learning, we have a label which may be categorical-like (high, middle, and low) or numerical-like (2.56, 7.83, and 3.32). When predicting a category, this is known as classification modeling and we can compare the predicted category to the actual result. A regression model focuses on predicting a numerical value. For example, predicting a compounds' activity in nanomolar concentration would be a regression task. Predicting whether a compound was considered active would be a classification task. Categories in classification tasks can be nominal such as a 0 or a 1 to represent true or false. Or they can be ordinal such as 1, 2, and 3 to represent category 1, category 2, or category 3. It is important to make sure you are aware of what the data look like and how it is represented as this can affect type of task you want to perform.

To demonstrate an example, imagine there are 400 compound-target associations. Each association has a measured activity (IC_{50} value in nanomolars). There are 8 different target classes and 20 inactive and 20 active examples for each target class. Various compound descriptors have also been calculated.

As mentioned, in classification modeling, we are predicting a categorical value. In this compound-target association dataset, we could ask the question, is it possible to predict whether a compound is active, whereas regression modeling is focused on predicting numeric values. For example, we may be asked to predict the activity value in nanomolars for a particular compound-target association.

Unsupervised learning

Unsupervised learning involves training a model without labels being available to help guide the model. To give an example using the previous datasets, imagine that you have a large set of compound-target associations and you needed to sort them into active and inactive. In a supervised manner you would have each compound-target association labeled as either active or inactive. However, how would you group these if you did not have the labels? You would need to rely on the compound properties, such as the molecular weight and the predicted log$P$. However, if you have no label, you cannot teach yourself which properties are associated with the compound being active or inactive. This is known as clustering. Clustering is a key task used within unsupervised learning.

This may be complicated by the compounds being associated with different targets. The properties that mean a compound is active on a target of one particular target class, may not be suitable to activity on a different target class. At the same time, compounds can be associated with different targets and target classes and therefore, this clustering can be difficult to fully interpret. In addition, it is often unclear how to best interpret the proposed clusterings, interpretation of clustering is often unclear. Much research assumes that the clusters seen are objective and meaningful, while neglecting how we know this or what "meaningful" means.[2, 3] Furthermore, cluster results are also strongly affected by the selected method and its interplay with the characteristics of a particular dataset,[4] making it possible to find clusters in homogenous or random data.[5, 6] In summary, every dataset contains clusters, with different clusters being revealed by different methods but not all these clusters are interesting or even real.

Semisupervised learning

As we have previous seen, supervised learning contains labeled data and unsupervised learning does not require labels to learn. Semisupervised allows you to use a dataset that is partially labeled. This is beneficial when labeled data are harder to acquire and unlabeled data are cheaper.[7] The semisupervised algorithms try to improve their performance by utilizing knowledge that is associated with either supervised or unsupervised algorithms. For example, an unlabeled approach may benefit from having some labels on some data and learning that it comes from the same class.[8]

Semisupervised learning to exploit unlabeled data has been combined with various approaches, as explained by Hady and Schwenker,[9] such as, semisupervised learning with graphs,[10] generative models (a type of model that includes the distribution of the data and gives a probability for a given example[11]), support vector machines (a type of ML algorithm[12, 13]), and by disagreement (multiple learners are trained and the disagreements between the learners are exploited during the learning process).[9, 14]

Model selection

Several models might be useful for the same task. It is also important to note that some models can be used for different types of ML. Such an example is that of decision tree which can be used for both classification and regression situations. Decision trees follow a leaf and branch to reach a conclusion as shown in Fig. 2 and are created by building a tree in a step known as induction. To make a decision tree generalize better on new data, and avoid being too specific for the training data, some bottom sections of a tree, which are not critical to the prediction task, can be removed. This process is called pruning, and it helps to reduce unnecessary complexity.[15]

This trade-off between model's complexity and its ability to describe the training data well is common for all ML algorithms and is referred to as *bias-variance trade-off*. Some models will be very flexible and would be able to match the training data exactly (this is described as high variance, because observation-to-observation predictions would have high variability), but would not generalize well to unseen data. This phenomenon is known as overfitting. Some other models might be not flexible

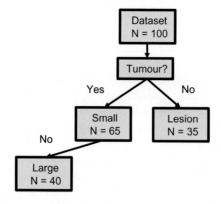

FIG. 2

Example of a decision tree.

enough and introduce systematic shifts in the predictions. This phenomenon is referred to as *bias*. Depending on the ML task (e.g., regression or classification), evaluation metrics need to be chosen to assess model performance. For regression, root mean squared error (RMSE) is a standard choice. It is calculated as a root of the averaged squared difference between the predicted and observed values. The smaller the RMSE, the smaller is the error made by a model. Overfitting can be diagnosed by small RMSE on training set, and high RMSE on test set. In the context of classification, the most widely spread metric is accuracy. It is calculated as a ratio of the number of correctly predicted examples to the total number of examples. Models with high variance would have high accuracy on the training set, but low accuracy on the test set. A metric to optimize for should be chosen based on a problem, for example, is sensitivity or specificity of a test, able to diagnose a disease more important? In case of binary classification, precision or recall might be preferred to accuracy. Precision measures how many observations predicted as positive are in fact positive. For example, we would only like to diagnose patients to have a certain condition, when they in fact have the condition. Recall, or sensitivity, measures how many observations out of all positive observations have been classified as positive. Optimizing for several metrics or their combination is also common: F1-score combines precision and recall in one metric, and area under receiver operation characteristics combines true-positive rate and false-positive rate. An example of a nontask-specific evaluation measures for model comparison are information criteria. An information criterion allows choosing a model based on the goodness-of-fit and the number of used parameters. If two models produce similar predictions but use different number of parameters to do so, a model with fewer parameters will be chosen. Two common techniques to avoid overfitting are cross-validation and regularization. Regularization techniques suggest that no one single parameter should have too much influence on the prediction; this goal can be achieved by adding a penalty, depending on the values of parameters, to the objective function. In case of neural networks, dropout represents another regularization technique: it switches off nodes of a network at random with a given probability. Cross-validation allows assessing the ability of a model to generalize on new data. For this, the whole training set is being subdivided into k subsets of equal size, and the model is being trained on all data but one subset (the hold-out set). The hold-out set is then used as a test set for evaluation. Of course, we can decide that these variables are therefore not informative and remove them, but this may not be true, this variable could be informative. This can depend on the question being asked.

Types of data

Understanding the type and nature of data correctly is crucial for the choice of a model. Such, in the supervised learning context, the type of outcome data defines which type of model can be applied. For instance, when the modeled value is binary, logistic regression could be used; continuous numerical values, linear regression is appropriate; count data can be modeled with a log-linear model. When the data are categorical and the task is the class prediction, the "one hot encoding" technique can be used. An example of a type of model that can use categorical values is decision trees. Simply converting to an integer may cause problems because naturally numbers follow an order and so the algorithm might pick up on this and try to learn something from that. For example, labeling a cancerous tumor as 1 and a benign tumor as 2 does not mean that a cancerous tumor is smaller than a benign one or that it is a higher priority. One hot encoding results in a binary representation of the categorical values (now the columns) where 1 represents presence and 0 represents absence. At this stage the modeling of binary data can be applied, where for each category the probability of being predicted is computed. If the categories are ordered, however, one hot encoding is inappropriate since it loses the notion of the correlation between neighboring categories. This type of data often comes in the form of "likert" scores, that is, a rating scale which is used to rank entities by groups, and can measure, for instance, various levels of response of an animal to a treatment in a clinical trial, or toxicity level of a drug in a safety study. Certain studies, due to their design, might contain a high number of the same observation.

Other key considerations

Feature generation and selection

Selection of features to be used in a predictive model is a crucial step alongside the choice of a model itself. Sometimes a set of features is provided in a dataset, such as, for instance, in vitro assay readouts in a toxicity study. Further features can be generated from these data by calculating products (to account for interactions), ratios or further transformations. In image analysis features are not readily available and need to be designed.

As with all model building, there is never going to be a one shoe fits all and every model will need to be treated individually to decide what the best way to handle the data is. This principle of each case must be considered separately, is considered often during feature elimination as well as when dealing with highly correlated values as well as descriptors that have zero or near-zero variance. For example, with regards to interpretability—the fewer descriptors the easier it can be to explain the output of a model. Therefore, the removal of highly correlated values may be beneficial to reduce the number of descriptors. However, highly correlated variables may carry valuable information about the investigated process, and hence could be modeled by causal models or by introducing correlation structures. Generally, for feature selection, the process of selecting the most relevant features for a prediction, is split into three method categories including filter, wrapper and embedded methodologies.[16] These categories also can be combined to form hybrid versions of these methods.[16] Filtering methods score features as a preprocessing step that is independent of the predictor, wrapper methods will combine different sets of the features, evaluate, compare and score between the different subsets.[17] Embedded methods are those that are able to learn which features contribute best to the prediction accuracy.[17] Overall, feature selection is used to try to identify the most useful features in a dataset and use those for prediction.

Feature selection is unlikely to be important in very small datasets, unless highly correlated, but what about when you have hundreds of descriptors? Are all of those descriptors really going to be equally important as one another?

A popular feature elimination method is known as recursive feature elimination. Essentially, this method goes in two possible directions. One way to use it is to start with a full set of features, and then remove features recursively to build a model assessing the model performance. It can be done for the combinations of features and determined which combination of features lead to the best accuracy. The opposite direction is to start with one feature only, and then iteratively add features observing improvement in performance, until performance ceases to improve.

With all ML and prediction tasks, it is appreciated that the predictions can only ever be as good as the data that is used to learn from. Essentially, if the data are poor or biased, the learning algorithm can only learn from poor or biased data and therefore, this will be reflected in quality of the output.

Censored and missing data

In certain cases, data can only be partially known due to censoring. Device measuring limits represent a typical example of censoring: true observed values below the upper censoring threshold would be known, but for the observations above the threshold only the upper censoring limit would be recorded as a value. Hence, censoring might bias the analysis. When censoring is taking place by design, treatment of these data might not be needed because censoring is informative on its own. But sometimes it is better to treat the censored values and impute them by modeling. A common approach is to make an assumption about the overall distribution of the variable and then model the values which are to be imputed using this distribution.

Missingness can occur due to various reasons: failure of a measuring device, or it can be informative itself, such as, it could reflect a measurement limit. During in vivo studies, animals can die, leading to missing observations. One way to treat missing values, is to remove records containing them. Another approach is to impute the values. To treat missing entries, it is important to understand the cause of absent data, that is, whether the entries are missing at random or there is a pattern. In case data are missing at random and imputation is needed, several methods can be applied: imputing with the average value or imputing with the mean or max values (if missingness is due to the sensitivity limit). Multiple imputations are another common approach. If data are missing not at random and the cause is known, this knowledge should be incorporated in the model, wherever possible.

Dependencies in the data: Time series or sequences, spatial dependence

In many settings the observed data points are not independent from each other and might display correlations. The nature of correlations can be structured and unstructured. Unstructured correlations are observed in data sets, where no relationship, other than association, can be assumed. Examples of structured correlations are dependency in space and time. Time series represent a collection of observations stamped with a time tag. In this case it can be assumed that neighboring (in time) observations are stringer correlated than distant observations and an observation at the current time point can be predicted from observations at previous time points. Same principle applies to spatial correlations, such as when an experiment is being run in a lab, the distribution of temperature and humidity in an incubator may play a role. Observations at near locations would be correlated higher than observations which are further apart. The form of dependencies in the data suggests a list of models that can be applied.

Deep learning

Deep learning is a subfield of ML that has been inspired by the structure and function of the brain. To understand deep learning, we need to break it down into its components. Neurons take weighted input and apply a nonlinear transformation, that is, an activation function, to them, returning an output (Fig. 3). The computation can be represented as an equation as shown in Eq. (1). Neural networks are made up of many neurons, grouped into layers. Different neural networks differ by the structure of the neurons and layers. Typically, a neural network will contain an input layer, output layer, and then hidden layers in between.

$$\gamma = \Sigma\left(\text{weight} \times \text{input}\right) + \text{bias} \tag{1}$$

Eq. (1) represents a neuron.

The weights determine the influence that the input has on the output and represent the connection between units and a bias which is applied to improve the fit of the model by shifting the activation function. The optimal value for the weights is one that helps the network map the example input into its associated output. The optimal set of values is achieved by measuring the distance between the output, based on the current values of weights and biases, and its true output. This is what the loss function is used for. The loss score is used to adjust the weights by an optimizer in an iterative procedure called backpropagation. At the start, all parameters (weights and biases) are initialized randomly, resulting in a high loss value but over the course of the training loop, as the network gets to observe more samples, the weights are adjusted and a reduction in the loss can be observed.

There are several types of activation functions that can be broadly categorized into two types, namely, linear activation functions and nonlinear activation functions. These functions determine the output of the neural network. Nonlinear activation functions are common at both hidden and output layers, whereas the linear activation function is mostly used at the output node. The nonlinear activation functions allow the model to generalize and adapt given a variety of data. Examples include the sigmoidal function where the output values are between 0 and 1 and can be utilized for predicting probability; the same approach is used to predict binary outcomes (i.e., 0 or 1). The hyperbolic tangent function (tanh) places values between -1 and 1 and the output is centered around zero. It is therefore also appropriate for classification tasks between two groups. One of the most popular nonlinear activation functions is ReLU (Rectified Linear Unit) and the values range between 0 and $+\infty$ but it should only be used in hidden layers.

One pass of the entire dataset through the neural network and back is known as an epoch. Multiple epochs are used to pass the dataset multiple times. To ease computational pressure, the epoch can be

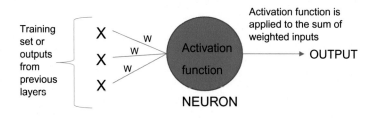

FIG. 3

Diagram of a deep learning neuron.

split into batches which dictate the number of iterations needed to complete one epoch. For example, 2000 examples that are split into 500 batches would require four iterations to complete one epoch.

Forward propagation is the running of a neural network from inputs to the outputs and the predicted values (the outputs) are used in the process of backpropagation. In backpropagation, for the output values (the predicted values) the error is calculated using a loss function. In addition the error is calculated with respect to each weight of the last layer and then these derivatives are used to calculate the derivatives of the second to last layer. This is repeated for all layers. The gradient value is subtracted from the weight value to reduce the error value. The derivatives of error are also known as gradients and the aim is to move closer (descent) to the local minima or the min loss. Example loss functions include the *F1/f* score which uses precision and recall to measure accuracy. Other loss functions include the mean absolute error and mean squared error—the equations are shown in Table 1. In general, the type of the loss function and appropriate evaluation measures depend on the task and the type of data.

Model optimizations are applied to improve the loss by updating bias and the weights in the model. Such examples are stochastic gradient decent which has been shown to have good performance for large scale problems[18] and Adam[19] of which is an extension of stochastic gradient decent. Normalization steps are used to aid in preventing over or underfitting and batch normalization methods help with dealing with internal covariant shift (where the distribution of activations changes at each layer during training). The use of batch normalization allows the use of higher learning rates.[20]

The rise of interest in Deep Learning began in 2009 when Google Brain and Nvidia cooperation led to the development of deep neural networks using high-performance computing capabilities of graphics processing units (GPU). This reduced time required for ML computations from weeks to days. Nowadays, there even exists a dedicated piece of hardware, named Tensor Processor Units (TPU) designed specifically to cope with training deep neural networks. Development in this field accelerated even more after 2010 when vanishing gradient problem (weights do not update due to too small of a gradient) was solved by ReLU activation.[21]

Artificial neural networks are applied for large and unstructured datasets in, for example, natural language processing, computer vision, medical image analysis, speech recognition, recommendation systems, financial fraud detection and bioinformatics in gene ontologies prediction.

Table 1 Example loss functions.

Loss function	Equation	Notes
F1/f score	$F1 = 2 * \dfrac{\text{Precision} \times \text{Recall}}{\text{Precision} + \text{Recall}}$	$\text{Precision} = \dfrac{TP}{TP + FP}$ $\text{Recall} = \dfrac{TP}{TP + FN}$ TP = true positive FP = false positive FN = False negative
Mean absolute error	$\dfrac{1}{n} \sum_{i=1}^{n} \lvert y_i - \hat{y}_i \rvert$	n represents the number of errors $\lvert y_i - \hat{y}_i \rvert$ represents the absolute errors
Mean squared error	$\dfrac{1}{n} \sum_{i=1}^{n} (y_i - \hat{y}_i)^2$	$(y_i - \hat{y}_i)^2$ represents the square of the errors

A deep neural network has more than one hidden layer, while shallow neural network has exactly one. A convolutional neural network is frequently used for extracting image features and for image classification problems. A recurrent neural network (RNN) performs calculations on sequence data. As they do not explicitly calculate features, they may be sensitive to short-term memory problems for long sequences.[22] In such cases, solutions such as long short-term memory units or gated recurrent units might be the right solution. Similarly, to nonneural-network architectures such as kNN, LG, SVM, Naive Bayes, multiple ANN can be combined into an ensemble.

Uncertainty quantification

Uncertainty quantification is a major concern for predictive models. Statistical tests are performed routinely and conclusions, such as, to accept or reject the null hypothesis (e.g., that there is no difference between treated and control groups) are being done solely on the basis of one single value (i.e., the p-value). Results of such analysis are rarely communicated together with the uncertainty. Confidence intervals (CIs) are a widely accepted uncertainty measure; however, their applications are limited: CIs show a range of plausible values for a quantity, but they do not describe how the values are distributed within the interval, that is, whether some values within the interval are more likely than the other.

Some ML models are able to naturally provide information about uncertainty: for instance, multiclass classification computes the probability of each class using the softmax function. This function assigns an individual probability to each class, by distributing the total probability of 1 between the classes. Even though the individual probabilities are being further used to predict the actual outcome (the prediction is the class with the highest probability), it could be also used to assess the confidence of the prediction. More widely, however, uncertainty quantification and communication is difficult and rare. Some further known approaches to uncertainty quantification include bootstrapping, conformal predictions, and Bayesian inference.

In the Bayesian approach the observed data are being used to update prior beliefs, which allows a researcher to work even with relatively small datasets, where classical statistical and ML methods may be not robust. Most of the models in the literature, however, make only point predictions, which mean that they provide no or little information about the uncertainty of the predictions. Such results might mislead decision-makers and lead to improvident choices. Bayesian inference allows to incorporate the prior knowledge and expert opinion into models and to take multiple sources of uncertainty into account.

Bayesian inference

Bayesian inference uses the Bayes rule to update prior beliefs with the information from the observed data. The prior knowledge or belief, as well as the inference method, is what distinguishes Bayesian method from the classical (a.k.a frequentist) approach. The two main schools of thought when it comes to uncertainty quantification—classical and Bayesian inference—interpret the unknown parameters and their estimates differently.

In classical statistics a parameter's value is considered to be constant, while in the Bayesian approach the unknown parameter is being treated as random. Regardless of the available information before the data collection process, the classical method only relies on the data obtained in one given experiment. Assumptions are being made in the form of the choice of the estimation method, for example, maximum likelihood or the method of moments. Both these methods can be used to estimate model parameters. Under the maximum likelihood method, parameters are estimated by maximizing model's likelihood (or minimizing its log-likelihood), so that the prediction of the model and observed outputs match well. Under the method of moments, analytical expressions of population moments are derived to be matched with the data, where moments are expected values of powers of a variable in question.

This step is rarely articulated in the literature when results of data analysis are being described.

Bayesian approach can be criticized for its subjectivity, that is, the role of priors in the final results. However, the assumptions made during the inference procedure are very explicit. Model formulation in the Bayesian context is flexible and can incorporate a multitude of knowledge about the system: biological or chemical laws, or causal relationships can be coded up as related to each other and governed by certain parameters.

The parameters are being estimated from the data and information about them is represented in the form of a distribution (called the posterior distribution), rather than a point estimate. Bayesian approach offers two ways to treat missing values. Missingness can be caused by, for instance, failure of measurement devices when working with in vitro data, or death of an animal when dealing with in vivo data. Traditional imputation approaches would be used as follows: when the missing value is in the response variable, it can be predicted from available covariates; if data in one of the predictors are missing, correlation between covariates can be used to construct a model to predict the variable of interest from the remaining predictors. The difference from the frequentist approach is that uncertainty in the predicted value can be carried forward into the uncertainty of main model. Alternatively—and this treatment of missing values is not available under the frequentist paradigm—each missing value can be viewed as an additional model parameter, that is, provided with a prior distribution and described by its posterior distribution as the result of model fitting. As in the previous case, uncertainty in the estimated value of the missing parameter will be naturally taken into account and propagated into the uncertainty of the whole model. Bayesian inference has been recently applied across various steps of the drug discovery pipeline: from early development and toxicology to design and analysis of clinical trials. Toxicity is a major cause of attrition and predicting it preclinically is hard. That is why safety evaluation is a crucial step in the drug development pipeline. It has been shown that Bayesian in silico toxicity profiling is useful for cardiotoxicity[23] and drug-induced liver injury (DILI) predictions.[24] Bayesian neural networks is a rapidly growing field and has been recently applied to model DILI from in vitro data.[25] In the clinical trial design setting, information from previous trials and research can be used in the form of priors. As the data is being collected, the model can be updated with the data newly acquired via the trial, to make efficient and timely inferences about the safety and/or efficacy of a treatment or therapy.[23]

References

1. *J. Brownlee, Your first machine learning project in python step-by-step, 2019.* Available at: https://machinelearningmastery.com/machine-learning-in-python-step-by-step/ [Accessed 14 September 2020].
2. Hennig C. What are the true clusters? *Pattern Recogn Lett* 2015;**64**:53–62.
3. Von Luxburg U, Williamson RC, Guyon I. Clustering: science or art? *Proceedings of ICML workshop on unsupervised and transfer learning JMLR Workshop and Conference Proceedings* 2012.
4. Rodriguez MZ, et al. Clustering algorithms: a comparative approach. *PLoS One* 2019;**14**:e0210236.
5. Ultsch A, Lötsch J. Machine-learned cluster identification in high-dimensional data. *J Biomed Inform* 2017;**66**:95–104.

6. Handl J, Knowles J, Kell DB. Computational cluster validation in post-genomic data analysis. *Bioinformatics* 2005;**21**:3201–12.

7. Zhu X. *Semi-supervised learning tutorial, International Conference on Machine Learning (ICML)*; 2007.

8. van Engelen JE, Hoos HH. A survey on semi-supervised learning. *Mach Learn* 2020;**109**:373–440.

9. Hady MFA, Schwenker F. Semi-supervised learning. In: *Handbook on neural information processing. Part of the intelligent systems reference library book series 49*, Berlin, Heidelberg: Springer; 2013. p. 215–39.

10. Sawant SS, Prabukumar M. A review on graph-based semi-supervised learning methods for hyperspectral image classification. *Egypt J Remote Sens Space Sci* 2020;**23**:243–8.

11. *Google Developers, Background: what is a generative model? 2019.* Available at: https://developers.google.com/machine-learning/gan/generative [Accessed 3 September 2020].

12. Boser BE, Guyon IM, Vapnik VN. Training algorithm for optimal margin classifiers. In: *Proceedings of the fifth annual ACM workshop on computational learning theory*. ACM; 1992. p. 144–52. https://doi.org/10.1145/130385.130401.

13. Noble WS. What is a support vector machine? *Nat Biotechnol* 2006;**24**:1565–7.

14. Zhou ZH, Li M. Semi-supervised learning by disagreement. *Knowl Inf Syst* 2010;**24**:415–39.

15. Osei-Bryson KM. Post-pruning in decision tree induction using multiple performance measures. *Comput Oper Res* 2007;**34**:3331–45.

16. Jović A, Brkić K, Bogunović N. A review of feature selection methods with applications. In: *2015 38th International convention on information and communication technology, electronics and microelectronics, MIPRO 2015—proceedings*; 2015. https://doi.org/10.1109/MIPRO.2015.7160458.

17. Iguyon I, Elisseeff A. An introduction to variable and feature selection. *J Mach Learn Res* 2003;**3**:1157–82.

18. Bottou L. Large-scale machine learning with stochastic gradient descent. In: *Proceedings of COMPSTAT'2010*. Physica-Verlag HD; 2010. p. 177–86. https://doi.org/10.1007/978-3-7908-2604-3_16.

19. Kingma DP, Ba JL. Adam: a method for stochastic optimization. In: *3rd international conference on learning representations, ICLR 2015—conference track proceedings (international conference on learning representations ICLR)*; 2015.

20. Ioffe S, Szegedy C. Batch normalization: accelerating deep network training by reducing internal covariate shift. In: *32nd international conference on machine learning, ICML 2015*. vol. 1. International Machine Learning Society (IMLS); 2015. p. 448–56.

21. Alom MZ, et al. *The history began from AlexNet: a comprehensive survey on deep learning approaches.* arXiv:1803.01164; 2018.

22. Karim A, et al. *Toxicity prediction by multimodal deep learning.* arXiv:1907.08333 [physics stat]; 2019.

23. Lazic SE, Edmunds N, Pollard CE. Predicting drug safety and communicating risk: benefits of a bayesian approach. *Toxicol Sci* 2018;**162**:89–98.

24. Williams DP, Lazic SE, Foster AJ, Semenova E, Morgan P. Predicting drug-induced liver injury with bayesian machine learning. *Chem Res Toxicol* 2020;**33**:239–48.

25. Semenova E, Williams DP, Afzal AM, Lazic SEA. Bayesian neural network for toxicity prediction. *Comput Toxicol* 2020;**16**:100133.

Data types and resources

3

Stephanie Kay Ashenden[a], Sumit Deswal[b], Krishna C. Bulusu[c], Aleksandra Bartosik[d], and Khader Shameer[e]

[a]*Data Sciences and Quantitative Biology, Discovery Sciences, R&D, AstraZeneca, Cambridge, United Kingdom,* [b]*Genome Engineering, Discovery Sciences, BioPharmaceuticals R&D, AstraZeneca, Gothenburg, Sweden,* [c]*Bioinformatics and Data Science, Translational Medicine, Oncology R&D, AstraZeneca, Cambridge, United Kingdom,* [d]*Clinical Data and Insights, Biopharmaceuticals R&D, AstraZeneca, Warsaw, Poland,* [e]*AI and Analytics, Data Science and Artificial Intelligence, Biopharma R&D, AstraZeneca, Gaithersburg, MD, United States*

Notes on data

Recent innovation in the field of machine learning has been enabled by the confluence of three advances: rapid expansion of affordable computing power in the form of cloud computing environments, the accelerating pace of infrastructure associated with large-scale data collection and rapid methodological advancements, particularly neural network architecture improvements. Development and adoption of these advances have lagged in the health care domain largely due to restrictions around public use of data and siloed nature of these datasets with respect to providers, payers, and clinical trial sponsors.

There are many different types of data that are relevant to drug discovery and development, each with its own uses, advantages and disadvantages. The type of data needed for a task will rely on an understanding and clarity of the task at hand. With an increasing amount of data being made available, new challenges continue to arise to be able to integrate (with a purpose), use and compare these data. Comparing data is important to capture a more complete picture of a disease, of which is often complex in nature.[1] One approach is to ensure that the data is FAIR, meaning that it is Findable, Accessible, Interoperable, and Reusable. A generic workflow for data FAIRification has been previously published[2] and discusses seven-core steps. These steps are to identify the objective, analyze the data and the metadata, define a semantic model for the data (and metadata), make the data (and the metadata) linkable, host the data somewhere, and then assess the data.[2] Other key considerations include assigning licenses and combining with other FAIR data.[3]

Data integration has been discussed by Zitnik and co-authors.[1] There are different integration stages such as early, Intermediate, and late integration.[1] These stages involve the transformation of the datasets into a single representation. This representation can then be used as input in a machine learning algorithm. In intermediate integration, many datasets are analyzed and representations that are shared between them are learnt. In late stage integration, each dataset has its own model built and these models are combined by building a model on the predictions of the previous models.[1]

Zitnik and co-authors[1] also discuss the fact that there are many challenges in integrating data such as the sparseness of biomedical data and its complexity. The authors note that the data are often biased and/or incomplete. For example, databases containing manually created data from papers may be limited to a certain

The Era of Artificial Intelligence, Machine Learning, and Data Science in the Pharmaceutical Industry. https://doi.org/10.1016/B978-0-12-820045-2.00004-0

number of journals. ChEMBL[4, 5] routinely extracts data from seven journals but does also include other journals not included in the seven.[6] The authors note that machine learning can be used for data integration.

However, there are other concerns with integrating data beyond the technical difficulties. Sharing and privacy concerns especially in relation to clinical data are a key consideration in the pharmaceutical industry. However, sharing clinical trial data is important in improving scientific innovation.[7] To this end, attempts have been made to improve clinical data sharing policies and practises[8] but only a small amount of companies met such measures with many failing to share data by a specified deadline and failed to report all data requests.[8] Other approaches include MELLODDY[9] which aims to bring together information to accelerate drug discovery by allowing pharmaceutical companies to collaborate. MELLODDY notes that huge amounts of data are generated during the drug discovery process and their hypothesis is that working across data types and partners will improve predictive power and understanding of models in the drug discovery process.[9] The large collection of small molecules with known activities can be used to enhance predictive machine learning models without exposure of proprietary information.[9]

Such large volumes of data are known as big data. In the medicinal field this may include omics data, clinical trials data and data collected in electronic health records. The data can be a combination of varying levels of structuredness and can be fantastic resources for information mining and machine learning projects.[10] Ishwarappa and Anuradha[11] discussed the five Vs of big data and explain that they correspond to:

- Volume (the amount of data)
- Velocity (how rapidly the data is generated and processed)
- Value (what the data can bring)
- Veracity (the quality of the data)
- Variety (the structure and types of data)

Different types of data will be used for different types of analysis and will enable for a variety of questions to be answered. Later, we discuss some of the key types of data that may be used.

Omics data

Omics studies aim to understand various organisms at the molecular level by studying specific components such as genes or proteins in both experimental and computational ways.[12] Such omics include genomics (study of genes), proteomics (study of proteins), metabolomics (study of metabolites), transcriptomics (concerned with mRNA) as well as more niche omics such as lipidomics and glycomics (Fig. 1). The rise of omics data gives thanks to technical advances in areas such as sequencing, microarray and mass spectrometry,[13] and omics data can be used throughout the drug discovery pipeline. For example, for identifying and validating novel drug targets[13] and understanding and interpreting genetic variations in patients for personalized medicine.[13, 14]

Bioinformatic techniques are used throughout the omics studies to analyze the resultant data, make sense of it and derive hypotheses and conclusions. There are a wide variety of omics data types available as well as databases that contain useful information that can be exploited throughout the drug discovery process. Later, we summarize the different omics methods and include some of the key databases.

Genomics

Genomics is concerned with understanding the genes that are within a genome (it is estimated that there are 20,000–25,000 genes in the human genome), it is also concerned with how those genes interact with

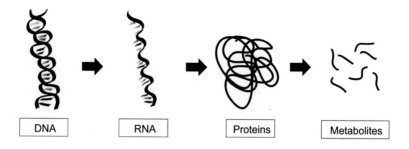

FIG. 1

Branches of the omics studies.

each other and other environmental factors.[15] Specifically, genomics is concerned with interactions between loci and alleles as well as considering other key interactions such as epistasis (effect of gene-gene interactions), pleiotrophy (effect of a gene on traits), and heterosis.

Libbrecht and Noble published an article on the applications of machine learning in genetics and genomics.[16] The authors discuss the different uses of supervised, semisupervised, unsupervised, generative, and discriminative approaches to modelling as well as the uses of machine learning using genetic data.[16] The authors explain that machine learning algorithms can use a wide variety of genomic data, as well as being able to learn to identifier particular elements and patterns in a genetic sequence.[16] Furthermore, it can be used to annotate genes in terms of their functions and understand the mechanisms behind gene expression.[16]

Transcriptomics

The transcriptome is the set of RNA transcripts that the genome produces in certain situations.[17] Transcriptomics signals aid in understanding drug target adverse effects.[18]

Methods such as RNA-Seq are used to profile the transcriptome. As a method it can detect transcripts from organisms where their genomic sequence is not currently profiled and has low background signal.[19] It can be used to understand differential gene expression,[20] RNA-Seq is supported with next generation sequencing of which allows for large numbers of read outs.[20]

Transcriptomic data have been used in machine learning algorithms in cases such as machine learning diagnostic pipeline for endometriosis where supervised learning approaches were used on RNA-seq as well as enrichment-based DNA methylation datasets.[21] Another use has been the development of GERAS (Genetic Reference for Age of Single-cell), which is based on their transcriptomes, the authors Singh and co-authors explain that it can assess individual cells to chronological stages which can help in understanding premature aging.[22] It has also been used alongside machine learning algorithms to aid in diagnostics and disease classification of growth hormone deficiency (random forest in this case).

Metabolomics and lipomics

Metabolomics and lipidomics are concerned with the metabolome and the lipidome, respectively. Metabolomics allows for the understanding of the metabolic status and biochemical events observed in a biological, or cellular, system.[13] Approaches in metabolomics includes the identification and quantification of known metabolites, profiling, or quantification of larger lists of metabolites (either identified or unknown compounds) or a method known as metabolic fingerprinting, of which is used to compare samples to a sample population to observe differences.[23] Metabolomics has been combined with machine learning to identify weight gain markers (again Random Forest algorithms were used).[24] Sen and co-authors have shown that deep learning has been applied to metabolomics in various areas such as biomarker discovery and metabolite identification (amongst others).[25]

Lipids are grouped into eight different categories including fatty acyls, glycerolipids, glycerophosolipids, sphingolipds, saccharolipds, polyketides, sterol, and prenol lipids.[26] They are important in cellular functions and are complex in nature, change under different conditions such as physiological, pathological and environmental.[27] Lipidomics has be used to show tissue-specific fingerprints in rat,[26] shown potential in risk prediction and therapeutic studies[28] and can be used through the drug discovery process.[27] Fan and co-authors used machine learning with lipidomics by developing SERRF (Systematic Error Removal using Random Forest) which aids in the normalization of large-scale untargeted lipidomics.[29]

Proteomics

Proteomics is concerned with the study of proteins. Proteomes can refer to the proteins at any level, for example, on the species level, such as all the proteins in the human species, or within a system or organ. In addition, one of the major difficulties with proteomics is its nature to change between cells and across time.[30] Questions may include understanding the protein expression level in the cell or identifying the proteins being modulated by a drug. Key areas of proteomic study involve, protein identification, protein structure, analysis of posttranslational modifications.

Typically a proteomic experiment is broken down into three key steps; the proteomics separation from its source such as a tissue. The acquisition of the protein structural information and finally, database utilization.[31] Experimental procedures to separate a protein from its source involve electrophoresis where the proteins appear as lines on a gel, separated by their molecular weight.[31] They are visualized by staining the gel and then preceded by acquiring an image of the gel. The proteins can be removed from the gel to be digested and put through a mass spectrometer. Sequencing is often completed by mass spectrometry methods, of which involves ionization of the sample, analysis of the mass, peptide fragmentation, and detection ultimately leads to database utilization. A typical global proteomics experiment involves profiling of several compounds to determine changes in particular proteins. By analyzing the observed abundance of the proteins across different treatment channels it is possible to observe treatment effects.

Swan and co-authors published applications of machine learning using proteomic data. The authors note that MS-derived proteomic data can be used in machine learning either directly using the mass spectral peaks or the identified proteins and can be used to identify biomarkers of disease as well as classifying samples.[32] Gessulat and co-authors developed Prosit, a deep neural network that predicts the chromatographic retention time as well as the fragment ion intensity of peptides.[33]

Chemical compounds

Compounds are often represented in a computer readable form. The ChemmineR[34] package for R[35] or RDKit[36] package in KNIME[37] or Python (https://www.python.org/) provides example compounds for analysis.

SDF format

SDF formats (structure data files formats) were developed by Molecular Design Limited (MDL) and are used to contain chemical information such as structure. The first section contains general information about the compound, including its name, its source and any relevant comments. The counts line has 12 fields that are of fixed length. The first two give the number of atoms and bonds described in the compound. Often Hydrogens are left implicit and can be included based on valence information.[38] The second block is known as the atom block (atom information encoded) and the third is known as the bond block where bond information is encoded. In the atom block, each line corresponds to each individual atom. The first three fields of each line correspond to the atoms position with its *x-y-z* coordinates.[38] Typically the atom symbol will be represented and the rest of the line relates to specific information such as charge information.[38] The bond blocks also have one line per individual block, and the first two fields index the atoms and the third field indicated the type of bond. The fourth refers to the stereoscopy.[38]

InChI and InChI Key format

InChI[39] is a nonproprietary line notation or 1D structural representation method of which aims to be canonical identifier for structures (and thus is suitable for cross database comparisons).[40] Owing to uniqueness of InChI, it has been used to derive canonical SMILES (described later) to create something called InChIfied SMILES.[41] InChI key is a hashed and condensed version of the full InChI string.

It was developed by the International Union of Pure and Applied Chemistry (IUPAC) along with the National Institute of Standards and Technology (NIST). It is continually updated by the InChI Trust. InChI captures a wide variety of compound information, not limited to its stereochemistry, charge and bond connectivity information.

InChI keys were developed to allow for searching of compounds as the full InChI is too long for this. It contains 27 characters, the first 14 corresponding to the connectivity information. Separated by a hyphen is the next eight characters that include other chemical information of the structure. The following characters (each separated by a hyphen) give information about the type of InChI, the version of it and finally, the protonation information of the compound.

It has an almost zero chance of two separate molecules having the same key. It was estimated that if 75 databases each had 1 billion structures, there would be one instance of two molecules having the same InChI key. Despite this, an example of a "collision" was identified with two compounds with different formulae and no stereochemistry.[42–44] This estimated rarity of collisions was experimentally tested and suggested that if uniqueness was desired it would probably need a longer hash.[45]

SMILES and SMARTS format

The simplified molecular-input line entry system also known as (SMILES) is one of the most commonly used.[46–48] SMILES are based on molecular graph theory where the nodes of a graph are the atoms and the edges are the bonds.[46, 47] Generic SMILES do not give details on the chirality or the isotopic nature of the structure (of which are known as isomeric SMILES).[49]

One problem with SMILES is that a single structure can be represented in multiple different SMILES strings and therefore, it is recommended to use canonicalized structures to prevent one compound being identified as multiple due to the different representations used. Daylight give an example of the ways that the SMILES string CCO can be written, including OCC, [Ch3][CH2][OH], C–C–O, and C(O)C.[49]

Daylight gives an in depth explanation of the rules for generating and understanding SMILES strings[49] and the common rules are summarized here. SMILES follow encoding rules, namely, the use of atomic symbols for atoms with aliphatic carbons being represented with a capital C and aromatic carbons being written with a lower case c. Brackets are used to describe abnormal valences and must include any attached hydrogens, as well as a number of + or − to indicate valance count. Absence of these will result in it being assumed there are zero hydrogens or charge. To indicate isotopic rules, the atomic symbol is preceded by its atomic mass such as [12C] or [13C]. On a side note, hydrogens are often omitted when writing SMILES strings and can be highlighted by either implicit nature (normal assumptions), explicit nature by either count (within brackets) or as explicit atoms themselves [H]. Bonds are represented by −, =, #, or : to depict single, double, triple, or aromatic bonds, respectively. Alternatively, atoms may be placed next to each other with the assumption that either a single or an aromatic bond separates them. To include direction, \ and / are used. Branching is dealt with within parentheses (of which can be nested) and cyclic structures contain a digit to indicate the breaking of a bond in the ring such as C1CCCCC1. Any disconnected structures are separated by a period. Dealing with tetrahedral centers can be represented by @ (neighbors are anticlockwise) or @@ (neighbors are clockwise) after the chiral atom. Many specific natures of compounds, such as tautomerization, chirality and shape, need to be explicitly specified in SMILES notation.

Extending on from SMILES is the SMARTS[49] notation which is designed to aid with substructure searching. SMARTS, extend atoms and bonds by including special symbols to allow for generalized identification, for example, the use of * to denote the identification of any atom or ~ to denote any bond. Many of these rules follow the rules of logical rule matching in coding languages such as the use of an explanation mark to denote NOT this, as an example, [!C] tells us to find not aliphatic carbons.

Daylight describes the difference between SMARTS and SMILES as SMARTS describing patterns and SMILES describing molecules. In addition, SMILES are valid SMARTS.

Fingerprint format

A molecular descriptor's role is to provide one and capture similarity and differences between compounds in a chosen dataset. There are multiple kinds of molecular descriptors that range in dimensionality (0D, 1D, 2D, 3D, and 4D). A molecular fingerprint is an example of a 1D-descriptor. It is a binary string with a list of substructures or other predefined patterns.[50] They are defined before a model is trained to avoid overfitting on sparse or small datasets. If a specified pattern is found in a molecule, the corresponding bit in the binary string is set to "1," otherwise it is set to "0."[51]

Example of fingerprints are ECFP4 (extended connectivity for high dimensional data, up to four bonds), FCFP4 (functional class-based, extended connectivity), MACCS (166 predefined MDL keys), MHFP6 (for circular structures) Bayes affinity fingerprints (bioactivity and similarity searching), PubChemFP (for existence of certain substructures), KRFP (from the 5-HT 5A dataset to classify between active or inactive compounds). Sometimes it is better to create custom fingerprints than rely on predefined ones.[52]

Essentially the features of the molecules (such as the presence of a particular atom) are extracted, hashed, and then the bits are set.[53] There are a wide host of available fingerprints that can be used as discussed in Table 1.

Performance of a machine learning model and prediction accuracy depends on the quality of data and descriptors and fingerprints chosen. For instance, fingerprint-based descriptors, for example, ECFP or MACCS, are recommended for active substances with functional groups located in meta or para positions.[65] For genotoxicity prediction, Support Vector Machines (SVM) models perform best with PubChemFPs. However, the authors recommend combining Random Forest (RF) and MACCS fingerprints.[66]

Table 1 Table of example of different types of molecular fingerprints.

Name	Notes
MACCS[54]	Substructure keys
Morgan[55, 56]	Circular fingerprints
Extended-Connectivity Fingerprints (ECFP)[55]	ECFP# where # is a number denoting the circle diameter. Typically, between 0 and 6
Daylight[57]	Path fingerprints that encode the substructure
Signature[58, 59]	Topological descriptor
MHFP6[60, 61]	For circular structures
Bayes affinity fingerprints[62]	Bioactivity and similarity searching
PubChemFP[63]	For existence of certain substructures
KRFP (Klekota Roth fingerprint)[64]	Substructure keys

Extended-connectivity fingerprints (ECFPs) were designed for structure-activity modeling of which are topological and circular.[55] They are related to Morgan fingerprints, but differ in their algorithm. The ECFP algorithm is well documented[55] and summarized here. Each atom is assigned an identifier of which is updated to capture neighboring atom information. Finally, any duplicate identifiers are removed (so the same feature is only represented once). Rather than a bit vector, ECFC derive a count of features.[67]

In comparison, to the ECFP algorithm of which has a predetermined set of iterations, Morgan fingerprints and their algorithm[56] continue to have iterative generations until uniqueness is achieved. This process is described by Rodgers and Hahn[55] in their extended-connectivity fingerprints paper where they explain that for Morgan fingerprints, their atom identifiers are not dependent on the atoms original numbering and uses identifiers from previous iterations after encoding invariant atom information into an initial identifier. Essentially the Morgan algorithm iterates through each atom and captures information about all possible paths through the atom, given a predetermined radius size.[68] Morgan fingerprints were designed to address molecular isomorphism[55] and are often used for comparing molecular similarity. These are hashed into a bit vector length (also predetermined). The iterative process involves each atom identifier in a compound and updating the information about it. For example, at iteration 0, only information about the atom is captured (as well as related bonds) whereas as the iterations increase, so does the information about the atom's neighbors, and so on.

Two other popular fingerprints are MACCS keys and Daylight fingerprints. The Molecular ACCess System (MACCS) keys is a predefined set of 116 substructures.[54] A problem with the MACCS keys is that there is no publication that defines what each of the 116 substructures are. Generally, when citing, individuals refer to a paper discussing the re-optimization of MDL keys.[54, 69] Daylight fingerprints are a form of path fingerprints which enumerate across the paths of a graph and translate them into a bit vector.[70] Signature fingerprints are not binary and are based on extended valence sequence.[58] They are topological descriptors that also describe the connectivity of the atoms within a compound.[71]

Other descriptors

A molecular descriptor can be derived from experimental data or calculated theoretically. Examples of nonfingerprint molecular descriptors include reactivity, shape, binding properties, atomic charges, molecular orbital energies, frontier orbital densities, molar refractivity, polarization, charge transfer, dipole moment, electrophilicity,[72] molecular and quantum dynamics.[73]

Molecular descriptors are generated with the use of tools, for example, PaDEL-Descriptor,[74] OpenBabel,[75] RDKit,[36, 53] CDKit,[76] and E-Dragon.[77]

Structural 2D descriptors perform well in models handling binary information such as classification and class probability estimation models[78] and in association rules learning.[79–81] There exists no universal descriptor that works best with every prediction model. However, various descriptor types can be combined as input data for a model to achieve higher performance.

There are various commercial and open-source software, databases, and servers that use molecular descriptors to predict toxic endpoints: OECD QSAR Toolbox,[82] Derek Nexus,[83] FAF-Drugs4,[84] eTOXsys,[85] TOXAlerts,[86] Schrödinger's CombiGlide[87–89] Predictor, Leadscope Hazard Expert,[90] VEGA,[91] METEOR.[83] ChemBench,[92] ChemSAR,[93] ToxTree,[94] Lazar,[95] admetSAR,[96] Discovery Studio[97] and Pipeline Pilot[98] are ML-based tools. For more detailed information, please refer to review on computational methods in HTC by Hevener, 2018.[99]

Furthermore, descriptors can also be calculated for protein structures. Local descriptors have been shown to aid in the characterization of amino acid neighborhoods.[100] The tool ProtDCal calculates numerical sequence and structure based descriptors of proteins.[101] Another publication had the authors develop a sequence descriptor (in matrix form) alongside a deep neural network that could be used for predicting protein-protein interactions.[102]

Similarity measures

It is often a requested task to compare the similarity of two compounds. Different similarity metrics are summarized in Table 2. Similarity can be rephrased as comparing the distance between the compounds to evaluate how different two compounds are. For fingerprint-based similarity calculations, Tanimoto index is a popular method.[110] A study compared several of these metrics comparing molecular finger-

Table 2 Table of different similarity metrics.

Name	Equation	Equation information
Tanimoto/Jaccard[103, 104]	$T(a,b) = \dfrac{N_c}{N_a + N_b - N_c}$	N = number of attributes in objects a and b C = intersection set
Tversky[105]	$similarity(A,B) = \dfrac{AB}{aA + bB + AB}$	α = weighs the contribution of the first reference molecule The similarity measure is asymmetric[106]
Dice[107]	$similarity(A,B) = \dfrac{2*AB}{A + B + AB}$	AB is bits present in both A and B
Manhattan[106]	$similarity(A,B) = \dfrac{A+B}{A + B + AB + !A!B}$	The more similar the fingerprint the lower the similarity score (acting more like a distance measure)[106]
Euclidean distance[106]	$Dist(A,B) = \sqrt{\dfrac{AB + !A!B}{A + B + AB + !A!B}}$	$!A!B$ represents the bits that are absent in both A and B
Cosine[108, 109]	$similarity(x,y) = \cos(\theta) = \dfrac{x.y}{\|x\| * \|y\|}$	x = compound x y = compound y

prints. They identified that the Tanimoto index, Dice index, Cosine coefficient and Soergel distance to be best and recommended that Euclidean and Manhattan distances not be used on their own.[110]

The reason for comparing the similarity of compounds is that, in combinatorial library design, chemists may reject compounds that have a Tanimoto coefficient ≥ 0.85 similar to another compound already chosen from the library.[111] This is for the purpose of ensuring structural diversity within the library. A study showed that by using Daylight fingerprints, and Tanimoto similarity, found that there was only a 30% chance that two compounds that were highly similar were both active, likely due to differences in target interactions.[111]

$$Similarity = \frac{1}{1 + distance} \qquad (1)$$

Eq. (1) is used for calculating the similarity of two compounds.

QSAR with regards to safety

QSAR studies involve pattern discovery, predictive analysis, association analysis, regression, and classification models that integrate information from various biological, physical, and chemical predictors. It relies on the assumption that chemical molecules sharing similar properties possess similar safety profile.[81] QSAR model establishes a relationship between a set of predictors and biological activity (e.g., binding affinity or toxicity). Biological properties correlate with the size and shape of a molecule, presence of specific bonds or chemical groups, lipophilicity, and electronic properties. Biological activity can be quantified, for example, as minimal concentration of a drug required to cause the response. According to the Organization for Economic Co-operation and Development (OECD) guidelines QSAR model should have (a) a defined endpoint; (b) an unambiguous algorithm; (c) a defined domain of applicability; (d) appropriate measures for goodness-of-fit, robustness, and predictivity; and (e) mechanistic interpretation.[112]

The largest advantage of QSAR modeling is feature interpretability, high predictability, and diversity of available molecular descriptors. QSAR enables calculation of biological activity and reduces significantly the number of molecules that need to be synthesized and tested in vitro. QSAR method has some limitations, though. To develop a model of high prediction power and high statistical significance, large datasets are necessary as well as a preselection of predictors. Additionally, it is not always possible to deduce human dose, duration of treatment or exposure without the use of animal data. Furthermore, not all structurally similar molecules exert a similar influence in vivo. Thus an experienced human expert should define the applicability domain and scope of interpretation of QSAR prediction.

QSAR approach dates to 1962, when Hansch assumed independence of features that influenced bioactivity and developed a linear regression model. In the Hansch model (Eq. 2), authors estimated logarithm of the reciprocal of the concentration (C) using the octanol/water partition coefficient (π) and the Hammett constant (σ):

$$\log\left(\frac{1}{C}\right) = 4.08\pi - 2.14\pi^2 + 2.78\sigma + 3.36 \qquad (2)$$

Eq. (2) represents Hansch Model.

A positive coefficient of a descriptor suggests a positive correlation between specific toxicity endpoint and that descriptor; negative coefficient is linked to negative correlation.[113] Two years later, in 1964, the Free-Wilson method basing on regression analysis was developed, and the chemical structure has been used as a single variable.[114] In the 1980s and 1990s, linear regression has been applied to develop toxicity prediction models with both single and multiple molecular properties as variables.

Approaches such as linear regression analysis and multivariate analysis perform well for single molecular properties prediction. However, currently, it is possible to generate many more types of molecular descriptors (1D to 4D) than it was 40 years ago, which leads to more high-dimensional datasets.[81] Hence, nowadays advanced nonlinear techniques have become more popular in toxicity prediction.

In certain cases, large numbers of input features, that is, dimensions (e.g., molecular descriptors) may result in decreased machine learning model performance. This phenomenon is referred as a *curse of* dimensionality because sample density decreases. The data set becomes sparse. As a result, the model may overfit, which means it learns too much about each data point. To assure that model's level of generalization is just right, preselection of preferably most relevant features may be indispensable. This process is called dimensionality reduction of *n*-dimensional feature space. Most common dimensionality reduction methods include: Least Absolute Shrinkage and Selection Operator (LASSO), Principal Component Analysis (PCA), Kernel Principal Component Analysis (KPCA), Linear Discriminant Analysis (LDA),[115] Multidimensional Scaling (MDS), Recursive Feature Elimination (RFE), Distributed Stochastic Neighbor Embedding (t-SNE), and Sequential floating forward selection (SFFS).[116]

Data resources

There are many resources available for data analytics, both commercial and open. Many of these resources can be used for multiple tasks. Below contains many of the key resources used in drug discovery, however, it is worth noting that as more data is created, and gaps are identified in available resources, new resources will be developed.

Toxicity related databases

As a result of the application of high throughput screening (HTS) and development of novel chemical and biological research techniques in the 21st century, a number of publicly available repositories is rapidly growing. This enables integration of siloed information and prediction of less evident side effects resulting from synergistic effects, and complex drug-drug interactions can be discovered. In this section, we present an overview of existing data sources related to toxicogenomics, organ toxicity, binding affinity, biochemical pathways, bioactivity, molecular interactions, gene-disease linkage, histopathology, oxidative stress, protein-protein interactions, metabolomics, transcriptomics, proteomics, and epigenomics (Table 3).

TOXNET[187] is an aggregator of other toxicity-related databases on breastfeeding and drugs, developmental toxicology literature, drug-induced liver injury, household product safety, and animal testing alternatives. TOXNET is available via PubMed since December 2019. ToxCast[158] and ECOTOX[161] are two databases created by the US Environmental Protection Agency. They contain high-throughput and high-level cell response data related to toxicity and environmental impact of over 1800 chemicals, consumer products, food and cosmetic additives. Tox21[156] is a collaborative database between some of the US Federal Agencies that aggregates toxicology data on commercial chemicals, pesticides,

Table 3 An overview of toxicity, chemical and multiomics databases useful in the computational evaluation of safety.

Database name	Data type	Description	Source publication DOI
CEBS (Chemical Effects in Biological Systems)	Adverse events	Biology-focused database of chemical effects that focuses on systems toxicology	[117]
IntSide	Adverse events	Chemical and biological side effects database	[118]
MetaADEDB	Adverse events	Integrates CTD, OFFSIDES, and SIDER, focusing on ADE-drug occurrences	[119]
OFFSIDES	Adverse events	ADRs reported during clinical trials before drug approval	[119]
SIDER	Adverse events	Contains information on marketed medicines and their recorded adverse drug reactions	[120]
BioGRID	Molecular interactions	Genetic and protein interactions	[121]
Biomodels	Molecular interactions	Rate-related interactions	[122]
Bioplex	Molecular interactions	Immunopurification and mass spectrometry-based protein interaction database	[123]
HAPPI-2	Molecular interactions	Protein-protein interactions with a confidence score	[124]
HPRD	Molecular interactions	Historic, no longer updated database of manually curated	[125]
IntAct	Molecular interactions	Open-source database system and analysis tools for molecular protein interaction data	[126]
InWeb_IM	Molecular interactions	Protein-protein interaction datasets with orthological predictions	[127]
mentha	Molecular interactions	A taxonomy browser of interactions from publications and databases	[128]
NRF2Ome	Molecular interactions	Manually curated human oxidative stress and NRF2 response specific database	[129]
OmniPath	Molecular interactions	Manually curated human signaling database	[130]
SignaLink2	Molecular interactions	Manually curated signaling database with regulations and predicted interactions	[131]
Signor	Molecular interactions	Manually curated pathway interactions with directions and signs	[132]
STRING	Molecular interactions	Curated databases using text mining interactions in different species	[133]
KEGG (Kyoto Encyclopedia of Genes and Genomes)	Pathways	The database focuses on high-level functions of biological systems from molecular-level information	[134]
MsigDB (Molecular Signature Database)	Pathways	A collection of annotated gene sets for use with GSEA (Gene Set Enrichment Analysis) software	[135]
Pathway Commons	Pathways	Free online database of pathways, bundled with open-source data analysis tools	[136, 137]

Continued

Table 3 An overview of toxicity, chemical and multiomics databases useful in the computational evaluation of safety—cont'd

Database name	Data type	Description	Source publication DOI
PharmGKB (The Pharmacogenomics Knowledgebase)	Pathways	Manually curated collection of PGx information from the primary literature	138, 139
Qiagen IPA (Ingenuity Pathway Analysis)	Pathways	A commercial pathway analysis tool, capable of complex analysis and prediction of downstream effects	140
Reactome	Pathways	Free online database of pathways, mostly focused on human biology	141, 142
The Gene Ontology Resource	Pathways	Database of pathways from molecular to organism-level for multiple species, focusing on the function of the genes and gene products. Datapoints have annotations on multiple levels of specificity	143
WikiPathways	Pathways	Community-curated collection of pathways with links to other sources and pathway databases	144
admetSAR	Toxicity-molecule associations	An online tool for the prediction of chemical ADMET properties	96, 145
BindingDB	Toxicity-molecule associations	A database of measured binding affinities, interactions of protein drug targets and small	146
CDT (Comparative Toxicogenomics Database)	Toxicity-molecule associations	Literature-based, manually curated associations between chemicals, gene products, phenotypes, diseases, and environmental exposures	147
ChEMBLdb	Toxicity-molecule associations	An EMBL manually curated chemical database with bioactivity data	5
ChemProt	Toxicity-molecule associations	A compilation of chemical-protein-disease annotation resources for studying systems pharmacology of a small molecule from molecular to clinical levels	148, 149
DSSTox	Toxicity-molecule associations	A subset of ACToR related to toxicity	150
eChemPortal	Toxicity-molecule associations	An aggregator of chemical hazard and risk information	151
PKKB (Pharmaco Kinetics Knowledge Base)	Toxicity-molecule associations	High-quality data for experimental ADMET properties	152
PubChem	Toxicity-molecule associations	An aggregator of chemical and physical properties, biological activities, safety and toxicity information, patents, literature citations	153
SuperToxic	Toxicity-molecule associations	Compounds and toxicity information	154

Table 3 An overview of toxicity, chemical and multiomics databases useful in the computational evaluation of safety—cont'd

Database name	Data type	Description	Source publication DOI
T3DB (Toxic Exposome Database)	Toxicity-molecule associations	Toxins data combined with target information	[155]
Tox21 (Toxicology in the 21st century)	Toxicity-molecule associations	Toxicity data for commercial chemicals, pesticides, food additives/contaminants, and medical compounds	[156]
ToxBank Data Warehouse	Toxicity-molecule associations	An aggregator of data for systemic toxicity	[157]
ToxCast Database (invitroDB)	Toxicity-molecule associations	HTS assay target information, study design information and quality	[158]
TOXNET	Toxicity-molecule associations	An aggregator of several toxicity databases, Integrated into PubMed in 2019	[159]
TTD (Therapeutic Targets Database)	Toxicity-molecule associations	Protein and nucleic acid targets, diseases, pathways	[160]
ECOTOX (Ecotoxicology Database)	Toxicity-molecule associations, adverse events	Adverse effects of single chemical stressors related to aquatoxicity	[161]
DrugBank	Toxicity-molecule associations, biological activity	A bioinformatics and cheminformatics resource on drug targets and properties of drugs	[162]
STITCH	Toxicity-molecule associations, pathways	Metabolic pathways, binding experiments, crystal structures, and drug-target relationships	[163]
Connectivity Map	Transcriptomics	Human cancer cell lines treated with various perturbants, Affymetrix GeneChip Human Genome	[164]
Drug Matrix	Transcriptomics	Rat Liver, kidney, heart and thigh muscle from Affymetrix GeneChip Rat Genome	[165]
LINCS L1000	Transcriptomics	Microscopy data, transcripts from L1000 database	[166]
Open TG-GATEs	Transcriptomics	Histopathology and clinical chemistry rat's liver, kidneys, hear and thigh muscle data, Affymetrix GeneChip Rat Genome	[167]
GEO (Gene Expression Omnibus)	Functional genomics data	Contains array and sequence-based data	[168]
ArrayExpress	Functional genomics data	Experimental data from high-throughput functional genomic tests	[169]
UniProt KnowledgeBase	Protein sequences and functional information	Database is split into two sections including UniProtKB/Swiss-Prot and UniProtKB/TrEMBL which respectively reflect whether the data are manually annotated and reviewed or not	[170–172]
Protein Databank	Protein information	Contains information about the "3D shapes of proteins, nucleic acids and complex assemblies"	[173, 174]

Continued

Table 3 An overview of toxicity, chemical and multiomics databases useful in the computational evaluation of safety—cont'd

Database name	Data type	Description	Source publication DOI
PRIDE	Proteomics	Repository of MS derived proteomics data	175–177
ProteomeDB	Proteomics	Aim to aid in the identification of the proteome	178
GnomAD (Genome Aggregation Database)	Sequencing data	Exome and genome sequencing data that has been combined from large-scale sequencing projects	179
WITHDRAWN	Withdrawn drugs	Contains withdrawn and discontinued drugs	180
DISGeNET	Target-disease information	Target-disease relationships	181
Open Targets	Target-disease information	Target-disease relationships	182
Clinical Pharmacology and British Pharmacology Society Guide to Pharmacology Database	Target and ligand information	Resource on targets and ligands	183
SuperTarget	Target-drug information	Target-drug information	184
GOSTAR	Target compound database	Manually curated target-compound database from literature and patents	185
SureChembl	Patent data	Open-source patent data	186

food additives, contaminants, and medical compounds? ToxBank Data Warehouse[157] stores systemic pharmacology information and additionally integrates into models predicting repeated-dose toxicity. PubChem[153] and DrugBank[162] are not purely toxicology databases; however, they collect bioactivity and biomolecular interactions data as well as clinical and patent information, respectively. ChEMBL[188] and CTD[147] databases contain manually curated data on chemical molecule and gene or protein interactions, chemical molecule and disease as well as gene and disease relationships. There exist various online public resources devoted to drug side effects: SIDER,[120] OFF-SIDES,[189] and CEBS.[66] These data are integrated with pathway-focused sites, for example, KEGG,[190] PharmGKB,[138] and Reactome,[141] which are curated and peer-reviewed pathway databases. The following table contains the main ones, however, it is not exhaustive.

A large number of molecular-omics data is present in the public domain and allow for reusing and exchange data from between experiments. High-dimensional and noisy biological signals used in, for example, differential gene expression, gene co-expression networks, compound protein-protein interaction networks, signature matching and organ toxicity analysis, often require a standardized ontology as well as manual data curation before they can be used to train a model.[18] However, the following public databases offer relatively high-quality data. DrugMatrix[191] contains in vivo rat liver, kidney, heart and thigh muscle from Affymetrix GeneChip Rat Genome 230 2.0 Array GE Codelink and Open TG-GATEs[167] contain rat liver and kidney data. The latter also contains human and rat in vitro hepatocytes histopathology, blood chemistry and clinical chemistry data. Toxicity data for five human cancer cell lines derived from the

Affymetrix GeneChip Human Genome U133A Array are stored in the Connectivity Map.[164] Microscopy images of up to 77 cell lines treated with various chemical compounds and gene expression data can be found in the Library of Integrated Network-based signatures L1000 (LINCS dataset).[164]

Many of the resources above have multiple applications. A wide variety of resources are available for proteomic studies from the EBI including UniProt KnowledgeBase (UniProtKB) and PRIDE.[192] Uniprot provides freely accessible resources of protein data such as protein sequences and functional information. UniProtKB is included in these resources.[170–172] It is split into two sections namely, the manually annotated and reviewed section known as UniProtKB/Swiss-Prot. The second section, UniProtKB/TrEMBL refers to the computationally annotated and nonreviewed section of the data. Owing to be computationally annotated, EBI states that there is high annotation coverage of the proteome.[172] These data can be used to find evidence for protein function or subcellular location.[172] Finally, PRIDE incudes protein and peptide identifications (such as details of posttranslational modifications) alongside evidence from mass spectrometry.[175–177]

This growth in the number of data repositories and databases has been fueled by the large amount of proteomic data generated.[175] The Protein DataBank is concerned with structural protein information such as the 3D shape if the protein and is maintained by the RCSB.[173, 174]

To deal with this, The HUPO Proteomics Standards Initiative,[193] or HUPO-PSI for short, was developed to ensure the universal adoption of stable data formats that has resulted in aggregation of proteomic data.[175] The HUPO-PSI's about section states that these standards were developed "to facilitate data comparison, exchange and verification".[193] However, it does not deal with the quality of data and the issues that brings.

Other key resources include WITHDRAWN[180] of which contains information about withdrawn and discontinued drug, DISGeNET[181] and Open Targets[182] for target-disease relationships. The Clinical Pharmacology and British Pharmacology Society Guide to Pharmacology Database[183] also contains target information and information about a variety of ligands. GOSTAR[185] and SureChEMBL[194] both contain information on target-compound information from patents with GOSTAR also containing that information available from literature.

Drug safety databases

To monitor, systematically review, and enable data-driven decisions on drug safety, WHO Collaborating Monitoring Centre in Uppsala[195] and National Competent Authorities (NCAs) maintain several databases dedicated to safety signals collection (Fouretier et al., 2016). The largest and the oldest ones are WHO VigiBase (1968), EU Eudravigilance, FDA FAERS, and VAERS, but most countries have established their own databases supported by Geographical Information Systems (GIS).[196] Geolocalization allows using these databases to detect both global and local trends. Table 4 presents an overview of the largest publicly accessible databases related both to postmarketing surveillance, unsolicited reporting, and solicited reporting from clinical trials.

Majority of the unsolicited resources is unstructured, fragmentary, unstandardized and suffering from the presence of confounders. Although WHO, ICH, and NCAs have taken a considerable standardization effort, the quality of ADR, reports vary across countries.[197] Additional curation of the data in indispensable as databases contains duplicates, missing data points, and it has high sample variance.

Furthermore, cases when patients were administered drugs as intended and no ADR occurred, are naturally not reported. From the perspective of data analysis and developing machine learning models lack their presence in a dataset results in class imbalance, survivorship bias and high numbers of false positives in predictions.[198] Thus, one cannot calculate the rate of occurrence for the whole population

Table 4 An overview of publicly available resources on adverse drug reactions maintained by WHO, competent national authorities (NCAs), WHO and scientific institutions.

Database name	Organization	Reporters	Content
VigiBase	Uppsala Monitoring Centre, WHO	MAHs, HCPs, consumers or any regional center	Twenty million ICSRs from 125 member states and 28 associate members on medicinal product-related suspected adverse events; postmarketing spontaneous severe and nonserious cases ICSRs, sometimes clinical trials, literature
			Related tools: <u>WHO-ART</u>, <u>MedDRA</u>, <u>WHO ICD</u>, <u>WHODrug</u>, VigiSearch VigiLyz, VigiMin, ICD, VigiAccess
Eudravigilance	EMA	MAH, NCAs, EEA sponsors of clinical trials	14.5 million ICSRs; Clinical Trial Module (EVCTM); Post-Authorization Module (EVPM) Related tools: EVDAS, Addreports.eu, MedDRA
FAERS	FDA	MAH, HCPs, consumers	Over 19 million postmarketing surveillance adverse event reports related to medications. Causality analysis not required for submission
			Related tools: Sentinel Initiative, FAERS Public Dashboard, AERSMIne, Open Vigil
VAERS	FDA, CDC	MAH, HCPs, consumers	700,000 postmarketing surveillance adverse event reports related to vaccines including <u>unverified</u> reports, misattribution, <u>underreporting</u>, and inconsistent data quality, Related tools: empirical Bayes and data mining tools built-in
Adverse Event Reports for Animal Drugs and Devices	FDA, Center for Veterinary Medicine	Veterinary professionals, consumers	Voluntary AE submission, database contain postmarketing surveillance adverse event reports related to animal drugs including drugs, supplements, vitamins
Yellow Card	MHRA, Commission on Human Medicines	HCP, hospital and community pharmacists, members of the public	ICSRs on medicines, OTCs, vaccines, herbal preparations and unlicensed medicines, e-cigarettes, counterfeit drug reports, defective medicinal products. Interactive Drug Analysis Profile IDAPs) can be downloaded for each drug
			Related tools: Android app, built-in analytics
<u>Canada Vigilance Adverse Reaction</u>	Health Canada	HCP, MAHs	Clinical and postmarket surveillance SAE reports prescription and nonprescription medications; natural health products; biologics (includes biotechnology products, vaccines, fractionated blood products, human blood and blood components, as well as human cells, tissues and organs); radiopharmaceuticals; and disinfectants and sanitizers with disinfectant claims

Table 4 An overview of publicly available resources on adverse drug reactions maintained by WHO, competent national authorities (NCAs), WHO and scientific institutions—cont'd

Database name	Organization	Reporters	Content
DAEN—medicines	Australian Department of Health TGA	HCPs, MAHs, members of public, therapeutic goods industry	ADR reports on adverse events related to medicines and vaccines used in Australia
LAREB	Netherlands Pharmacovigilance Centre Lareb	HCPs, community pharmacists, members of the public	Downloadable reports with preprocessed data and literature related to ADR reporting in the Netherlands
PROTECT	EMA and partners	None	PROTECT ADR database is a downloadable Excel file listing of all MedDRA preferred terms or low-level terms adverse drug reactions (ADRs), text mined Summary of Product Characteristics (SPC) of medicinal products authorized in the EU, automated mapping of ADR terms, fuzzy text matching, expert review
SIDER	EMBL	None	Postmarket surveillance, extracted from public documents and package leaflets and Summary of Product Characteristics include side effect frequency, drug and side effect classifications, links to drug target relations, top-down database

Database contain both solicited and unsolicited data.

basing on spontaneous resources only. Otherwise, the risk of false-positive reporting for certain medicines may be artificially elevated.[199] Finally, statistical significance in a model does not always mean clinical relevance. A majority of patients might be likely to respond better to certain medications statistically. However, some atypical side effects may occur that lower the quality of life of a small number of patients and hence outweigh the benefits.

Finally, longitudinal patient medical history may not always be easily retrieved, and thus it is challenging to verify reported information as well as establish causality understood as in ICH-E2A guideline.[200] Reports submitted to SRS databases are subjective and often contain inconsistent records when compared with original medical documentation.

Key public data-resources for precision medicine

This section describes many completed and ongoing efforts to generate large-scale datasets from cell lines, patients and healthy volunteers. These datasets are a necessary asset that will be used to generate novel AI/ML-based models to guide precision medicine.

Resources for enabling the development of computational models in oncology

Beginning with the characterization of NCI60 cell lines for predicting drug sensitivity, there has been enormous number of large-scale studies to generate genomics, proteomics, functional genomics, or drug sensitivity datasets that can be utilized to predict cancer cells sensitivity to a targeted agent (Table 5). Among them Cancer Cell Line Encyclopedia (CCLE) project by the BROAD Institute is one of the most

Table 5 Key public data resources in the oncology for enabling the development of computational models and knowledge discovery.

Resource	Biological material	Omics readout (#cell lines)	Weblink	Last update	Reference
NCI-60	60 cancer cell lines	Drug sensitivity (> 100,000 compounds), SNV, CNV, RNAseq, DNA methylation	https://discover.nci.nih.gov/cellminer/		201
McDermott et al	500 cell lines	Drug sensitivity (14 kinase inhibitors)		2007	202
GSK	311 cell lines	Drug sensitivity (19 compounds)		2010	
CCLE	~1000 cancer cell lines	WES (326), WGS (329) RNAseq (1019), Methylation (RRBS, 843), RPPA (899), microRNA profiling (954), global histone modifications (897), drug sensitivity (24 compounds, 479) and metabolic profiling for 225 metabolites (928)	https://portals.broadinstitute.org/ccle	May 2019	203–205
GDSC	1001 cell lines, 453 compounds	Transcription (microarray) Methylation (Infinium HumanMethylation450 BeadChip arrays) Drug sensitivity	https://www.cancerrxgene.org/	July 2019	206, 207
CTRP	481 Compounds across 860 cancer cell lines	Drug sensitivity	http://portals.broadinstitute.org/ctrp.v2/		208
Genentech	675 human cancer cell lines	RNA-seq and SNP array analysis	https://www.nature.com/articles/nbt.3080	2015	209
Connectivity Map	Nine cell lines	1,319,138 L1000 profiles from 42,080 perturbagens (19,811 small molecule compounds, 18,493 shRNAs, 3,462 cDNAs, and 314 biologics), corresponding to 25,200 biological entities (19,811 compounds, shRNA and/or cDNA against 5075 genes, and 314 biologics) for a total of 473,647 signatures	http://www.lincsproject.org/ https://clue.io/cmap	2017	164, 210
MCLP	Cell lines	RPPA	https://tcpaportal.org/mclp/#/		211
Cheng et a.l	15 HPV and 11 HPV + HNSCC cell lines	Whole exome sequencing and RNA-seq		Oct 2018	212

Name	Description	Data types	URL	Date	Reference
TCGA	> 11,000 primary cancer and matched normal samples spanning 33 cancer types	Genomic, methylation (Infinium HumanMethylation450 BeadChip arrays), transcriptomic and proteomics (RPPA)	https://www.cancer.gov/about-nci/organization/ccg/research/structural-genomics/tcga		213
ICGC	86 cancer projects across 22 sites, ~ 25,000 patients	Genome sequencing	https://icgc.org/		214
TCPA	8167 tumor samples	RPPA	https://tcpaportal.org/tcpa/		215, 216
CPTAC	Cancer patients, primary tumor and the adjacent tissue. 45 total studies for 10 tissue types, resulting in a total of 2696 samples	Phosphoproteomics, proteomics, transcriptomics, SCNA, mutations	https://cptac-data-portal.georgetown.edu/	May, 2019	
Dong-Gi mun et al.	Paired tumor and adjacent normal tissues, as well as blood samples, from 80 patients with EOGCs under 45 years of age	Exome sequencing, RNA-seq, global proteome, 26 ummariz-proteome and glycoproteome			217

Abbreviations: NCI60, National Cancer Institute collection of 60 cell lines; CCLE, Cancer Cell Line Encyclopedia; GDSC, genomics of drug sensitivity in cancer; COSMIC, Catalogue of Somatic Mutations in Cancer; TCGA, the cancer genome atlas; MCLP, MD Anderson Cell Lines Project; CPTAC, Clinical Proteomic Tumor Analysis Consortium; RPPA, reverse phase protein array.

comprehensive. In its first round in 2012, CCLE included gene expression, copy number and mutation profile data for 947 cell lines, and pharmacological profile for 24 anticancer drugs in 479 of the cell lines. In 2019, project extended to include data on RNA sequencing (RNAseq; 1019 cell lines), whole-exome sequencing (WES; 326 cell lines), whole-genome sequencing (WGS; 329 cell lines), reverse-phase protein, array (RPPA; 899 cell lines), reduced representation bisulfite sequencing (RRBS; 843 cell lines), microRNA expression profiling (954 cell lines), and global histone modification profiling (897 cell lines) for CCLE cell lines. In addition, abundance of 225 metabolites was measured for 928 cell lines. An additional project from Genentech profiles gene expression, mutations, gene fusions and expression of nonhuman sequences in 675 human cancer cell lines. MLCP project characterized the proteome of the human cancer cell lines. Two resources that include the drug sensitivity data are Genomics of Drug Sensitivity (GDSC) from the Sanger Institute and the Cancer Therapeutics Response Portal (CTRP) from the BROAD institute.[208, 218, 219] By generating expression data (and making it public) that indicates how cells respond to various genetic and environmental stressors, the LINCS project from the NIH helps to gain a more detailed understanding of cell pathways.[164, 220]

Although cancer cell line data is crucial for many insights and some of the large-scale experiments such as CRISPR functional genomics screens can only be done in cell lines, primary data on patients is vital to understand and modeling of human disease.

Several large consortiums/projects took this challenge of characterizing tumor samples in various genomics, epigenomics, and proteomics aspects. Prominent among them is the cancer genome atlas (TCGA) which has sequenced and characterized more than 11,000 patient samples in 33 cancer types.[213] International cancer genome consortium (ICGC) is another consortium of several national projects to sequence the cancer samples.[214] The Cancer Proteome Atlas (TCPA) performed RPPA analysis on more than 800 samples and Clinical Proteomic Tumor Analysis Consortium (CPTAC) launched in 2011 by NCI pioneered the integrated proteogenomic analysis of colorectal, breast, and ovarian cancer.[215] These efforts revealed new insights into these cancer types, such as identification of proteomic-centric subtypes, prioritization of driver mutations, and understanding cancer-relevant pathways through posttranslational modifications. The CPTAC has produced proteomics data sets for tumor samples previously analyzed by TCGA program.

Key genomic/epigenomic resources for therapeutic areas other than oncology
There are multitudes of ongoing projects outside oncology domain for large-scale data generation. Some of them are summarized in Table 6.

Resources for accessing metadata and analysis tools
Accessing and analyzing raw sequencing data can be quite cumbersome for most biologists. Resources that present analyzed or easy to grasp data on genetic alterations as well as pathway level analysis are very helpful. Several such resources that can be used directly for hypothesis generation/verification exist. Some of these are listed in Table 7.

Fig. 2 recapitulates progress on data generation frontier that include drug screening in cell lines, functional genomics (RNAi and CRISPR) screens, detailed characterization of cell lines and finally exome or whole genome sequencing of patients and healthy volunteers. Some of these data were already used employing AI/ML-based approaches to identify novel synthetic lethality pairs, predict drug IC50, or even clinical outcome prediction.[207, 242, 243] By designing an AI algorithm to analyze CT scan images, researchers have created a radiomic signature that defines the level of lymphocyte infiltration

Table 6 Large genomic datasets for nononcology applications.

Resource	Biological material	Omics readout	Weblink	Last update	Reference
GWAS Catalog	Human primary tissue/cells	DNA	https://www.ebi.ac.uk/gwas/	Every week	221
Expression Atlas	Multiple species and tissues	RNA	https://www.ebi.ac.uk/gxa/home	August 2020	222
ClinVar	Human primary tissue/cells	DNA	https://www.ncbi.nlm.nih.gov/clinvar/		223
OMIM	Human primary tissue/cells	DNA	https://www.omim.org/	Everyday	
ENCODE	Cell lines, primary cells, cell free samples, tissue	Epigenetic profiling	https://www.encodeproject.org/	August2019	224
Human Cell Atlas	Human primary tissue/cells	Single cell sequencing	https://www.humancellatlas.org/		225, 226
Single Cell Portal	Collection of studies on single cells (288 so far)	Single cell sequencing	https://portals.broadinstitute.org/single_cell	August 2020	
GTEx Portal	54 nondiseased tissue sites across nearly 1000 individuals	Primarily for molecular assays including WGS, WES, and RNA-Seq. Remaining samples are available from the GTEx Biobank. The GTEx Portal provides open access to data including gene expression, QTLs, and histology images	https://gtexportal.org/home/	August 2019	227
PsychENCODE	Human brain samples and organoids	DNA, RNA and epigenetics profiling	http://www.psychencode.org/	December 2018	228
NIAGADS	61 datasets, >59,000 samples	Genotypic data for the study of genetics of late-onset Alzheimer's disease	https://www.niagads.org/	February 2019	
ADSP	The Alzheimer's Disease Sequencing Project	DNA	https://www.niagads.org/adsp/	November 2018	229
ADNI	Alzheimer's Disease Neuroimaging Initiative, >800 subjects	Clinical, genetic, MRI image, PET image, Biospecimen	http://adni.loni.usc.edu/		230
AutDB	resource for exploring the impact of genetic variations associated with autism spectrum disorders (ASD)	Human Gene, which annotates all ASD-linked genes and their variants; Animal Model, which catalogs behavioral, anatomical and physiological data from rodent models of ASD; Protein Interaction (PIN), which builds interactomes from direct relationships of protein products of ASD genes; and Copy Number Variant (CNV), which catalogs deletions and duplications of chromosomal loci identified in ASD	http://autism.mindspec.org/autdb	Quarterly	231
NDAR	National Database for Autism Research	Genetics, behavioral data	https://nda.nih.gov/	November 2018	232
NIMH Dara Archive (NDA)		NDA is a collection of data repositories including the Research Domain Criteria Database (RdoCdb). The National database for Clinical trials related to mental illness (NDCT) and the NIH pediatric MRI Repository (PedsMRI)	https://nda.nih.gov/	August 2019	

Continued

Table 7 Resources for accessing metadata and analysis tools.

Database	Content	Omics readout	Weblink	Last update	Reference
COSMIC	Tumor samples and >1000 cell lines	Expert curated database of somatic mutations	https://cancer.sanger.ac.uk/cosmic	V92, August 2020	233
Cancer DepMap	Data from CCLE, TCGA, GDSC, RNAi and CRISPR screens	Genomics, proteomics, RNAi/CRISPR screens and drug sensitivity	https://depmap.org/portal/	Every 90 days	
Cell Model Passports	Cell lines and organoids	Mutations, expression, CNV, methylation, fusions, drug response, CRISPR score	https://cellmodelpassports.sanger.ac.uk/passports		234
cBioPortal	The portal hosts a total of 263 cancer studies including CCLE and TCGA data	Mutations, CNV, RNAseq, RPPA	http://www.cbioportal.org/		235
TCGA-CDR	Clinical data resource for high quality survival outcome analytics	Survival data	See reference		236
mSigDB	Annotated gene sets for use with GSEA	Gene sets	http://software.broadinstitute.org/gsea/msigdb/index.jsp		237
Enricher	Annotated gene sets for use with GSEA	Gene sets	https://amp.pharm.mssm.edu/Enrichr/		238
GeneMANIA	Hypothesis generation regarding function of a gene	Multiple omics-based data	https://genemania.org/		239
L1000CDS2	LINCS L1000 characteristic direction signature search engine	Finds consensus L1000 small molecule signatures that match user input signatures	https://amp.pharm.mssm.edu/L1000CDS2/#/index		240
Geneshot	Ranking genes based on text mining	Literature, expression data	https://amp.pharm.mssm.edu/geneshot/		241

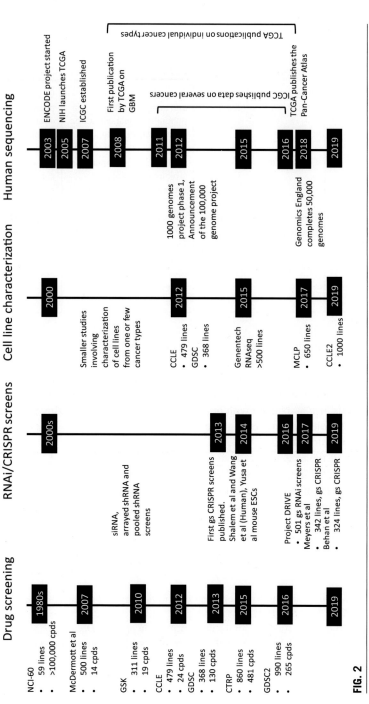

FIG. 2

Historic resources for clinical trials.

Table 8 Selected examples of historical data sets, potential methods to analyze them and their respective applications in biopharma.

Examples	Data type	Data and methods	Applications in biopharma
National Biomedical Imaging Archive (NBIA); GenomeRNAi	Imaging data	Image preprocessing and analyses, data annotation, data extraction, segmentation, deep learning, computer vision	Clinical or cellular phenotyping, patient stratification and disease subclassification
TCGA; dbGAP	Genomic data	Variant calling, annotation, structural variants differential expression	Diagnosis, disease subtyping, therapeutic matching, clinical trial matching
UK Biobank; BioMe Biobank	Biobanks and electronic health records	Clinical trajectory estimation, biomarker-based modeling	Predict risk of diseases, real world evidence modeling
ClinicalTrials.gov; AACT Database	Clinical trials databases	Clinical trial protocols, performance metrics, patient population summaries	Predictive modeling of clinical trial metrics

of a tumor and provides a predictive score for the efficacy of immunotherapy in the patient.[244] Gene expression profile analysis of needle biopsy specimens was performed from the livers of 216 patients with hepatitis C-related early-stage cirrhosis who were prospectively followed up for a median of 10 years. Evaluation of 186-gene signature used to predict outcomes of patients with hepatocellular carcinoma showed this signature is also associated with outcomes of patients with hepatitis C-related early-stage cirrhosis.[245] Recently, whole-genome sequencing was used to accurately predict profiles of susceptibility to first-line antituberculosis drugs.[246]

Table 8 lists some of the examples of historical data sets, potential methods to analyze them, and their respective applications in biopharma. The recent innovation in the field of AI has been enabled primarily by the confluence of rapid advances in affordable computing power in the form of cloud computing, infrastructure to process and manage large-scale data sets and architectures and methodologies such as neural networks.

References

1. Zitnik M, et al. Machine learning for integrating data in biology and medicine: Principles, practice, and opportunities. *Inf Fus* 2019;**50**:71–91.
2. Jacobsen A, et al. A generic workflow for the data fairification process. *Data Intell* 2020;**2**:56–65.
3. FAIRification process—GO FAIR. Available at: https://www.go-fair.org/fair-principles/fairification-process/ [Accessed 11 August 2020].
4. ChEMBL. Available at: https://www.ebi.ac.uk/chembl/ [Accessed 5 September 2018].
5. Gaulton A, et al. ChEMBL: a large-scale bioactivity database for drug discovery. *Nucleic Acids Res* 2012;**40**:D1100–7.

6. ChEMBL data questions—ChEMBL interface documentation. Available at: https://chembl.gitbook.io/chembl-interface-documentation/frequently-asked-questions/chembl-data-questions [Accessed 11 August 2020].

7. The evolving role of clinical trial data sharing. Available at: https://pharmaphorum.com/views-and-analysis/clinical-trial-data-sharing/ [Accessed 4 September 2020].

8. Miller J, Ross JS, Wilenzick M, Mello MM. Sharing of clinical trial data and results reporting practices among large pharmaceutical companies: cross sectional descriptive study and pilot of a tool to improve company practices. *BMJ* 2019;**366**:l4127.

9. MELLODDY. Available at: https://www.melloddy.eu/ [Accessed 4 September 2020].

10. Rouse M, Botelho B, Bigelow S. *Big data. Search Data Management*; 2020. Available at: https://searchdatamanagement.techtarget.com/definition/big-data. [Accessed 7 September 2020].

11. Ishwarappa, Anuradha J. A brief introduction on big data 5Vs characteristics and hadoop technology. *Procedia Comput Sci* 2015;**48**:319–24.

12. Horgan RP, Kenny LC. 'Omic' technologies: genomics, transcriptomics, proteomics and metabolomics. *Obstet Gynaecol* 2011;**13**:189–95.

13. Paananen J, Fortino V. An omics perspective on drug target discovery platforms. *Brief Bioinform* 2019. bbx122.

14. Simon R, Roychowdhury S. Implementing personalized cancer genomics in clinical trials. *Nat Rev Drug Discov* 2013;**12**:358–69.

15. A brief guide to genomics. Available at: https://www.genome.gov/about-genomics/fact-sheets/A-Brief-Guide-to-Genomics [Accessed 14 October 2019].

16. Libbrecht MW, Noble WS. Machine learning applications in genetics and genomics. *Nat Rev Genet* 2015;**16**:321–32.

17. Transcriptomics—Latest research and news | Nature. Available at: https://www.nature.com/subjects/transcriptomics [Accessed 14 July 2020].

18. Alexander-Dann B, et al. Developments in toxicogenomics: understanding and predicting compound-induced toxicity from gene expression data. *Mol Omics* 2018;**14**:218–36.

19. Wang Z, Gerstein M, Snyder M. RNA-Seq: a revolutionary tool for transcriptomics. *Nat Rev Genet* 2009;**10**:57–63.

20. Transcriptomics today: Microarrays, RNA-seq, and more | Science | AAAS. Available at: https://www.sciencemag.org/features/2015/07/transcriptomics-today-microarrays-rna-seq-and-more [Accessed 14th July 2020].

21. Akter S, et al. Machine learning classifiers for endometriosis using transcriptomics and methylomics data. *Front Genet* 2019;**10**:766.

22. Singh SP, et al. Machine learning based classification of cells into chronological stages using single-cell transcriptomics. *Sci Rep* 2018;**8**:17156.

23. Roessner U, Bowne J. What is metabolomics all about? *BioTechniques* 2009;**46**:363–5.

24. Dias-Audibert FL, et al. Combining machine learning and metabolomics to identify weight gain biomarkers. *Front Bioeng Biotechnol* 2020;**8**.

25. Sen P, et al. Deep learning meets metabolomics: a methodological perspective. *Brief Bioinform* 2020. https://doi.org/10.1093/bib/bbaa204.

26. Pradas I, et al. Lipidomics reveals a tissue-specific fingerprint. *Front Physiol* 2018;**9**:1165.

27. Yang K, Han X. Lipidomics: techniques, applications, and outcomes related to biomedical sciences. *Trends Biochem Sci* 2016;**41**:954–69.

28. Meikle PJ, Wong G, Barlow CK, Kingwell BA. Lipidomics: potential role in risk prediction and therapeutic monitoring for diabetes and cardiovascular disease. *Pharmacol Ther* 2014;**143**:12–23.

29. Fan S, et al. Systematic error removal using random forest for normalizing large-scale untargeted lipidomics data. *Anal Chem* 2019;**91**:3590–6.

30. What is proteomics? I EMBL-EBI Train online. Available at: https://www.ebi.ac.uk/training/online/course/proteomics-introduction-ebi-resources/what-proteomics [Accessed 8 October 2019].

31. Graves PR, Haystead TAJ. Molecular biologist's guide to proteomics. *Microbiol Mol Biol Rev* 2002;**66**:39–63.

32. Swan AL, Mobasheri A, Allaway D, Liddell S, Bacardit J. Application of machine learning to proteomics data: classification and biomarker identification in postgenomics biology. *Omi A J Integr Biol* 2013;**17**:595–610.

33. Gessulat S, et al. Prosit: proteome-wide prediction of peptide tandem mass spectra by deep learning. *Nat Methods* 2019;**16**:509–18.

34. Cao Y, Charisi A, Cheng L-C, Jiang T, Girke T. ChemmineR: a compound mining framework for R. *Bioinformatics* 2008;**24**:1733–4.

35. R Core Team. *R: A language and environment for statistical computing*. Vienna, Austria: R Foundation for Statistical Computing; 2020. https://www.R-project.org/.

36. Landrum G. *RDKit: open-source cheminformatics*. https://www.rdkit.org.

37. Berthold MR, Cebron N, Dill F, Gabriel TR, Kötter T, Meinl T, Ohl P, Sieb C, Thiel K, Wiswedel B. KNIME: The Konstanz Information Miner. In: *Studies in Classification, Data Analysis, and Knowledge Organization*. Springer; 2007.

38. What is the correct format for compounds in SDF or MOL files?—Progenesis SDF studio. Available at: http://www.nonlinear.com/progenesis/sdf-studio/v0.9/faq/sdf-file-format-guidance.aspx [Accessed 18 October 2019].

39. Heller SR, McNaught A, Pletnev I, Stein S, Tchekhovskoi D. InChI, the IUPAC international chemical identifier. *J Cheminform* 2015;**7**.

40. Heller S, McNaught A, Stein S, Tchekhovskoi D, Pletnev I. InChI—the worldwide chemical structure identifier standard. *J Cheminform* 2013;**5**.

41. O'Boyle NM. Towards a Universal SMILES representation—a standard method to generate canonical SMILES based on the InChI. *J Cheminform* 2012;**4**:22.

42. chem-bla-ics: InChIKey collision: the DIY copy/pastables. Available at: https://chem-bla-ics.blogspot.com/2011/09/inchikey-collision-diy-copypastables.html?_sm_au_=iHHRkrfFZLWsZNV6 [Accessed 16 September 2019].

43. An InChIkey collision is discovered and NOT based on stereochemistry ChemConnector blog. Available at: http://www.chemconnector.com/2011/09/01/an-inchikey-collision-is-discovered-and-not-based-on-stereochemistry/ [Accessed 16 September 2019].

44. Willighagen EL. *InChIKey collision: the DIY copy/pastables*; 2011.

45. Pletnev I, et al. InChIKey collision resistance: an experimental testing. *J Cheminform* 2012;**4**.

46. Weininger D. SMILES, a chemical language and information system. 1. Introduction to methodology and encoding rules. *J Chem Inf Comput Sci* 1988;**28**:31–6.

47. Weininger D, Weininger A, Weininger JL. SMILES. 2. algorithm for generation of unique SMILES notation. *J Chem Inf Comput Sci* 1989;**29**:97–101.

48. Weininger D. Smiles. 3. Depict. Graphical depiction of chemical structures. *J Chem Inf Comput Sci* 1990;**30**:237–43.

49. Daylight theory: SMARTS—a language for describing molecular patterns. Daylight Chemical Information Systems, Inc; 2012. Available at: http://www.daylight.com/dayhtml/doc/theory/theory.smarts.html [Accessed 8 September 2018].

50. Yang H, et al. Evaluation of different methods for identification of structural alerts using chemical ames mutagenicity data set as a benchmark. *Chem Res Toxicol* 2017;**30**:1355–64.

51. Cammarata A, Menon GK. Pattern recognition. Classification of therapeutic agents according to pharmacophores. *J Med Chem* 1976;**19**:739–48.

52. Wu Y, Wang G. Machine learning based toxicity prediction: from chemical structural description to transcriptome analysis. *Int J Mol Sci* 2018;**19**:2358.

53. Landrum G. *Fingerprints in the RDKit. RDKit UGM 2012: fingerprints in the RDKit*; 2012. Available at: https://www.rdkit.org/UGM/2012/Landrum_RDKit_UGM.Fingerprints.Final.pptx.pdf. [Accessed 16 September 2019].

54. Durant JL, Leland BA, Henry DR, Nourse JG. Reoptimization of MDL keys for use in drug discovery. *J Chem Inf Comput Sci* 2002;**42**:1273–80.

55. Rogers D, Hahn M. Extended-connectivity fingerprints. *J Chem Inf Model* 2010;**50**:742–54.

56. Morgan H, The L. Generation of a unique machine description for chemical structures—a technique developed at chemical abstracts service. *J Chem Doc* 1965;**5**:107–13.

57. Daylight theory: fingerprints. Available at: https://www.daylight.com/dayhtml/doc/theory/theory.finger.html [Accessed 16 September 2019].

58. Faulon JL, Visco DP, Pophale RS. The signature molecular descriptor. 1. Using extended valence sequences in QSAR and QSPR studies. *J Chem Inf Comput Sci* 2003;**43**:707–20.

59. Faulon JL, Churchwell CJ, Visco DP. The signature molecular descriptor. 2. Enumerating molecules from their extended valence sequences. *J Chem Inf Comput Sci* 2003;**43**:721–34.

60. GitHub—reymond-group/mhfp: Molecular MHFP fingerprints for cheminformatics applications. Available at: https://github.com/reymond-group/mhfp [Accessed 9 October 2020].

61. Probst D, Reymond JL. A probabilistic molecular fingerprint for big data settings. *J Cheminform* 2018;**10**.

62. Bender A, et al. 'Bayes affinity fingerprints' Improve retrieval rates in virtual screening and define orthogonal bioactivity space: when are multitarget drugs a feasible concept? *J Chem Inf Model* 2006;**46**:2445–56.

63. Wang Y, et al. PubChem BioAssay: 2017 update. *Nucleic Acids Res* 2017;**45**:D955–63.

64. Klekota J, Roth FP. Chemical substructures that enrich for biological activity. *Bioinformatics* 2008;**24**:2518–25.

65. Banerjee P, Siramshetty VB, Drwal MN, Preissner R. Computational methods for prediction of in vitro effects of new chemical structures. *J Cheminform* 2016;**8**.

66. Fan D, et al. In silico prediction of chemical genotoxicity using machine learning methods and structural alerts. *Toxicol Res (Camb)* 2018;**7**:211–20.

67. O'Boyle NM, Sayle RA. Comparing structural fingerprints using a literature-based similarity benchmark. *J Cheminform* 2016;**8**.

68. How to choose bits and radius during circular fingerprint calculation in RDKit? Available at: https://www.researchgate.net/post/How_to_choose_bits_and_radius_during_circular_fingerprint_calculation_in_RDKit [Accessed 18 September 2019].

69. Dalke A. *No title*; 2019. Available at: http://www.dalkescientific.com/writings/diary/archive/2014/10/17/maccs_key_44.html. [Accessed 18 September 2019].

70. Fingerprint generation—Toolkits—Python. Available at: https://docs.eyesopen.com/toolkits/python/graphsimtk/fingerprint.html#section-fingerprint-path [Accessed 5 February 2020].

71. Alvarsson J, et al. Ligand-based target prediction with signature fingerprints. *J Chem Inf Model* 2014;**54**:2647–53.

72. Dhawan A, Kwon S. In vitro toxicology. *Int J Toxicol* 2017. https://doi.org/10.1080/10915810305079.

73. Yang H, Sun L, Li W, Liu G, Tang Y. Identification of nontoxic substructures: a new strategy to avoid potential toxicity risk. *Toxicol Sci* 2018;**165**:396–407.

74. Yap C, PaDEL-descriptor W. An open source software to calculate molecular descriptors and fingerprints. *J Comput Chem* 2011;**32**:1466–74.

75. O'Boyle NM, et al. Open Babel: An Open chemical toolbox. *J Cheminform* 2011;**3**.

76. Steinbeck C, et al. The Chemistry Development Kit (CDK): an open-source Java library for chemo- and bioinformatics. *J Chem Inf Comput Sci* 2003;**43**:493–500.

77. Tetko IV, et al. Virtual computational chemistry laboratory—design and description. *J Comput Aided Mol Des* 2005;**19**:453–63.

78. Hewitt M, Enoch SJ, Madden JC, Przybylak KR, Cronin MTD. Hepatotoxicity: a scheme for generating chemical categories for read-across, structural alerts and insights into mechanism(s) of action. *Crit Rev Toxicol* 2013;**43**:537–58.

79. Borgelt C, Berthold MR. Mining molecular fragments: finding relevant substructures of molecules. In: *2002 IEEE International Conference on Data Mining. ICDM*; 2002. p. 51–8. IEEE Comput. Soc, 2002. https://doi.org/10.1109/ICDM.2002.1183885.

80. Venkatapathy R, Wang NCY. Developmental toxicity prediction. In: Reisfeld B, Mayeno AN, editors. *Computational toxicology*. vol. 930. Humana Press; 2013. p. 305–40.

81. Raies AB, Bajic VB. In silico toxicology: computational methods for the prediction of chemical toxicity. *Wiley Interdiscip Rev Comput Mol Sci* 2016;**6**:147–72.

82. Gómez-Jiménez G, et al. The OECD principles for (Q)SAR models in the context of knowledge discovery in databases (KDD). *Adv Protein Chem Struct Biol* 2018;**113**:85–117.

83. Marchant CA, Briggs KA, Long A. In silico tools for sharing data and knowledge on toxicity and metabolism: derek for windows, meteor, and vitic. *Toxicol Mech Methods* 2008;**18**:177–87.

84. Lagorce D, Sperandio O, Baell JB, Miteva MA, Villoutreix BO. FAF-Drugs3: a web server for compound property calculation and chemical library design. *Nucleic Acids Res* 2015;**43**:W200–7.

85. Sanz F, et al. Integrative modeling strategies for predicting drug toxicities at the eTOX project. *Mol Inform* 2015;**34**.

86. Sushko I, Salmina E, Potemkin VA, Poda G, Tetko IV. ToxAlerts: a web server of structural alerts for toxic chemicals and compounds with potential adverse reactions. *J Chem Inf Model* 2012;**52**:2310–6.

87. CombiGlide 2.5 User Manual. Library; 2009.

88. Friesner RA, et al. Glide: a new approach for rapid, accurate docking and scoring. 1. Method and assessment of docking accuracy. *J Med Chem* 2004;**47**:1739–49.

89. Halgren TA, et al. Glide: a new approach for rapid, accurate docking and scoring. 2. Enrichment factors in database screening. *J Med Chem* 2004;**47**:1750–9.

90. Amberg A, et al. Principles and procedures for handling out-of-domain and indeterminate results as part of ICH M7 recommended (Q)SAR analyses. *Regul Toxicol Pharmacol* 2019;**102**:53–64.

91. Benfenati E, Manganaro A, Gini G. VEGA-QSAR: AI inside a platform for predictive toxicology. In: *CEUR workshop proceedings, vol. 1107*. CEUR-WS; 2013. p. 21–8.

92. Capuzzi SJ, et al. Chembench: a publicly accessible, integrated cheminformatics portal. *J Chem Inf Model* 2017;**57**:105–8.

93. Dong J, et al. ChemSAR: an online pipelining platform for molecular SAR modeling. *J Cheminform* 2017;**9**.

94. Patlewicz G, Jeliazkova N, Safford RJ, Worth AP, Aleksiev B. An evaluation of the implementation of the Cramer classification scheme in the Toxtree software. *SAR QSAR Environ Res* 2008;**19**:495–524.

95. Maunz A, et al. Lazar: a modular predictive toxicology framework. *Front Pharmacol* 2013;**4**.

96. Cheng F, et al. AdmetSAR: a comprehensive source and free tool for assessment of chemical ADMET properties. *J Chem Inf Model* 2012;**52**:3099–105.

97. Kemmish H, Fasnacht M, Yan L. Fully automated antibody structure prediction using BIOVIA tools: validation study. *PLoS One* 2017;**12**, e0177923.

98. Vellay SGP, Latimer NEM, Paillard G. Interactive text mining with Pipeline Pilot: a bibliographic web-based tool for PubMed. *Infect Disord Drug Targets* 2009;**9**:366–74.

99. Hevener KE. Computational toxicology methods in chemical library design and high-throughput screening hit validation. *Methods Mol Biol* 2018;**1800**:275–85.

100. Hvidsten TR, Kryshtafovych A, Fidelis K. Local descriptors of protein structure: a systematic analysis of the sequence-structure relationship in proteins using short- and long-range interactions. *Proteins Struct Funct Bioinform* 2009;**75**:870–84.

101. Ruiz-Blanco YB, Paz W, Green J, Marrero-Ponce Y. ProtDCal: a program to compute general-purpose-numerical descriptors for sequences and 3D-structures of proteins. *BMC Bioinform* 2015;**16**.
102. Wang X, Wu Y, Wang R, Wei Y, Gui Y. A novel matrix of sequence descriptors for predicting protein-protein interactions from amino acid sequences. *PLoS One* 2019;**14**, e0217312.
103. Segaran T. *Programming collective intelligence: building smart Web 2.0 applications.* Sebastopol, CA: O'Reilly Media; 2007.
104. Discussion of SImilarity metrics—Jaccard/Tanimoto coefficient. Available at: http://mines.humanoriented. com/classes/2010/fall/csci568/portfolio_exports/sphilip/tani.html [Accessed 19 September 2019].
105. Tversky A. Features of similarity. *Psychol Rev* 1977;**84**:327–52.
106. Similarity measures—Toolkits—Python. Available at: https://docs.eyesopen.com/toolkits/python/ graphsimtk/measure.html [Accessed 6 February 2020].
107. Dice LR. Measures of the amount of ecologic association between species. *Ecology* 1945;**26**:297–302.
108. Tan P-N, Steinbach M, Karpatne A, Kumar V. *Introduction to data mining. in introduction to data mining.* Pearson Addison Wesley; 2006.
109. Discussion of SImilarity Metrics—Cosine Similarity.
110. Bajusz D, Rácz A, Héberger K. Why is Tanimoto index an appropriate choice for fingerprint-based similarity calculations? *J Cheminform* 2015;**7**.
111. Martin YC, Kofron JL, Traphagen LM. Do structurally similar molecules have similar biological activity? *J Med Chem* 2002;**45**:4350–8.
112. Burello E. Review of (Q)SAR models for regulatory assessment of nanomaterials risks. *NanoImpact* 2017;**8**:48–58.
113. Topliss JG. A manual method for applying the Hansch approach to drug design. *J Med Chem* 1977;**20**:463–9.
114. Craig PN. Comparison of the Hansch and Free-Wilson approaches to structure-activity correlation. In: Van Valkenburg W, editor. *Biological correlations—the Hansch approach.* vol. 114. American Chemical Society; 1974. p. 115–29.
115. Cover T, Hart P. Nearest neighbor pattern classification. *IEEE Trans Inf Theory* 1967;**13**:21–7.
116. Idakwo G, et al. A review of feature reduction methods for QSAR-based toxicity prediction. In: Hong H, editor. *Advances in computational toxicology.* vol. 30. Springer International Publishing; 2019. p. 119–39.
117. Waters M, et al. CEBS—chemical effects in biological systems: a public data repository integrating study design and toxicity data with microarray and proteomics data. *Nucleic Acids Res* 2008;**36**:D892–900.
118. Juan-Blanco T, Duran-Frigola M, Aloy P. IntSide: a web server for the chemical and biological examination of drug side effects. *Bioinformatics* 2015;**31**:612–3.
119. Cheng F, et al. Adverse drug events: database construction and in silico prediction. *J Chem Inf Model* 2013;**53**:744–52.
120. Kuhn M, Letunic I, Jensen LJ, Bork P. The SIDER database of drugs and side effects. *Nucleic Acids Res* 2016;**44**:D1075–9.
121. Stark C, et al. BioGRID: a general repository for interaction datasets. *Nucleic Acids Res* 2006;**34**:D535–9.
122. Juty N, et al. BioModels: content, features, functionality, and use. *CPT Pharmacometr Syst Pharmacol* 2015;**4**, e3.
123. Huttlin EL, et al. The BioPlex network: a systematic exploration of the human interactome. *Cell* 2015;**162**:425–40.
124. Chen JY, Pandey R, Nguyen TM. HAPPI-2: a comprehensive and high-quality map of human annotated and predicted protein interactions. *BMC Genomics* 2017;**18**:182.
125. Peri S, et al. Development of human protein reference database as an initial platform for approaching systems biology in humans. *Genome Res* 2003;**13**:2363–71.
126. Hermjakob H, et al. IntAct: an open source molecular interaction database. *Nucleic Acids Res* 2004;**1**:D452–5.
127. Li T, et al. A scored human protein-protein interaction network to catalyze genomic interpretation. *Nat Methods* 2016;**14**:61–4.

128. Calderone A, Castagnoli L, Cesareni G. Mentha: a resource for browsing integrated protein-interaction networks. *Nat Methods* 2013;**10**:690–1.
129. Türei D, et al. NRF2-ome: an integrated web resource to discover protein interaction and regulatory networks of NRF2. *Oxidative Med Cell Longev* 2013;**2013**.
130. Türei D, Korcsmáros T, Saez-Rodriguez J. OmniPath: guidelines and gateway for literature-curated signaling pathway resources. *Nat Methods* 2016;**13**:966–7.
131. Fazekas D, et al. SignaLink 2—a signaling pathway resource with multi-layered regulatory networks. *BMC Syst Biol* 2013;**7**.
132. Perfetto L, et al. SIGNOR: a database of causal relationships between biological entities. *Nucleic Acids Res* 2016;**44**:D548–54.
133. Szklarczyk D, et al. STRING v11: protein-protein association networks with increased coverage, supporting functional discovery in genome-wide experimental datasets. *Nucleic Acids Res* 2019;**47**:D607–13.
134. Kanehisa M, Goto S. KEGG: Kyoto encyclopedia of genes and genomes. *Nucleic Acids Res* 2000;**28**:27–30.
135. Liberzon A, et al. The Molecular Signatures Database (MSigDB) hallmark gene set collection. *Cell Syst* 2015;**1**:417–25.
136. Rodchenkov I, et al. Pathway commons 2019 update: integration, analysis and exploration of pathway data. *Nucleic Acids Res* 2020;**48**:D489–97.
137. Cerami EG, et al. Pathway commons, a web resource for biological pathway data. *Nucleic Acids Res* 2011;**39**:D685–90.
138. Barbarino JM, Whirl-Carrillo M, Altman RB, Klein TE. PharmGKB: a worldwide resource for pharmacogenomic information. *Wiley Interdiscip Rev Syst Biol Med* 2018;**10**, e1417.
139. Thorn CF, Klein TE, Altman RB. PharmGKB: the pharmacogenomics knowledge base. *Methods Mol Biol* 2013;**1015**:311–20.
140. Yu J, Gu X, Yi S. Ingenuity pathway analysis of gene expression profiles in distal nerve stump following nerve injury: Insights into wallerian degeneration. *Front Cell Neurosci* 2016;**10**.
141. Croft D, et al. Reactome: a database of reactions, pathways and biological processes. *Nucleic Acids Res* 2011;**39**:D691–7.
142. Reactome | EMBL-EBI Train online. Available at: https://www.ebi.ac.uk/training/online/course/proteomics-introduction-ebi-resources/proteomics-resources-ebi/reactome [Accessed 10 October 2019].
143. Carbon S, et al. The Gene Ontology Resource: 20 years and still GOing strong. *Nucleic Acids Res* 2019;**47**:D330–8.
144. Slenter DN, et al. WikiPathways: a multifaceted pathway database bridging metabolomics to other omics research. *Nucleic Acids Res* 2018;**46**:D661–7.
145. Yang H, et al. AdmetSAR 2.0: web-service for prediction and optimization of chemical ADMET properties. *Bioinformatics* 2019;**35**:1067–9.
146. Gilson MK, et al. BindingDB in 2015: a public database for medicinal chemistry, computational chemistry and systems pharmacology. *Nucleic Acids Res* 2016;**44**:D1045–53.
147. Davis AP, et al. The Comparative Toxicogenomics Database: update 2019. *Nucleic Acids Res* 2019;**47**:D948–54.
148. Taboureau O, et al. ChemProt: a disease chemical biology database. *Nucleic Acids Res* 2011;**39**:D367–72.
149. Kringelum J, et al. ChemProt-3.0: a global chemical biology diseases mapping. *Database (Oxford)* 2016. bav123.
150. Richard AM, Williams CLR. Distributed structure-searchable toxicity (DSSTox) public database network: a proposal. *Mutat Res Fundam Mol Mech Mutagen* 2002;**499**:27–52.
151. Austin T, Denoyelle M, Chaudry A, Stradling S, Eadsforth C. European chemicals agency dossier submissions as an experimental data source: refinement of a fish toxicity model for predicting acute LC50 values. *Environ Toxicol Chem* 2015;**34**:369–78.

152. Douguet D. Data sets representative of the structures and experimental properties of FDA-approved drugs. *ACS Med Chem Lett* 2018;**9**:204–9.
153. Kim S, et al. PubChem substance and compound databases. *Nucleic Acids Res* 2016;**44**:D1202–13.
154. Schmidt U, et al. SuperToxic: a comprehensive database of toxic compounds. *Nucleic Acids Res* 2009;**37**:D295–9.
155. Wishart D, et al. T3DB: the toxic exposome database. *Nucleic Acids Res* 2015;**43**:D928–34.
156. Thomas RS, et al. The US Federal Tox21 Program: a strategic and operational plan for continued leadership. *ALTEX* 2018;**35**:163–8.
157. Kohonen P, et al. The ToxBank data warehouse: supporting the replacement of in vivo repeated dose systemic toxicity testing. *Mol Inform* 2013;**32**:47–63.
158. Richard AM, et al. ToxCast chemical landscape: paving the road to 21st century toxicology. *Chem Res Toxicol* 2016;**29**:1225–51.
159. Wexler P. TOXNET: an evolving web resource for toxicology and environmental health information. *Toxicology* 2001;**157**:3–10.
160. Chen X, Ji ZL, Chen YZ. TTD: therapeutic target database. *Nucleic Acids Res* 2002;**30**:412–5.
161. Kostich MS, et al. Aquatic concentrations of chemical analytes compared to ecotoxicity estimates. *Sci Total Environ* 2017;**579**.
162. Wishart DS, et al. DrugBank 5.0: a major update to the DrugBank database for 2018. *Nucleic Acids Res* 2018;**46**:D1074–82.
163. Kuhn M, von Mering C, Campillos M, Jensen LJ, Bork P. STITCH: interaction networks of chemicals and proteins. *Nucleic Acids Res* 2008;**36**:D684–8.
164. Subramanian A, et al. A next generation connectivity map: L1000 platform and the first 1,000,000 profiles. *Cell* 2017;**171**. 1437–1452.e17.
165. Barel G, Herwig R. Network and pathway analysis of toxicogenomics data. *Front Genet* 2018;**9**.
166. Musa A, Tripathi S, Dehmer M, Emmert-Streib F. L1000 viewer: a search engine and Web interface for the LINCS data repository. *Front Genet* 2019;**10**.
167. Igarashi Y, et al. Open TG-GATEs: a large-scale toxicogenomics database. *Nucleic Acids Res* 2015;**43**:D921–7.
168. Clough E, Barrett T. The gene expression omnibus database. *Methods Mol Biol* 2016;**1418**:93–110.
169. Athar A, et al. ArrayExpress update—from bulk to single-cell expression data. *Nucleic Acids Res* 2019;**47**:D711–5.
170. Apweiler R, et al. Ongoing and future developments at the Universal Protein Resource. *Nucleic Acids Res* 2011;**39**:D214–9.
171. UniProt. Available at: https://www.uniprot.org/ [Accessed 10 October 2019].
172. UniProtKB | EMBL-EBI Train online. Available at: https://www.ebi.ac.uk/training/online/course/proteomics-introduction-ebi-resources/proteomics-resources-ebi/uniprotkb [Accessed 10 October 2019].
173. RCSB PDB: homepage. Available at: http://www.rcsb.org/ [Accessed 10 October 2019].
174. Berman HM, et al. The protein data bank. *Nucleic Acids Res* 2000;**28**:235–42.
175. Vizcaíno JA, et al. A guide to the Proteomics Identifications Database proteomics data repository. *Proteomics* 2009;**9**:4276–83.
176. PRIDE | EMBL-EBI Train online. Available at: https://www.ebi.ac.uk/training/online/course/proteomics-introduction-ebi-resources/proteomics-resources-ebi/pride [Accessed 10 October 2019].
177. PRIDE archive. Available at: https://www.ebi.ac.uk/pride/archive/ [Accessed 10 October 2019].
178. Schmidt T, et al. ProteomicsDB. *Nucleic Acids Res* 2018;**46**:D1271–81.
179. gnomAD. Available at: https://gnomad.broadinstitute.org/ [Accessed 5 August 2020].
180. Siramshetty VB, et al. WITHDRAWN—a resource for withdrawn and discontinued drugs. *Nucleic Acids Res* 2016;**44**:D1080–6.

181. DisGeNET—a database of gene-disease associations. Available at: https://www.disgenet.org/ [Accessed 26 July 2020].

182. Home—open targets. Available at: https://www.opentargets.org/ [Accessed 26 July 2020].

183. Home | IUPHAR/BPS Guide to PHARMACOLOGY. (2015). Available at: https://www.guidetopharmacology.org/ [Accessed 31 July 2020].

184. SuperTarget. Available at: http://insilico.charite.de/supertarget/ [Accessed 26 July 2020].

185. Excelra | Data science to empower life science innovation. Available at: https://www.gostardb.com/about-gostar.jsp [Accessed 5 April 2018].

186. Search—SureChEMBL. Available at: https://www.surechembl.org/search/ [Accessed 31 July 2020].

187. Fonger GC, Stroup D, Thomas PL, Wexler P. Toxnet: a computerized collection of toxicological and environmental health information. *Toxicol Ind Health* 2000;**16**:4–6.

188. Gaulton A, et al. The ChEMBL database in 2017. *Nucleic Acids Res* 2017;**45**:D945–54.

189. Tatonetti NP, Ye PP, Daneshjou R, Altman RB. Data-driven prediction of drug effects and interactions. *Sci Transl Med* 2012;**4**. 125ra31.

190. Kanehisa M. The KEGG database. *Novartis Found Symp* 2002;**247**. 91–103, 119–128, 244–252.

191. Römer M, Backert L, Eichner J, Zell A. ToxDBScan: large-scale similarity screening of toxicological databases for drug candidates. *Int J Mol Sci* 2014;**15**:19037–55.

192. Proteomics resources at the EBI | EMBL-EBI Train online. Available at: https://www.ebi.ac.uk/training/online/course/proteomics-introduction-ebi-resources/proteomics-resources-ebi [Accessed 10 October 2019].

193. HUPO-PSI Working groups and Outputs | HUPO proteomics standards initiative. Available at: http://www.psidev.info/ [Accessed 10 October 2019].

194. Search—SureChEMBL. Available at: https://www.surechembl.org/search/ [Accessed 4 August 2017].

195. Wilson AM, Thabane L, Holbrook A. Application of data mining techniques in pharmacovigilance. *Br J Clin Pharmacol* 2004;**57**:127–34.

196. Duggirala HJ, et al. Use of data mining at the Food and Drug Administration. *J Am Med Inform Assoc* 2016;**23**:428–34.

197. Xu Z, Kass-Hout T, Anderson-Smits C, Gray G. Signal detection using change point analysis in postmarket surveillance: CHANGE POINT ANALYSIS. *Pharmacoepidemiol Drug Saf* 2015;**24**:663–8.

198. Perner P, Bichindaritz I, Salvetti O. Advances in data mining applications in medicine, web mining, marketing, image and signal mining; proceedings. In: *Industrial conference on data mining <6 Leipzig>, Springer*; 2006.

199. Ventola C, Big L. Data and pharmacovigilance: data mining for adverse drug events and interactions. *P T A Peer-Review J Formul Manag* 2018;**43**:340–51.

200. Basile AO, Yahi A, Tatonetti NP. Artificial intelligence for drug toxicity and safety. *Trends Pharmacol Sci* 2019;**40**:624–35.

201. Reinhold WC, et al. CellMiner: a web-based suite of genomic and pharmacologic tools to explore transcript and drug patterns in the NCI-60 cell line set. *Cancer Res* 2012;**72**:3499–511.

202. McDermott U, et al. Identification of genotype-correlated sensitivity to selective kinase inhibitors by using high-throughput tumor cell line profiling. *Proc Natl Acad Sci U S A* 2007;**104**:19936–41.

203. Barretina J, et al. The Cancer Cell Line Encyclopedia enables predictive modelling of anticancer drug sensitivity. *Nature* 2012;**483**:603–7.

204. Ghandi M, et al. Next-generation characterization of the Cancer Cell Line Encyclopedia. *Nature* 2019;**569**:503–8.

205. Li H, et al. The landscape of cancer cell line metabolism. *Nat Med* 2019;**25**:850–60.

206. Garnett MJ, et al. Systematic identification of genomic markers of drug sensitivity in cancer cells. *Nature* 2012;**483**:570–5.

207. Iorio F, et al. A landscape of pharmacogenomic interactions in cancer. *Cell* 2016;**166**:740–54.

208. Basu A, et al. An interactive resource to identify cancer genetic and lineage dependencies targeted by small molecules. *Cell* 2013;**154**:1151–61.
209. Klijn C, et al. A comprehensive transcriptional portrait of human cancer cell lines. *Nat Biotechnol* 2015;**33**:306–12.
210. Lamb J. The Connectivity Map: a new tool for biomedical research. *Nat Rev Cancer* 2007;**7**:54–60.
211. Li J, et al. Characterization of human cancer cell lines by reverse-phase protein arrays. *Cancer Cell* 2017;**31**:225–39.
212. Cheng H, et al. Genomic and transcriptomic characterization links cell lines with aggressive head and neck cancers. *Cell Rep* 2018;**25**. 1332–1345.e5.
213. Hutter C, Zenklusen JC. The cancer genome atlas: creating lasting value beyond its data. *Cell* 2018;**173**:283–5.
214. International Cancer Genome, C, et al. International network of cancer genome projects. *Nature* 2010;**464**:993–8.
215. Rudnick PA, et al. A description of the clinical proteomic tumor analysis consortium (CPTAC) common data analysis pipeline. *J Proteome Res* 2016;**15**:1023–32.
216. Zhang H, et al. Integrated proteogenomic characterization of human high-grade serous ovarian cancer. *Cell* 2016;**166**:755–65.
217. Mun DG, et al. Proteogenomic characterization of human early-onset gastric cancer. *Cancer Cell* 2019;**35**. 111–124.e10.
218. Rees MG, et al. Correlating chemical sensitivity and basal gene expression reveals mechanism of action. *Nat Chem Biol* 2016;**12**:109–16.
219. Seashore-Ludlow B, et al. Harnessing connectivity in a large-scale small-molecule sensitivity dataset. *Cancer Discov* 2015;**5**.
220. Stathias V, et al. LINCS Data Portal 2.0: next generation access point for perturbation-response signatures. *Nucleic Acids Res* 2020;**48**:D431–9.
221. Buniello A, et al. The NHGRI-EBI GWAS Catalog of published genome-wide association studies, targeted arrays and summary statistics 2019. *Nucleic Acids Res* 2019;**47**:D1005–12.
222. Papatheodorou I, et al. Expression Atlas update: from tissues to single cells. *Nucleic Acids Res* 2020;**48**:D77–83.
223. Landrum MJ, et al. ClinVar: Public archive of relationships among sequence variation and human phenotype. *Nucleic Acids Res* 2014;**42**:D980–5.
224. Sloan CA, et al. ENCODE data at the ENCODE portal. *Nucleic Acids Res* 2016;**44**:D726–32.
225. Regev A, et al. The Human Cell Atlas. *elife* 2017;**6**.
226. Rozenblatt-Rosen O, Stubbington MJT, Regev A, Teichmann SA. The Human Cell Atlas: from vision to reality. *Nature* 2017;**550**:451–3.
227. Mele M, et al. Human genomics. The human transcriptome across tissues and individuals. *Science (80-)* 2015;**348**:660–5.
228. Sestan E. Revealing the brain's molecular architecture. *Science (80)* 2018;**362**:1262–3.
229. Beecham GW, et al. The Alzheimer's Disease Sequencing Project: Study design and sample selection. *Neurol Genet* 2017;**3**, e194.
230. Lambert JC, et al. Meta-analysis of 74,046 individuals identifies 11 new susceptibility loci for Alzheimer's disease. *Nat Genet* 2013;**45**:1452–8.
231. Pereanu W, et al. AutDB: a platform to decode the genetic architecture of autism. *Nucleic Acids Res* 2018;**46**:D1049–54.
232. Hall D, Huerta MF, McAuliffe MJ, Farber GK. Sharing heterogeneous data: the national database for autism research. *Neuroinformatics* 2012;**10**:331–9.
233. Forbes SA, et al. COSMIC: somatic cancer genetics at high-resolution. *Nucleic Acids Res* 2017;**45**:D777–83.

234. van der Meer D, et al. Cell Model Passports—a hub for clinical, genetic and functional datasets of preclinical cancer models. *Nucleic Acids Res* 2019;**47**:D923–9.

235. Gao J, et al. Integrative analysis of complex cancer genomics and clinical profiles using the cBioPortal. *Sci Signal* 2013;**6**. pl1.

236. Liu J, et al. An integrated TCGA pan-cancer clinical data resource to drive high-quality survival outcome analytics. *Cell* 2018;**173**. 400–416.e11.

237. Liberzon A, et al. Molecular signatures database (MSigDB) 3.0. *Bioinformatics* 2011;**27**:1739–40.

238. Chen EY, et al. Enrichr: interactive and collaborative HTML5 gene list enrichment analysis tool. *BMC Bioinform* 2013;**14**.

239. Warde-Farley D, et al. The GeneMANIA prediction server: biological network integration for gene prioritization and predicting gene function. *Nucleic Acids Res* 2010;**38**:W214–20.

240. Duan Q, et al. L1000CDS(2): LINCS L1000 characteristic direction signatures search engine. *NPJ Syst Biol Appl* 2016;**2**.

241. Lachmann A, et al. Geneshot: search engine for ranking genes from arbitrary text queries. *Nucleic Acids Res* 2019;**47**:W571–7.

242. Jerby-Arnon L, et al. Predicting cancer-specific vulnerability via data-driven detection of synthetic lethality. *Cell* 2014;**158**:1199–209.

243. Behan FM, et al. Prioritization of cancer therapeutic targets using CRISPR-Cas9 screens. *Nature* 2019;**568**:511–6.

244. Sun R, et al. A radiomics approach to assess tumour-infiltrating CD8 cells and response to anti-PD-1 or anti-PD-L1 immunotherapy: an imaging biomarker, retrospective multicohort study. *Lancet Oncol* 2018;**19**:1180–91.

245. Hoshida Y, et al. Prognostic gene expression signature for patients with hepatitis C-related early-stage cirrhosis. *Gastroenterology* 2013;**144**:1024–30.

246. Allix-Beguec C, et al. Prediction of susceptibility to first-line tuberculosis drugs by DNA sequencing. *N Engl J Med* 2018;**379**:1403–15.

Target identification and validation

4

Stephanie Kay Ashenden[a], Natalie Kurbatova[b], and Aleksandra Bartosik[c]

[a]Data Sciences and Quantitative Biology, Discovery Sciences, R&D, AstraZeneca, Cambridge, United Kingdom, [b]Data Infrastructure and Tools, Data Science and Artificial Intelligence, R&D, AstraZeneca, Cambridge, United Kingdom, [c]Clinical Data and Insights, Biopharmaceuticals R&D, AstraZeneca, Warsaw, Poland

Introduction

Target identification and target validation are involved with the identification and validation of a target that can be modulated for a therapeutic purpose. Whilst there is much variation on what a target is and how to define one, we can consider a therapeutic target to be one that is involved with the mechanisms of a particular disease and modulation of it will affect the disease in a beneficial way. It is important to also note that when a drug fails, it tends to be due to safety concerns or lack of efficacy.[1]

There are several computational methods that are applied within the target identification stages of drug discovery, including cheminformatic methods, such as similarity searching and network-based methods.[2] However, much information is also contained within databases and scientific publications.

Whilst here we primarily focus on small molecules as the main drug modality, it must be noted, that there are wide variety of modalities beyond small molecules and antibodies, and therefore, it remains to be seen how much of the genome can be drugged effectively. Efforts to expand further into and utilize more of the druggable genome have been undertaken such as the initiative called Illuminating the Druggable Genome of which started in 2013.[3] However, it is important to note the vast array of modalities that are now available. The druggable genome is the genes that encode proteins that can potentially be modulated. It is thought that of the approximately 20,000 genes that make up the human genome, approximately 3000 of these are considered part of the druggable genome despite only a few hundred targets being currently exploited.[3] The program aims to create a resource for the study of protein families.[3] The program along with Nature Reviews Drug Discovery publish content on understudied protein targets frequently and have included examples like $K_{na}1.1$ channels as potential therapeutic targets for early-onset epilepsy and PIP5K1A as a potential target for cancers that have KRAS or TP53 mutations.[4] Again, it is important to note that developments in technology have allowed for new modalities to be used as therapeutic agents.[5] Such an example is the approvals of ASOs (two) by the FDA in 2016.[6]

Animal models are another approach to understand human biology. A series of challenges named systems biology verification for Industrial Methodology for PROcess Verification in Research (sbv IMPROVER) was designed to address translatability between rodents and humans.[7] This initiative saw

The Era of Artificial Intelligence, Machine Learning, and Data Science in the Pharmaceutical Industry. https://doi.org/10.1016/B978-0-12-820045-2.00005-2

scientists apply computational methodologies on datasets containing phosphoproteomics, transcriptomics, and cytokine data from both human and rat tissues that were exposed to various stimuli.[7] A key finding was that even similar computational method had diverse performance and there was no clear "best practise".[7] Despite this, the computational methods were able to translate some of these stimuli and biological processes between the human and rat.[7]

Despite such successes in translatability, animal studies have been considered poor predictors of human responses for a very long time. Bracken explored this idea of why animal studies are often poor predictors of human reactions to exposure.[8] Bracken summarized the reasoning as to why there is often a lack of translatability between animal studies and human trials is that many animal experiments do not appear to be designed and performed well.[8] There may be other issues such as limitations in methods and biases.[8] Despite this, Denayer and Van Roy explain that animal models allow for a greater understanding of biology, disease, and medicine.[9]

Targeting the wrong target can lead to failure of the development of the drug and therefore result in cost, not just financially, but in time as well.[1] Failli and co-authors discuss that to improve this process, many developed platforms rely on estimates of efficacy that have been derived from computational methods that create scores of target-disease associations.[1] Failli and co-authors also present novel methods that consider both efficacy and safety by using gene expression data, transcriptome based, as well as tissue disease-specific networks.[1]

Defining what a drug target is crucial but is not well defined.[10] Santos and co-authors[10] highlighted the need to keep accurate and current record of approved drugs and the target in which the drugs modulation results in therapeutic effects.[10] The authors note several resources including databases and papers that contain this information but note that it is still difficult to find a complete picture.[10] Owing to this, the authors attempt to annotate the current list of FDA-approved drugs (up to when published).

The understanding of what makes a good drug target is important in target identification and the drug discovery process as a whole. Gashaw and co-authors[11] have previously discussed what a good drug target is and also detail what the target evaluation at Bayer HealthCare involved at time of writing. The authors highlight that first a disease is identified that does not currently have treatment and then the identification of a drug target follows. Target assessment follows comprising of experimental target assessment methods, theoretical drugability assessment, and thoughts on target-related/stratification biomarkers. Alongside this adverse event evaluation and intellectual property is carefully considered.[11] These steps are important and investigate key areas of a target such as whether the target actually has a role within the disease of interest as well as whether the target can be modulated without adverse events.

Predicting drug target interactions is an area of great interest in the drug discovery process. Predicting such interactions can help to save time and money. The term drug prediction can be an ambiguous term. It can refer to the identification of a target of which will modulate the disease of interest (related to target focused approaches), but also can refer to the identification of a target that a drug is binding to (more related to phenotypic approaches). Unlike in targeted drug discovery approaches, phenotypic approaches do not have prior knowledge of a specific target or what role that target places within a disease.

The large number of resources available for aiding in the discovery of a drug target have been summarized by an online article by BiopharmaTrend of which listed 36 different web resources for identifying targets.[12] This list includes databases such as WITHDRAWN,[13] which is a resource for withdrawn and discontinued drugs, DisGeNET,[14] which is focused on target-disease relationships as is Open Targets.[15, 16]

Open Targets uses human genetic and genomic data to aid in target identification and prioritization. The website describes that the Open Targets Platform has public data to aid in this task and the Open

Targets Genetics Portal uses GWAS (Genome Wide Association Studies of which aims to associate particular variations with diseases) and functional genomics to identify targets. The data integration is complemented with large-scale systematic experimental approaches.[15]

Drug-target databases mentioned by Santos and co-authors included the Therapeutic Targets Database,[17] of which focuses on known and explored therapeutic protein and nucleic acid targets, as well as the diseases, pathway information and any corresponding drugs. DrugBank,[18] contains bioinformatic and cheminformatic resources for combining drug data and drug target information. Next, the authors mentioned SuperTarget[19] which includes various drug information for target proteins. Finally, the authors mention the International Union of Basic and Clinical Pharmacology and British Pharmacology Society Guide to Pharmacology Database.[20] This resource aggregates target information and separates targets into eight different target classes. Namely, GPCRs, Ion channels, Nuclear hormone receptors, Kinases, Catalytic receptors, Transporters, Enzymes, and other protein targets. It has been argued that focusing on a class of targets that are known to have key roles can provide beneficial outcomes in early drug discovery phases.[21] The thought is that it will reduce changes for technical failure due to the diversification of the target space, and thus available binding pockets.[21] This in turn increases potential clinical applications.[21]

These resources are important for understand the mechanisms of disease. Understanding the mechanism of disease is an important step in identifying the targets involved. However, a difficulty with understanding disease is that regardless of the technology used, capturing the complete biological picture, with all of its complexities, is not possible in most human diseases.[22] The use of integrating multiple technologies such as multiomics is used to capture a more complete picture. For example, genetic data and transcriptomic data have been combined for the purpose of using gene expression profiles to identify disease-related genes for prioritization and drug repositioning.[23, 24]

Target identification predictions

As previously mentioned, target identification can relate to two distinct questions. First, we have a compound, what is the protein it is associated with where we have a linkage between a compound and a protein. Second, we have a disease what is the protein involved where there is a linkage between protein and disease.

Gene-disease association data have been used to predict therapeutic targets using a variety of classifiers and validating the results by mining scientific literature for the proposed therapeutic targets.[25] Protein-protein interactions have also been used to predict genes involved in disease.[26] Predicting mutations that are associated with disease is also of interest such as is demonstrated with the tool BorodaTM (Boosted RegressiOn trees for Disease-Associated mutations in TransMembrane) proteins.[27]

Several tools have been developed to aid in drug-target prediction with many different strategies in place. For example, Bayesian approaches, such as the tool BANDIT for predicting drug binding targets[28] as well as deep learning approaches, such as the tool deepDTnet that is network based.[29] Many methods have been summarized by Mathai and co-authors in their paper discussing validation strategies for target prediction methods.[30] Here we have picked out a handful of key methods to discuss.

The SwissTargetPrediction tool uses multiple logistic regression to combine the different similarity values from shape and fingerprint compound similarities.[31] HitPick is another online service[32] for target prediction where a user uploads their compounds in SMILES format and uses 2D molecular similarity-based methods such as nearest neighbor similarity searching and machine learning methods that are based on naive Bayesian models (Laplacian modified).[32] The chemical similarity is measured using Tanimoto coefficients. The tool SEA[33] stands for the similarity ensemble approach and performs the predictions based on the chemical similarity of the compounds. The compounds have been annotated into sets (based on the receptor that the ligand modulates) and then the Tanimoto coefficient similarity score between these sets was calculated. Another method named SPiDER[34] use self-organizing maps to identify targets. SLAP is a tool that uses a network approach and is able to make links between related data and these data can be of differing types.[35, 36] Finally, PIDGIN,[37, 38] which stands for Prediction Including Inactivity uses nearly 200 million bioactivity data points containing both active and inactive compounds from ChEMBL and PubChem for target prediction.

The validation strategies that are used in target prediction have been previously discussed and note that most strategies that are used are not accounting for biases in the data. The authors[30] suggest external validation of the parameterized models and compared with the results from the internal validation.[30]

Mathai and co-authors[30] summarized such validation strategies for target prediction methods. The authors note that it is important to understand how the data were validated, based on the data partitioning schemes (how the data is split into train and test sets) used, as well as how the performance was measured, based on the metrics used). The authors note that performance assessment for target prediction methods do not often take into account biases in the data and that new metrics are needed. However, they do suggest several steps that can be taken to represent a more realistic view on the prediction performance. Examples of their suggestions include using a combination of metrics and methods should be used, the minima, maxima, and distributions should be reported. The use of stratified sampling and external data should be used.[30] The authors also discuss the validation strategies used by various tools including most of the examples shown above in Table 1.

Table 1 List of some key drug-target prediction tools.

Tool	Notes	Website	Validation strategy	References
SwissTargetPrediction	Online tool using both 2D and 3D similarity	http://www.swisstargetprediction.ch/	Cross validation	*31,39,40*
HitPick	Online tool molecular similarity-based methods	http://mips.helmholtz-muenchen.de/hitpick/	Large data set	*32*
SEA	Online tool using chemical similarity to a set of reference molecules	http://sea.bkslab.org/	Experimental validation	*33*
SPiDER	Uses self-organizing maps along with CATS pharmacophore and molecular operating environment descriptors	http://modlabcadd.ethz.ch/software/spider/	Cross validation	*34*
SLAP	Network-based approach links different but related types of information	http://cheminfov.informatics.indiana.edu:8080/slap/	Large data set	*35,36*
PIDGIN	Bernoulli Naive Bayes algorithm	https://github.com/lhm30/PIDGIN	Cross validation and other datasets	*37,38*

When validating targets, methods are required to identify the most complexing targets that have been identified. MANTIS_ML[41] is a tool that uses genetic screens from HTS that uses machine learning to rank disease-associated genes where the authors showed good predictive performance when applied to chronic kidney disease and amyotrophic lateral sclerosis. The tool works by compiling features from multiple sources, followed by an automated machine learning preprocessing step with exploratory data analysis and finally stochastic semisupervised learning.[41] These features are derived from sources such as disease-specific resources such as CKDdb,[42] a database containing multiomic studies of chronic kidney disease, data filtered by tissue and disease such as from the GWAS catalog[43] and resources that are more generic like the Genome Aggregation Database (GnomAD)[44] and the Molecular Signatures database (MSigDB),[45] of which has the purpose of combining exome and genome sequencing data from large-scale sequencing projects.[41]

In some cases, a compound will have been discovered to modulate a protein for a particular disease, but the actual target is not known. For example, the observed change in protein expression that is the result of a target upstream interacting with a compound. The structure of the target can also be predicted, and deep learning has been applied to such effect. The distance between pairs of residues can be predicted via neural networks even with fewer homologous sequences.[46]

Gene prioritization methods

After identifying a list of genes, it is often useful to prioritize this list to focus the search, ideally starting with the most likely target gene. Methods for gene prioritization vary in their approaches from network-based methods to matrixes. Moreau and Tranchevent[47] previously reviewed computational tools for prioritizing candidate genes and noted that the overall aim is to ingrate data for the purpose of identifying the most promising genes for validation. It often utilizes prior knowledge such as in the form of keywords that describe the phenotype or known gene associations to a particular process or phenotype. Furthermore, with a number of gene prioritization tools available, methods to benchmark such gene prioritization tools have been undertaken with the authors[48] (Guala and Sonnhammer) recommending the use of Gene Ontology[49] data alongside the tool FunCoup.[50, 51] Because of the cost of identifying and validating new genes that are involved in diseases, many of the algorithms have had the performance measured by using known disease genes.[52]

Many of these tools rely on different types of data and algorithms to perform gene prioritization. For example, the use of genomic data fusion, such as is the case with the gene prioritization tool ENDEAVOUR.[53, 54] Other approaches include Beegle,[55] of which mines the literature to find genes that have a link to the search query. It then uses ENDEAVOUR to perform the gene prioritization step. Cross-species approaches have been proposed such as in the case of GPSy.[56] The use of Bayesian matrix factorization using both genomic and phenotypic information has also been implemented.[57] Network based approaches are also used in gene prioritization tasks.

Such network-based tools included Rosalind which uses knowledge graphs to predict linkages between disease and genes that uses data integration (in graph format) and relational inference.[58] MaxLink, which is a guilt-by-association network which aims to identify and rank genes in relation to a given query list.[59] NetRank is based on the algorithm used in the ToppGene Suite[60, 61] and uses the Page Rank with priors algorithm where a random surfer is more likely to end up in nodes that are initially relevant.[60, 62] Other algorithms include, NetWalk, which is based on the random walk with restarts algorithm which operates by walking from a node to another randomly but at any time the walk can be restarted (based on probability).[52]Using not only the shortest path interaction but also taking into the global interactome structure has been shown to have performance advantages.[52] NetProp is based on the network propagation algorithm of which was designed to address the issue of many approaches using only local network information.

GUILD[63] (Genes Underlying Inheritance Linked Disorders), uses four network-based disease-gene prioritization algorithms, namely, NetShort (considers node importance for a given phenotype will have a shorter distance to other seed nodes within the network), NetZcore (normalizes the scores of nodes in the network in relation to a set of random networks with similar topology and therefore considers the relevance of the node for a given phenotype), NetScore (considers multiple shortest paths from source to target) and NetCombo (averaged normalization score for each of the other three prioritization algorithms). GUILD also includes NetRank, NetWalk, NetProp, and fFlow of which is based on the FunctionalFlow algorithm.[60, 64] This algorithm generalizes the guilt-by-association method by considering groups of proteins that may or may not physically interact with each other.[64] In fFlow, the scores annotated from nodes with higher scores "flow" toward nodes with lower scores depending on the edge capacity.[60] The authors who describe the FunctionalFlow algorithm compare it with three others, Majority[64, 65] (consider all neighboring proteins and calculate how many times each annotate occurs for each protein), Neighborhood[64, 66] (considers all proteins within a radius and then for each function consider if it is over-represented), and GenMultiCut. The authors note there are different considerations of GenMultiCut with two instances agreeing that functional annotations on interaction networks should be made with the purpose of reducing the occurrences of different annotations that are associated with neighboring proteins.[64]

Machine learning is also used in gene prioritization tasks. In the case of Amyotrophic Lateral Sclerosis, Bean and co-authors published a knowledge-based machine learning approach that used a variety of available data to predict new genes involved in Amyotrophic Lateral Sclerosis.[67] Isakov and co-authors used machine learning gene prioritization tasks to identify risk genes for inflammatory bowel disease and the authors performed a literature search to identify whether particular genes had been published alongside inflammatory bowel disease.[68] Prioritizing loci from genome wide association studies (GWAS) studies is another important prioritization task that machine learning has aided in.[69] Nicholls and co-authors reviewed the machine learning approaches used in loci prioritization and noted that GWAS results require aid in sorting through noise as well and understanding which are the most likely causal genes and variants.

Machine learning and knowledge graphs in drug discovery
Introduction

The rapid accumulation of unstructured knowledge in the biomedical literature and high throughput technologies are primary sources of information for modern drug discovery.

Knowledge graph concept allows linking of vast amounts of biomedical data with ease. It provides many algorithmic approaches, including machine learning, to answer pharmaceutical questions of drug repositioning, new target discovery, target prioritization, patient stratification, and personalized medicine.[70–73]

There is no single definition of a knowledge graph. The term in its modern meaning appeared in 2012 when Google announced its new intelligent model of the search algorithm. From the list of definitions, the most descriptive one is the following:

"A knowledge graph acquires and integrates information into an ontology and applies a reasoner to derive new knowledge".[74]

A knowledge graph is a graph in its classical graph theory definition. The underlying mathematical graph concept distinguishes the knowledge graph from other knowledge-based concepts. Besides, it is a semantic graph when we encode the meaning of the data alongside the data in the form of ontology. The main idea here is that we are collecting data into the self-descriptive graph by integrating data and metadata. In all applications and domains, the knowledge graph is a flexible structure allowing to change, delete, and add new data. This dynamism requires an accurate gathering of provenance and versioning information.

Reasoning and the fact that knowledge graph being a graph in the mathematic sense, are fundamental characteristics of the knowledge graph.

Knowledge graphs are usually large structures. For example, Google's Knowledge graph by the end of 2016 contained around seventy thousand millions of connected facts.[75] Typical knowledge graph in biopharmaceutical domain consists of millions of nodes and a thousand-million of edges. Significant volumes of data are applicable for reasoning with the help of machine learning techniques. The bonus of the graph structure is the set of graph theory algorithms.

Graph theory algorithms

Graph theory is a well-established mathematical field that started in the 18th century with Leonhard Euler paper on the Seven Bridges of Königsberg.

Mathematically, we are defining a graph as a set of nodes connected by edges: $G = (N, E)$, where N is a set of nodes also called vertices, and E is a set of two-nodes, whose elements are called edges. An undirected graph is a graph where all the edges are bidirectional. In contrast, a graph where the edges point in a direction is called a directed graph.

Both nodes and edges can have properties that play the role of weights in different algorithms (Fig. 1).

In the context of knowledge graphs, the important graph theory algorithms that are used either directly or as feature extraction techniques for machine learning:

- connected components,
- shortest path,
- minimum spanning tree,
- variety of centrality measures, and
- spectral clustering.

"Connected components" is a collective name for algorithms which find clusters/islands. Within the biomedical domain, this class of algorithms operates to identify protein families, to detect protein complexes in protein-protein interactions networks, to identify technological artefacts.[76]

"Shortest path" is an algorithm to find the shortest path along edges between two nodes. The most famous example of this class is Dijkstra's algorithm.[77] In biological knowledge graphs, this algorithm is used to calculate network centralities.

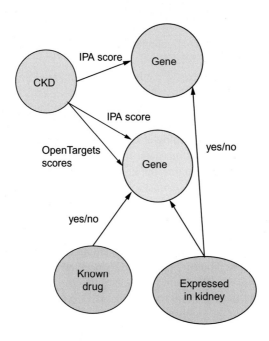

FIG. 1

Directed graph. Color code: *blue*, disease node; *green*, gene node; *pink*, knowledge node.

"Minimum Spanning tree" is an algorithm that helps to find a subset of the edges of a connected, edge-weighted undirected graph that connects all the nodes, without any cycles and with the minimum possible total edge weight. The greedy algorithm, commonly in use is Kruskal's algorithm.[78] In the drug discovery domain, this class of graph theory algorithms is used to define disease hierarchies, cluster nodes, compare biological pathways, and subgraphs.

A widely accepted fact in biological data analyses is that in most graphs, some nodes or edges are more important or influential than others. This importance can be quantified using centrality measures.[79] We often use centrality measures as features for machine learning models. Here is the list of commonly used centrality measures:

- Degree centrality is simply the number of connections for a node.
- Closeness centrality indicates how close a node is to all other nodes in the graph.[80]
- Betweenness centrality quantifies how many times a particular node comes in the shortest chosen path between two other nodes.[81]
- Eigenvector centrality is a measure of the influence of a node in a network. It assigns relative scores to all nodes in the network based on the concept that connections to high-scoring nodes contribute more to the score of the node in question than equal connections to low-scoring nodes.[82] Popular Google's Pagerank algorithm is a variant of the eigenvector centrality (Fig. 2).[83]

We apply centrality measures for nodes prioritization in drug discovery knowledge graphs, for instance, to find the essential proteins/genes for a particular disease, and to find the artefacts.

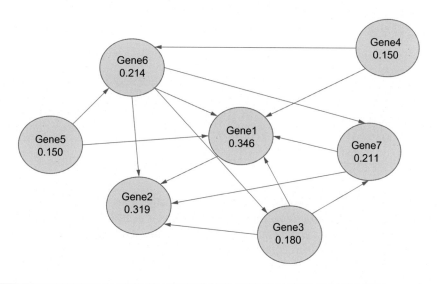

FIG. 2

ArticleRank (the variant of the PageRank algorithm) results. Gene1 has the highest centrality score.

Spectral clustering algorithm has three primary stages:

- construction of a matrix representation of the graph (Laplacian matrix),
- computation of eigenvalues and eigenvectors of the matrix and mapping of each point to a lower-dimensional representation based on one or more eigenvectors, and
- finally, clustering of points based on the new representation.

Spectral clustering is used as an unsupervised feature learning technique for machine learning models.

Graph-oriented machine learning approaches

In general, machine learning approaches on graph structures can be divided into two broad categories:

- feature extraction from the graph followed by an application of machine learning and
- graph-specific deep network architectures that directly minimize the loss function for a downstream prediction task.

Feature extraction from graph

Feature extraction or feature learning from the graph can be done in a supervised, semisupervised, and unsupervised manner using classic graph algorithms and machine learning techniques. Our ultimate goal in the feature learning from a graph is to extract the information encoded in graph structure to cover diverse patterns of graphs.

One of the examples of unsupervised feature learning for feature extraction from a graph is the application of spectral clustering algorithms. This explicit deterministic approach provides mathematically

Table 2 Centrality measures for gene nodes extracted as features.	
Gene	**Centrality measure (ArticleRank)**
Gene1	0.346
Gene2	0.319
Gene3	0.180
Gene4	0.150
Gene5	0.150
…	

defined features that cover graph patterns, but it is computationally expensive and hard to scale. Another example of the usage of graph algorithms for feature extraction is centrality measures (Table 2).

As an alternative to the direct application of graph algorithms, representation learning approaches started to win the race of feature extraction from graphs.[73, 84–86] This type of representation learning is a semisupervised method with an objective to represent local neighborhoods of nodes in the created features.

A broadly used representation learning algorithm is *node2vec*.[84] Given any graph, it can learn continuous feature representations for the nodes, which can then be used for various downstream machine learning tasks. The algorithm creates new features also called embeddings, or compressed representations by using random walks together with graph theory connected components, algorithmic ideas and simple neural network with one layer. The algorithm *node2vec* consists of three steps:

- Sampling. A graph is sampled by random walks with two parameters p and q that can be tuned specifically for the downstream analysis task. Parameter q controls the discovery of the broader neighborhood, and parameter p controls the search of the microscopic view around the node.
- Training. The method uses the skip-gram network that accepts a node from the sampling step as a one-hot vector as an input and maximizes the probability for predicting neighbor nodes using backpropagation on just single hidden-layer feedforward neural network.
- Computing embeddings. Embeddings are the output of a hidden layer of the network (Fig. 3).

Tensor decomposition is another embedding technique we are using in knowledge graphs since graph can be represented as a third-order binary tensor, where each element corresponds to a triple, 1 indicating a true fact and 0 indicating the unknown.

There are many approaches to factorize the third-order binary tensor,[87–89] amongst them relatively complex nonlinear convolutional models,[90, 91] and simple linear models like RESCAL[87] and TuckER.[92] The latest is based on Tucker decomposition[93] of the binary tensor of triples into a core tensor multiplied by a matrix along with each mode. Rows of the matrices contain entity and relation embeddings, while entries of the core tensor determine the level of interaction between them. This approach uses a mathematically principled widely studied model. TuckER's results for edge prediction are compatible with the nonlinear models. Besides, tensor decomposition model can serve as a baseline for more elaborate models, for instance by incorporating graph algorithms.

We use *node2vec*, tensor decomposition and other embedding techniques[94–96] in pharmaceutical knowledge graphs to predict the most probable labels of the nodes, for instance, functional labels of genes/proteins, and possible side effects.

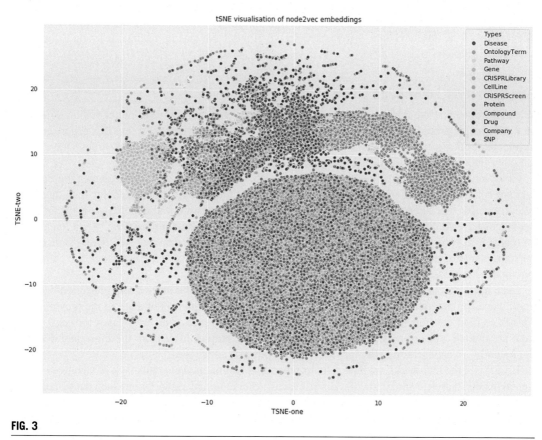

FIG. 3

node2vec embeddings used for tSNE plot. Shows the clustering of the node types in embedded space.

Another application of embedding techniques is edge prediction, when we wish to predict whether a pair of nodes in a graph should have an edge connecting them, for instance, to discover novel interactions between genes.

After the feature extraction, the standard preprocessing ML methods can be used to reduce dimensionality following by preferable ML classification or regression methodology for the downstream task.

Graph-specific deep network architectures

There are supervised machine learning approaches specially developed for graph structures.

These methods base on graph-specific deep network architectures that match the structure of the graphs and directly minimize the loss function for a downstream prediction task using several layers of nonlinear transformations. The basic idea is the creation of a special graph neural network layer that is able to interact with the knowledge graph directly. It takes a graph as input and performs computations over the structure.[97] By propagating information between nodes principally using the graph edges, it is possible to maintain the relational inductive biases present in the graph structure.[98] The relational inductive bias term means that we are forcing an algorithm to prioritize structural, relational solutions, independent of the observed data. A generic example of inductive bias in machine learning techniques is a regularization term.

Drug discovery knowledge graph challenges

As we mentioned before the volumes of structured data that are coming from -omics technologies and natural language processing give us an opportunity to link together masses of research papers and experimental results to build knowledge graphs that connect genes, diseases, cell lines, drugs, side effects, and other interconnected biomedical information in a systematic semantic manner.

However, there are a lot of challenges to deal with: trustworthiness of the resources the original data are coming from, different experimental designs and technological biases, noisy data, usage of different ontologies that do not always have cross-references, and ascertainment biases. Some of these challenges have solutions in the field of data science dealing with data unification.[99] Data integration for biomedical knowledge graph requires assigning of weights for the data that are coming from different sources and have different levels of trust. Both data unification and data integration are extremely time-consuming tasks.

Even having an ideal drug discovery knowledge graph with ideally unified and integrated data, we would have computational challenges as well. Most of the techniques mentioned in the previous sections have been developed for the social networks, paper references, and web pages ranking, where there are limited types of nodes and relations. Biomedical graphs differ from these original domains significantly—we are dealing with the diversity of node and edge types.

Besides, in the real world, we have a lot of noise amongst biomedical data, lack of negative experimental results, biases toward particular disease areas. All mentioned earlier biomedical data challenges requires careful, systematic testing of different methods, including the search processes of the most relevant subgraphs for a particular drug discovery task and methods for the evaluation and prioritization of the resources data are coming from.

Data, data mining, and natural language processing for information extraction

What is natural language processing

Natural language processing (NLP) is ultimately about accessing information fast and finding the relevant parts of the information. It differs from text mining in that if you have a large chunk of text, in text mining you could search for a specific location such as London. In text mining, you would be able to pull out all the examples of London being mentioned in the document. With NLP, rather than asking it to search for the word London, you could ask it to bring back all mentions of a location or ask intelligent questions such as where an individual lives or which English cities are mentioned in the document. It takes into consideration the surrounding information as well. To summarize, natural language processing is concerned with processing the interactions between source data, computers, and human beings.

How is it used for drug discovery and development

Implementing an NLP pipeline can be completed with packages such as NLTK (Natural Language Toolkit)[100] and spacy[101] in python[102] and there are many tutorial on line.[103] This tutorial is summarized here. The first step with natural language processing is to process sentences one word at a time. This is known as word tokenization. For example, take the sentence "What drugs were approved last year?", you would split this into individual word tokens; "what," "drugs," "were," "approved," "last," "year" and "?". We can then remove punctuation which is often useful as it is treated as a separate token and you may find punctuation being highlighted as the most common token. However, punctuation can really affect the meaning of a sentence. It is also possible to remove stop words which are those that occur frequently such as "and" or "the" and therefore can be beneficial to remove to focus on the more informative text. Next, we predict the parts of speech for example determining the type of word (noun, adjective, etc.) and how each word is separated. Following this, lemmatization is performed where we reduce a word to its simplest form. For example, run, runner and running would be reduced to the lemma formation run. These all are referring to the same concept and reducing to the lemma formation helps a computer interpret as such. The next stage is known as dependency parsing which works out how all the words within a sentence relate to each other. Following this, there is a range of analysis we can perform next such as named entity recognition where a model predicts what each word (or if you have done further processing steps to identify phrases rather than single words) refers to such as names, locations, organizations, and objects. Steps like coreference resolution (identifying expressions that are related to a particular entity)[104] can be used to identify what we are talking about even after the name has been mentioned for example: "Random forest is a popular machine learning algorithm. It can perform both classification and regression tasks" The "it" on its own may be difficult to determine what we are talking about but taking the whole two sentences we can understand that in this case "it" refers to the random forest algorithm. This step is difficult and as technology improves—we should expect to see this improve.

The use of ontologies (relationship structure within a concept) decrease linguistic ambiguity by mapping words to standardized terms and to establish a hierarchy of lower and higher-level terms[105] MedDRA is an example of the medical dictionary used by HCPs, pharmaceutical companies and regulators.[106] Another example includes SNOMED CT, which standardized clinical terminology with the use of concept codes, descriptions, relationships, and reference sets.[107] One can download more ontologies dedicated to drug safety-related NLP from online repositories, for example, BioPortal or via NCBO Ontology Recommender Service.

Several NLP techniques have become standard tools available in text processing tools. For example, a search engine such as Apache Solr has built-in functions that allow for: tokenization (splitting text into tokens), stemming (removing stems from words), ignoring diacritics (which sometimes is helpful for non-English texts), conversion to lowercase words, applying phonetic algorithms like Metaphone. Omics technologies generate a lot of data that needs computational approaches to pull out the useful information, by means of data mining.[23] These data can be used to help identify and validate potential drug targets.

Another essential NLP method in NLP is building the n-gram model of a corpus (a collection of texts). N-grams are tuples (2-gram is a doublet, 3-gram is triplet, and 4-gram is a quadruplet) of letters or words that appear consecutively in the original text, for example, when the text is "GAATTC," the 3-grams are "GAA," "AAT," "ATT," and "TTC". N-grams are most commonly used in machine translations and computational biology, especially in analyzing DNA and RNA sequences (e.g., finding sequences with similar n-gram profiles). A generalization of an n-gram is a skip-gram in which the elements constituting a group do not have to be next to each other in the original sequence [e.g., grouping Nth word with $(N+2)$th word]. Google OpenRefine free online tool allows user to apply n-gram fingerprints for noisy data curation, transformation, and mining.

N-gram is a special case of Markov models. A simple Markov model called Markov chain is a set of possible states together with probabilities of transitioning between two states (every pair A, B of states has attached probabilities of the transition A → B and B → A). Markov models identified high cumulative toxicity during a phase II trial ifosfamide plus doxorubicin and granulocyte colony-stimulating factor in soft tissue sarcoma in female patients.[108]

Another NLP technique which has become popular in recent years is word2vec. It relies on word n-grams or skip-grams to find which words have a similar meaning. word2vec relies on the distributional hypothesis: that words with similar distributions in sentences have similar meanings. The internal representation of every word in such word2vec model is a numerical vector, usually with hundreds or more dimensions (hence the "vec" part in the name of the method). An interesting property of this method is that the word vectors can be assessed for their similarity (e.g., by computing a dot product of two vectors), or even added or subtracted (e.g., a word2vec model can be trained to compute that "king − man + woman = queen" which in other words could be expressed as "king is to man what queen is to woman"). This model has lots of potential in natural language applications, but there are also other applications such as BioVec, an application of word2vec to proteins and gene vectors.[109] Extracting drug-drug interactions[110] and drug repurposing[111] are two more examples of word2vec application.

Where is it used in drug discovery and development (and thoughts on where it is going at the end)

NLP has applications in summarizing documents, speech recognition, and chatbots. A recent patent[112] has been filled which details a system and a method for delivering targeted advertising by receiving verbal requests, processing, and interpreting the utterance and selecting advertisement based on the determined context.

Apache Solr has used for indexing both structured and unstructured data from studies in DataCelerate platform as a part of BioCelerate's Toxicology Data Sharing Initiative[113] launched in May 2018. The platform aims to integrate high-quality toxicology data from various sources and different pharmaceutical companies and scientific institutions. Linguamatics IQVIA I2E AMP 2.0 tool also has been used to search and analyze ADRs and drug-drug interactions.[114] Chillbot[115] online NLP app integrates gene names with PubMed publications used to discover relations with other genes. Not only literature search can be automated but also the experimental design itself. Hybrow (Hypothesis Browser) developed in Nigam Shah lab is a system for computer-aided hypothesis evaluation, experimental design and model building.[116]

BioBERT (Bidirectional Encoder Representations from Transformers for Biomedical Text Mining)[117] is described by the authors as a "pretrained biomedical language representation model" that can be used for text mining of biomedical content. It is based on the pretrained BERT model (Bidirectional Encoder Representations from Transformers model)[118] which allows for tokens to be analyzed in a multidirectional (left to right and right to left) manner which can help with understanding context).

BeFree is a system that was developed to identify relationships between entities such as drugs and genes with diseases.[119] The authors found that only some of the gene-disease associations that were discovered were also collected in curated databases highlighting the need to find alternative ways of curating information rather than just manually. DigDee is another example of a tool using NLP for data extraction of disease–gene relationships. This method **also takes** into account biological events such as phosphorylation and mutations and the authors argue this may represent abnormal behavior caused by the disease.[120]

Adverse event detection is another area of use for NLP in the drug discovery and understanding pipeline. This has been an area of consideration for some time and in one study, Melton and Hripcsak used NLP to assess discharge summaries of patients and was found to be an effective method.[121]

Another proposed use for natural language processing relates to mining historic clinical narratives which could reveal important information. In the study,[122] the authors focus on chronic diseases, stating that globally there are growing incidences of chronic illnesses and electronic health records are used to analyze patient data that are used to aid chronic disease risk prediction. Historic data can be used to support these predictions.

References

1. Failli M, Paananen J, Fortino V. Prioritizing target-disease associations with novel safety and efficacy scoring methods. *Nat Sci Rep* 2019;**9**.
2. Katsila T, Spyroulias GA, Patrinos GP, Matsoukas MT. Computational approaches in target identification and drug discovery. *Comput Struct Biotechnol J* 2016;**14**:177–84.
3. Illuminating the Druggable Genome (IDG) | National Center for Advancing Translational Sciences. Available at: https://ncats.nih.gov/idg. [Accessed 4 September 2020].
4. IDG and Nature Reviews Drug Discovery Target Watch | NIH Common Fund. Available at: https://commonfund.nih.gov/IDG/NRDD [Accessed 4 September 2020].
5. Di Fusco D, et al. Antisense oligonucleotide: Basic concepts and therapeutic application in inflammatory bowel disease. *Front Pharmacol* 2019;**10**.
6. Goyal N, Narayanaswami P. Making sense of antisense oligonucleotides: a narrative review. *Muscle Nerve* 2018;**57**:356–70.
7. Rhrissorrakrai K, et al. Understanding the limits of animal models as predictors of human biology: Lessons learned from the sbv IMPROVER Species Translation Challenge. *Bioinformatics* 2015;**31**:471–83.
8. Bracken MB. Why animal studies are often poor predictors of human reactions to exposure. *J R Soc Med* 2009;**102**:120–2.
9. Denayer T, Stöhrn T, Van Roy M. Animal models in translational medicine: validation and prediction. *New Horizons Transl Med* 2014;**2**:5–11.
10. Santos R, et al. A comprehensive map of molecular drug targets Europe PMC Funders Group. *Nat Rev Drug Discov* 2017;**16**:19–34.
11. Gashaw I, Ellinghaus P, Sommer A, Asadullah K. What makes a good drug target? *Drug Discov Today* 2011;**16**:1037–43.
12. Ajami ABA. *36 online tools for target hunting in drug discovery | BioPharmaTrend*; 2017. Available at: https://www.biopharmatrend.com/post/45-27-web-resources-for-target-hunting-in-drug-discovery/. [Accessed 26 July 2020].

13. Siramshetty VB, et al. WITHDRAWN—a resource for withdrawn and discontinued drugs. *Nucleic Acids Res* 2016;**44**:D1080–6.

14. DisGeNET—a database of gene-disease associations. Available at: https://www.disgenet.org/ [Accessed 26 July 2020].

15. Home—open targets. Available at: https://www.opentargets.org/. [Accessed 26 July 2020].

16. Koscielny G, et al. Open Targets: a platform for therapeutic target identification and Validation. *Nucleic Acids Res* 2017;**45**:D985–94.

17. Chen X, Ji ZL, Chen YZ. TTD: therapeutic target database. *Nucleic Acids Res* 2002;**30**:412–5.

18. DrugBank. Available at: https://www.drugbank.ca/ [Accessed: 26th July 2020].

19. SuperTarget. Available at: http://insilico.charite.de/supertarget/. [Accessed 26 July 2020].

20. Home | IUPHAR/BPS Guide to PHARMACOLOGY. (2015). Available at: https://www.guidetopharmacology.org/ [Accessed 31 July 2020].

21. Barnash KD, James LI, Frye SV. Target class drug discovery. *Nat Chem Biol* 2017;**13**:1053–6.

22. Karczewski KJ, Snyder MP. Integrative omics for health and disease. *Nat Rev Genet* 2018;**19**:299–310.

23. Paananen J, Fortino V. An omics perspective on drug target discovery platforms. *Brief Bioinform* 2019. bbx122.

24. Ferrero E, Agarwal P. Connecting genetics and gene expression data for target prioritisation and drug repositioning. *BioData Min* 2018;**11**.

25. Ferrero E, Dunham I, Sanseau P. In silico prediction of novel therapeutic targets using gene-disease association data. *J Transl Med* 2017;**15**:182.

26. Oti M, Snel B, Huynen MA, Brunner HG. Predicting disease genes using protein-protein interactions. *J Med Genet* 2006;**43**:691–8.

27. Popov P, Bizin I, Gromiha M, Kulandaisamy A, Frishman D. Prediction of disease-associated mutations in the transmembrane regions of proteins with known 3D structure. *PLoS One* 2019;**14**, e0219452.

28. Madhukar NS, et al. A Bayesian machine learning approach for drug target identification using diverse data types. *Nat Commun* 2019;**10**.

29. Zeng X, et al. Target identification among known drugs by deep learning from heterogeneous networks. *Chem Sci* 2020;**11**:1775–97.

30. Mathai N, Chen Y, Kirchmair J. Validation strategies for target prediction methods. *Brief Bioinform* 2020;**21**:791–802.

31. Gfeller D, Michielin O, Zoete V. Shaping the interaction landscape of bioactive molecules. *Bioinformatics* 2013;**29**:3073–9.

32. HitPick: hit identification & target prediction. Available at: http://mips.helmholtz-muenchen.de/hitpick/cgi-bin/index.cgi?content=help.html [Accessed 16 January 2020].

33. Keiser MJ, et al. Relating protein pharmacology by ligand chemistry. *Nat Biotechnol* 2007;**25**:197–206.

34. Reker D, Rodrigues T, Schneider P, Schneider G. Identifying the macromolecular targets of de novo-designed chemical entities through self-organizing map consensus. *Proc Natl Acad Sci U S A* 2014;**111**:4067–72.

35. Chen B, Ding Y, Wild DJ. Assessing drug target association using semantic linked data. *PLoS Comput Biol* 2012;**8**, e1002574.

36. Semantic Link Association Prediction. Available at: http://cheminfov.informatics.indiana.edu:8080/slap/ [Accessed 21 January 2020].

37. Mervin LH, et al. Target prediction utilising negative bioactivity data covering large chemical space. *J Cheminform* 2015;**7**:51.

38. GitHub—lhm30/PIDGIN: prediction IncluDinG INactives (**OLD VERSION**) NEW VERSION: https://github.com/lhm30/PIDGINv2). Available at: https://github.com/lhm30/PIDGIN [Accessed 5 August 2020].

39. SwissTargetPrediction. Available at: http://www.swisstargetprediction.ch/ [Accessed 16 January 2020].

40. Daina A, Michielin O, Zoete V. SwissTargetPrediction: updated data and new features for efficient prediction of protein targets of small molecules. *Nucleic Acids Res* 2019;**47**:W357–64.

41. Vitsios D, Petrovski S. Mantis-ml: disease-agnostic gene prioritization from high-throughput genomic screens by stochastic semi-supervised learning. *Am J Hum Genet* 2020;**106**:659–78.
42. CKDdb. Available at: http://www.padb.org/ckddb/ [Accessed 5 August 2020].
43. GWAS Catalog. Available at: https://www.ebi.ac.uk/gwas/ [Accessed 5 August 2020].
44. gnomAD. Available at: https://gnomad.broadinstitute.org/ [Accessed 5 August 2020].
45. GSEA | MSigDB. Available at: https://www.gsea-msigdb.org/gsea/msigdb/index.jsp [Accessed 5 August 2020].
46. Senior AW, et al. Improved protein structure prediction using potentials from deep learning. *Nature* 2020;**577**:706–10.
47. Moreau Y, Tranchevent LC. Computational tools for prioritizing candidate genes: boosting disease gene discovery. *Nat Rev Genet* 2012;**13**:523–36.
48. Guala D, Sonnhammer ELL. A large-scale benchmark of gene prioritization methods. *Nat Sci Rep* 2017;**7**.
49. Ashburner M, et al. Gene ontology: tool for the unification of biology. *Nat Genet* 2000;**25**:25–9.
50. Networks of functional couplings/associations—FunCoup. Available at: http://funcoup.sbc.su.se/search/ [Accessed 7 August 2020].
51. Ogris C, Guala D, Mateusz K, Sonnhammer ELL. FunCoup 4: new species, data, and visualization. *Nucleic Acids Res* 2018;**46**:D601–7.
52. Köhler S, Bauer S, Horn D, Robinson PN. Walking the interactome for prioritization of candidate disease genes. *Am J Hum Genet* 2008;**82**:949–58.
53. Tranchevent LC, et al. Candidate gene prioritization with Endeavour. *Nucleic Acids Res* 2016;**44**:W117–21.
54. Tranchevent LC, et al. ENDEAVOUR update: a web resource for gene prioritization in multiple species. *Nucleic Acids Res* 2008;**36**:W377–84.
55. ElShal S, et al. Beegle: from literature mining to disease-gene discovery. *Nucleic Acids Res* 2016;**44**, e18.
56. Britto R, et al. GPSy: a cross-species gene prioritization system for conserved biological processes-application in male gamete development. *Nucleic Acids Res* 2012;**40**:W458–65.
57. Zakeri P, Simm J, Arany A, Elshal S, Moreau Y. Gene prioritization using Bayesian matrix factorization with genomic and phenotypic side information. *Bioinformatics* 2018;**34**:i447–56.
58. Paliwal S, de Giorgio A, Neil D, Michel JB, Lacoste AM. Preclinical validation of therapeutic targets predicted by tensor factorization on heterogeneous graphs. *Sci Rep* 2020;**10**.
59. Guala D, Sjölund E, Sonnhammer ELL. Maxlink: network-based prioritization of genes tightly linked to a disease seed set. *Bioinformatics* 2014;**30**:2689–90.
60. GUILD (software). Available at: http://sbi.imim.es/web/index.php/research/software/guildsoftware?page=guild.software#SECTION00041000000000000000 [Accessed 9 August 2020].
61. Chen J, Bardes EE, Aronow BJ, Jegga AG. ToppGene Suite for gene list enrichment analysis and candidate gene prioritization. *Nucleic Acids Res* 2009;**37**:W305–11.
62. Xiang B, et al. PageRank with priors: An influence propagation perspective. In: *IJCAI international joint conference on artificial intelligence*; 2013. p. 2740–6.
63. Guney E, Oliva B. Exploiting protein-protein interaction networks for genome-wide disease-gene prioritization. *PLoS One* 2012;**7**, e43557.
64. Nabieva E, Jim K, Agarwal A, Chazelle B, Singh M. Whole-proteome prediction of protein function via graph-theoretic analysis of interaction maps. *Bioinformatics Suppl* 2005;**1**:302–10.
65. Schwikowski B, Uetz P, Fields S. A network of protein-protein interactions in yeast. *Nat Biotechnol* 2000;**18**:1257–61.
66. Hishigaki H, Nakai K, Ono T, Tanigami A, Takagi T. Assessment of prediction accuracy of protein function from protein-protein interaction data. *Yeast* 2001;**18**:523–31.
67. Bean DM, Al-Chalabi A, Dobson RJB, Iacoangeli A. A knowledge-based machine learning approach to gene prioritisation in amyotrophic lateral sclerosis. *Genes (Basel)* 2020;**11**:668.

68. Isakov O, Dotan I, Ben-Shachar S. Machine learning-based gene prioritization identifies novel candidate risk genes for inflammatory bowel disease. *Inflamm Bowel Dis* 2017;**23**:1516–23.

69. Nicholls HL, et al. Reaching the end-game for GWAS: machine learning approaches for the prioritization of complex disease loci. *Front Genet* 2020;**11**.

70. Xue H, Li J, Xie H, Wang Y. Review of drug repositioning approaches and resources. *Int J Biol Sci* 2018;**14**:1232–44.

71. Malas TB, et al. Drug prioritization using the semantic properties of a knowledge graph. *Sci Rep* 2019;**9**.

72. Ping P, Watson K, Han J, Bui A. Individualized knowledge graph: a viable informatics path to precision medicine. *Circ Res* 2017;**120**:1078–80.

73. Mohamed SK, Nounu A, Nováček V. Drug target discovery using knowledge graph embeddings. In: *Proceedings of the ACM symposium on applied computing, Association for Computing Machinery*; 2019. p. 11–8. https://doi.org/10.1145/3297280.3297282.

74. Ehrlinger L, Wöß W. Towards a definition of knowledge graphs. In: *CEUR workshop proceedings*; 2016.

75. Vincent J. *Apple boasts about sales; Google boasts about how good its AI is. The Verge*; 2016. Available at: https://www.theverge.com/2016/10/4/13122406/google-phone-event-stats.

76. Pavlopoulos GA, et al. Using graph theory to analyze biological networks. *BioData Min* 2011;**4**.

77. Dijkstra EW. A note on two problems in connexion with graphs. *Numer Math* 1959;**1**:269–71.

78. Kruskal JB. On the shortest spanning subtree of a graph and the traveling salesman problem. *Proc Am Math Soc* 1956;**7**:48–50.

79. Ghasemi M, Seidkhani H, Tamimi F, Rahgozar M, Masoudi-Nejad A. Centrality measures in biological networks. *Curr Bioinforma* 2014;**9**:426–41.

80. Golbeck J. *Analyzing the social web*. Morgan Kaufmann—Elseiver; 2013.

81. Freeman LC. A set of measures of centrality based on betweenness. *Sociometry* 1977;**40**:35–41.

82. Newman M. *Networks: an introduction*. Oxford University Press; 2010.

83. Austin D. *How Google Finds Your Needle in the Web's Haystack*; 2006. p. 1–9.

84. Grover A, Leskovec J. *Node2Vec: scalable feature learning for networks*. ACM; 2016. p. 855–64.

85. Dong Y, Chawla NV, Swami A. Metapath2vec: Scalable representation learning for heterogeneous networks. In: *Proceedings of the ACM SIGKDD international conference on knowledge discovery and data mining, Association for Computing Machinery*; 2017. p. 135–44. https://doi.org/10.1145/3097983.3098036.

86. Hoff PD, Raftery AE, Handcock MS. Latent space approaches to social network analysis. *J Am Stat Assoc* 2002;**97**:1090–8.

87. Nickel M, Tresp V, Kriegel H-P. *A three-way model for collective learning on multi-relational data*; 2011. p. 809–16.

88. Trouillon T, et al. Knowledge graph completion via complex tensor factorization. *J Mach Learn Res* 2017;**18**:1–38.

89. Kazemi SM, Poole D. In: Bengio S, et al., editors. *SimplE embedding for link prediction in knowledge graphs*. Curran Associates, Inc.; 2018. p. 4284–95.

90. Dettmers T, Minervini P, Stenetorp P, Riedel S. *Convolutional 2D knowledge graph embeddings*. arXiv1707.01476 [cs]; 2018.

91. Balažević I, Allen C, Hospedales TM. *Hypernetwork knowledge graph embeddings*. arXiv:1808.07018 [cs, stat]; 2019. https://doi.org/10.1007/978-3-030-30493-5_52.

92. Balažević I, Allen C, Hospedales TM. TuckER: tensor factorization for knowledge graph completion. In: *Proc. 2019 conf empir methods nat lang process. 9th int jt conf nat lang process*; 2019. p. 5185–94. https://doi.org/10.18653/v1/D19-1522.

93. Tucker LR. Some mathematical notes on three-mode factor analysis. *Psychometrika* 1966;**31**:279–311.

94. Tang J, et al. LINE: large-scale information network embedding. In: *International World Wide Web conferences steering committee*; 2015. p. 1067–77. https://doi.org/10.1145/2736277.2741093.

95. Perozzi B, Al-Rfou R, Skiena S. *DeepWalk: online learning of social representations*. ACM; 2014. p. 701–10.

96. Qiu J, et al. *Network embedding as matrix factorization: unifying deepwalk, LINE, PTE, and Node2Vec.* ACM; 2018. p. 459–67.

97. Bronstein MM, Bruna J, LeCun Y, Szlam A, Vandergheynst P. Geometric deep learning: going beyond euclidean data. *IEEE Signal Process Mag* 2017;**34**:18–42.

98. Battaglia PW, et al. *Relational inductive biases, deep learning, and graph networks.* CoRR abs/1806.0; 2018.

99. Greene AC, Giffin KA, Greene CS, Moore JH. Adapting bioinformatics curricula for big data. *Brief Bioinform* 2016;**17**:43–50.

100. Natural Language Toolkit — NLTK 3.5 documentation. Available at: https://www.nltk.org/ [Accessed 23 October 2020].

101. spaCy—Industrial-strength Natural Language Processing in Python. Available at: https://spacy.io/ [Accessed 5 February 2020].

102. Python Software Foundation. Welcome to Python.org. (2016). Available at: https://www.python.org/ [Accessed 27 June 2018].

103. Natural Language Processing is Fun!—Adam Geitgey—Medium. Available at: https://medium.com/@ageitgey/natural-language-processing-is-fun-9a0bff37854e [Accessed 5 February 2020].

104. Zheng J, Chapman WW, Crowley RS, Savova GK. Coreference resolution: a review of general methodologies and applications in the clinical domain. *J Biomed Inform* 2011;**44**:1113–22.

105. Hu X. Natural language processing and ontology-enhanced biomedical literature mining for systems biology. In: *Computational systems biology.* Elsevier; 2006. p. 39–56.

106. Gunnar D, Cédric B, Marie-Christine J. Automatic generation of MedDRA terms groupings using an ontology. *Stud Health Technol Inform* 2012;**180**:73–7.

107. El-Sappagh S, Franda F, Ali F, Kwak K-S. SNOMED CT standard ontology based on the ontology for general medical science. *BMC Med Inform Decis Mak* 2018;**18**:76.

108. Fernandes LL, Murray S, Taylor JMG. Multivariate Markov models for the conditional probability of toxicity in phase II trials. *Biom J* 2016;**58**:186–205.

109. Asgari E, Mofrad MRK. Continuous distributed representation of biological sequences for deep proteomics and genomics. *PLoS One* 2015;**10**, e0141287.

110. Kavuluru R, Rios A, Tran T. Extracting drug-drug interactions with word and character-level recurrent neural networks. In: *IEEE int conf healthc informatics IEEE int conf healthc informatics 2017*; 2017. p. 5–12.

111. Patrick MT, et al. Drug repurposing prediction for immune-mediated cutaneous diseases using a word-embedding–based machine learning approach. *J Invest Dermatol* 2019;**139**:683–91.

112. US20190087866A1—system and method for delivering targeted advertisements and/or providing natural language processing based on advertisements—Google Patents; 2007.

113. Mangipudy WHR. *TOXICOLOGY DATA SHARING: leveraging data sharing to enable better decision making in research and development*; 2019. Available at: https://transceleratebiopharmainc.com/wp-content/uploads/2019/10/BioCelerate-TDS-Awareness-Manuscript_October2019.pdf.

114. Ventola C, Big L. Data and pharmacovigilance: data mining for adverse drug events and interactions. *P T A Peer-Reviewed J Formul Manag* 2018;**43**:340–51.

115. Chen H, Sharp BM. Content-rich biological network constructed by mining PubMed abstracts. *BMC Bioinform* 2004;**5**.

116. Racunas SA, Shah NH, Albert I, Fedoroff NV. HyBrow: a prototype system for computer-aided hypothesis evaluation. *Bioinformatics* 2004;**20**:i257–64.

117. Lee J, et al. BioBERT: a pre-trained biomedical language representation model for biomedical text mining. *Bioinformatics* 2020;**36**:1234–40.

118. Devlin J, Chang M-W, Lee K, Toutanova K. BERT: pre-training of deep bidirectional transformers for language understanding. In: *NAACL HLT 2019—2019 conf North Am chapter assoc comput linguist hum lang technol proc conf*; 2018. p. 4171–86.

119. Bravo À, Piñero J, Queralt-Rosinach N, Rautschka M, Furlong LI. Extraction of relations between genes and diseases from text and large-scale data analysis: implications for translational research. *BMC Bioinform* 2015;**16**.
120. Kim J, Kim JJ, Lee H. An analysis of disease-gene relationship from Medline abstracts by DigSee. *Sci Rep* 2017;**7**:1–13.
121. Melton GB, Hripcsak G. Automated detection of adverse events using natural language processing of discharge summaries. *J Am Med Inform Assoc* 2005;**12**:448–57.
122. Sheikhalishahi S, et al. Natural language processing of clinical notes on chronic diseases: systematic review. *J Med Internet Res* 2019;**7**, e12239.

Hit discovery

5

Hannes Whittingham and Stephanie Kay Ashenden

Data Sciences and Quantitative Biology, Discovery Sciences, R&D, AstraZeneca, Cambridge, United Kingdom

Hit discovery is concerned with the identification of a compound that "hits" or interacts with a target in a biological sense. However, the use of this term varies between researchers, and it can also be described as a compound showing desired activity in a screen, which is confirmed in retesting.[1] In this phase of the drug discovery process, screening assays are developed, and tools and methods to identify new compounds, such as high-throughput screening, play an important role.

Chemical space

Chemical space refers to the set of all molecules that could possibly exist.[2] The size of drug-like chemical space is subject to debate and has been estimated to lie anywhere between 10^{23} and 10^{180} compounds depending on the way it is calculated, as explained in Polishchuk et al.[3] However, an oft-cited middle ground for the number of synthetically accessible small organic compounds is 10^{60}.[4] This figure is based on a molecule that contains 30 atoms (C, N, O, or S), 4 rings and up to 10 branch points, but of course, the larger the structures one considers, the larger the number of potential compounds.

The immense size of chemical space makes the identification of therapeutic compounds an extremely difficult task. Huge swathes of this space are occupied by molecules of no medicinal interest, which could never be viable as drugs. Identifying a way to focus search methods is therefore essential in finding these rare potential leads. As an example, natural products have been shown to be a good source of therapeutically relevant leads.[5]

However, search methods need to remain broad to cover as much chemical space as possible. Part of the problem regarding the exploration of chemical space has been due to the lack of diversity in many compound libraries used today.[6] Chemical libraries containing the greatest diversity will likely be most promising. For example, a high success rate when screening natural products,[8] combined with the knowledge that they have been biosynthesized in nature to modulate biological processes, has led to an assumption that the chemical space where these natural products reside has good potential for the discovery of biologically relevant molecules.[7] However, it would be incorrect to assume that exclusively those areas of chemical space occupied by natural products contain all biologically active compounds.[7] Vast areas of chemical space that may have good drug-like properties remain unexplored, and so it is important to ask the question: how can we efficiently explore these regions and find the promising leads that they contain?[9]

The Era of Artificial Intelligence, Machine Learning, and Data Science in the Pharmaceutical Industry. https://doi.org/10.1016/B978-0-12-820045-2.00006-4

Techniques to reduce the number of compounds that need to be screened while increasing their relative diversity are crucial to making this task feasible, and good visualization techniques can provide aid in understanding chemical diversity and the associated changes in physicochemical properties. Both methods can be used to intelligently expand the diversity in chemical libraries to aid in the exploration of chemical space. Appropriate analysis of chemical space has enabled a broad scope of studies and applications, such as library design and the development of compound collections, in both the public and industrial domains. However, it is important to be aware that the choice of descriptors and parameters used to represent molecules can have a dramatic impact on their distribution over the chemical space they define.[10]

Screening methods

The work of Hughes et al.[1] on the early stages of the drug discovery process divides the different screening methods used in hit discovery into High Throughput Screening (HTS), focused screening, fragment screening, physiological screening, NMR screening, and virtual screening.

A main focus in screening library design is diversity.[11, 12] Several methods, and various combinations of these methods, have been explored[13] to increase diversity in libraries due to the concern it poses: in many cases, a screening process has not delivered a lead for a biologically relevant target, which can be attributed at least in part to the insufficient variation provided by libraries consisting of clusters of similar compounds.[14]

However, there is a need to identify which parameters are relevant in creating the diversity we need: often, a seemingly minor change to a compound such as an alternative functional group can have dramatic effects. As an example, tetrazole, a heterocycle, can be used as a substitute for carboxylic acid groups. While having some similar properties such as pK_a, it differs in that it is much more lipophilic and can therefore move through a cell membrane much more readily.[15, 16] Some discussion as to the true nature of the desired diversity here has suggested that diversity of structure itself should not be the goal, and that the breadth of a library should instead be judged by the variation in the behavior of its constituent compounds and other external criteria.[17] The design of adequate screening libraries must also ensure that they contain molecules with the properties associated with successful drug-like compounds.[18]

High-throughput screening

A part of the solution to the challenges posed by the scale of chemical space has been the development of the ability to screen compounds against targets as quickly and efficiently as possible. HTS utilizes robotics and automated technologies to allow pharmacological tests to be conducted rapidly and at scale. The capacity to screen large numbers of compounds in assays against biological targets at the same time has made HTS a powerful tool in the search for new medicines,[19] with a roughly 60% success rate in finding leads across targets.[20] Furthermore, in a study of 58 drugs approved between 1991 and 2008, 19 showed origins starting with HTS—impressive, given that the adoption of HTS only began in earnest in the mid-1990s.[20] Despite these successes, there is now broad consensus that HTS alone is insufficient to produce the leads that the industry requires.[20]

Computer-aided drug discovery

Computer-aided drug discovery can be split into structure-based and ligand-based approaches.[21] Structure-based drug design methods make use of target information, while ligand-based methods make use only of the known ligands with bioactivity. It has been suggested that the use of these methods to complement experimental approaches is a powerful tool for rational drug design.[22] Macalino and co-authors review computer-aided drug design and summarize its constituent methods: for example, structure-based methods involve binding site identification, docking and scoring, while ligand-based drug design comprises of quantitative structure-activity relationship (QSAR) modeling and pharmacophore modeling.[22]

De novo design

De novo design is a cost-effective and time-efficient method that attempts to generate novel pharmaceutically active agents (molecules, peptides, etc.) with desirable properties[23] and synthetic accessibility. It attempts to produce these structures from scratch computationally, prior to any real-world experimentation. De novo design software uses local optimization strategies: while full exploration of chemical space to find a global optimum is infeasible, a practical local solution can often be found based on existing chemical knowledge.[23] Methods used to assess the suitability of the generated structures for drug design includes structure-activity relationship models, relationships to existing ligands with known properties, macromolecular target structures and individually acquired domain knowledge.[24] This task is, however, made more challenging by the commonly observed sensitivity of ligands to small changes in structure; design efforts frequently encounter activity cliffs, where two compounds that are very similar structurally have large differences in potency.[25] Further to this, there are a very large number of such small changes that can be made to any given starting molecule, and each one may affect several important variables relevant to a potential drug. The complex, multifaceted nature of this problem means that methods like pattern recognition can be useful in the discovery of new chemical entities.[24, 26] Design methods can be split into positive design and negative design. Positive design ensuring that only the areas of chemical space that are more likely to find potential candidates with such described properties are explored whereas negative design aims to prevent adverse proteins and unwanted structures.[27] Various scoring methods are used to select the best compounds.[24, 28]

There are three core challenges in de novo design: structure generation, scoring and optimization.[23, 24] From a large pool of generated compounds, scoring identifies the best compounds to take forward into optimization. A variety of scoring methods are used, some based on the receptor (docking and pharmacophore methods) and others on known ligands (similarity methods).[27]

There are a wide variety of de novo drug design-based tools available; Suryanarayanan and co-authors provide a comprehensive list of both free to use and commercial software for performing de novo design,[29] and Hartenfeller and Schneider also provide such a list of software published since 2005.[27] Many early tools were atom-based, allowing for small, detailed changes in structure, and meaning that in theory the whole of chemical space could be accessed. In practice, however, such fine-grained methods are usually observed to generate a high proportion of unsuitable structures.[27] Fragment-based approaches are one approach to solving this, and reduce the search space by using fragments for molecular assembly, especially if those fragments already occur in drug molecules.[27]

Kawai et al. developed a fragment-based evolutionary approach to the de novo design of drug-like molecules, using an active reference molecule to aid in navigating chemical space and to seed fragments.[30] Starting with a first set of structures prepared using the seed fragments and others from a fragment library, offspring were generated by mutation (addition, removal, or replacement of a fragment), and cross over (exchange of fragment sets from two different parent molecules), subject to a set of chemistry-based fragment connection rules. The Tanimoto coefficient was then used to evaluate their fitness and determine which structures to take further, and these steps were repeated to generate a pool of candidate drug-like molecules.

A review by Suryanarayanan and co-authors lists six different computational methods for de novo design: fragment positioning, site point connection, fragment connection, library construction, molecule growth, and random connection.[29] The connection methods (fragment connection, site point connection, and random connection) relate to connecting fragments together in some form. For example, in random connection methods, fragments are connected in a random manner, whereas in fragment connection methods the fragments are connected at particular positions within a binding site.[29] Site point connections identify unique sites for placing the fragments within the binding site.[29]

Another approach is that of the de novo drug design tool FOG (Fragment Optimized Growth).[31] The algorithm aims to generate molecules that sit within drug-like chemical space by statistically biasing the growth of molecules to have desired features. The method uses a Markov Chain (transitions from one state to another according to probabilistic rules[32]) to add fragments in a biased manner to the new molecule.[31] This is dependent on the frequencies of specific fragment-fragment connections within the training database.[31]

DOGS (Design of Genuine Structures) is another de novo drug design method that is instead reaction-based.[33] The algorithm places emphasis on the synthetic accessibility of the designed molecules by limiting the program to construction pathways that are possible synthesis routes.[33] This is done by basing the assembly process on available molecular building blocks and established chemical reactions.[33] An iterative process of fragment assembly is used to generate new molecules: each intermediate product has a chemical reaction applied to it, chosen based on a favorability score, and then a new reactant is selected to grow the molecule.[33] Tools have been built to help to enhance the reaction framework, such as the multilabel Reaction Class Recommender.[34] This works as a filter, which, for a given starting material, identifies a set of recommended reaction classes and removes molecules that are less synthetically feasible.[34]

Chemical space-based methods have also been applied to generate drug-like molecules.[35] DAECS (de novo algorithm for exploring chemical space) was developed to first visualize the distribution of properties across chemical space on a 2D plane. Initial structures are then selected by the user to seed the generation of new ones.[35] The new structures are filtered, and those passing are added to a pool from which the seeds for the next generation are selected, in an iterative process.[35] REINVENT is another publicly available tool that makes use of generative models to explore chemical space, aided by the use of diversity filters.[36]

Examples of atom-based de novo drug design include DLD,[37] GenStar,[38] and LEGEND.[39] DLD stands for Dynamic Ligand Design,[37] in which molecules are generated using Monte-Carlo simulated annealing. This technique was shown to successfully construct ligands that complemented the shape and charge distribution of the binding site.[37] GenStar also suggests structures that can fill the active sites of enzymes. In this case, the structures are composed of sequentially-grown sp3 carbons (rings and branch points are also permitted). Seed atoms are selected to determine where in the structure the generation begins, or a predocked inhibitor core can be used as a base on which to add atoms. An enzyme contact model is then used to score the generated candidates.[38] Finally, LEGEND uses stochastic search methods to achieve the same goal. LEGEND works by first selecting an anchor atom within the protein to which a hydrogen bond can be formed, and positioning the first suggested atom within the generated ligand so as to form this bond. The program then adds new atoms in an iterative process, which involves the following steps: (i) random selection of a root atom from those already placed; (ii) random positioning of a new atom, subject to rules on bond length and angle from the chosen root; and (iii) a check that the new position does not violate the van der Waals radii of existing atoms or have an excessively high van der Waals interaction energy. If this check is failed, the three-step process is repeated. Finally, the generation of the new structure is terminated by the completion of fragmentary aromatic rings, the addition of hydrogens to fulfil valence rules, and Simplex optimization of the finished structure.[39]

Virtual screening

The preceding sections have emphasized some of the core problems faced in the discovery of new hits, which stem from the scale of chemical space and the difficulty of exploring it effectively. "Virtual Screening" (VS) refers to a collection of computational methods which help to address these problems by identifying likely hits via in silico experimentation. In the foreseeable future, these techniques can never replace real-life screening—but they can offer considerable value in focusing the search, increasing hit rates by intelligent selection of compounds and driving down the time and investment required to reach a satisfactory lead.

Machine learning techniques for VS have increased rapidly in performance and in popularity over recent years in accordance with the general trend, but non-ML techniques are still more widely used, and three main categories of non-ML techniques are summarized below. Traditionally, different VS methods have been categorized depending primarily on whether they require structural information on the target as input: techniques using this information are referred to as structure-based virtual screening methods (SBVS) and those using the ligand only as Ligand-based (LBVS).

- **Docking** is a well-established and heavily used SBVS method which effectively fits a simulated ligand into a simulated active site.[40] It can model totally new ligands and scaffolds in the absence of preexisting activity data and offer in-depth understanding of the nature of ligand-target interactions, revealing clear physical reasons why particular features of the compound result in effective binding. However, it also requires knowledge of the target structure which may not be available, and necessarily depends on approximations of the underlying physics and thermodynamics: in particular, the results can be very sensitive to the *scoring function* used to evaluate the entropic favorability of a given pose of the ligand and target.[41]
- **Pharmacophore modeling** refers to techniques which focus on pharmacophores—partial structures within a drug molecule responsible for biological activity.[42] These may fit into the structure- or ligand-based categories[43]: some examine the shape and properties of unoccupied spaces within known protein structures to suggest matching pharmacophores[44]; others look for common themes among sets of known actives to infer the crucial features. It is hard to generalize about the strengths and weaknesses of these techniques as they are extremely diverse and have been in development for a century, but a 2010 review by Yang suggests that they are often found to suffer from a high false positive rate—that is, a low proportion of virtual hits hold up on experimental verification.[43]

- **Similarity methods** are a purely ligand-based technique and work on the principle that similar molecules have similar properties. In general, they require (1) a way of encoding a molecule into a descriptor and (2) a metric describing the similarity between different compound descriptors, which together allow compounds in a virtual library to be ranked according to their similarity to experimentally verified actives. Suggestions can then be made about likely activity on this basis. A broad range of 1D, 2D, and 3D descriptors have been used in such methods, ranging from simple physicochemical descriptors to representations of full 3D volumes.[45] Although various distance metrics have been used, the Tanimoto coefficient is by far the most common and found to perform well where comparisons are made.[46] As one of the simplest VS tools available, similarity-based methods have been widely used, but they extrapolate poorly beyond a very short distance in chemical space and are in most cases used just to identify a few potential actives before further VS steps using more advanced techniques.[47]

Since our goal in this book is primarily to introduce the reader to comparatively new methods based on data science, machine learning, and AI, we will focus for the remainder of this section on supervised machine learning methods for VS. While the body of research on this topic is large, we aim to provide the reader with some sense of the key directions in this work in the recent past, up to the time of writing and point out and discuss the main practical steps and challenges for practitioners. To this end, the next few sections in this chapter discuss: (1) data collection and curation for this purpose; (2) the representation of compounds to machine learning algorithms; (3) a selection of the most commonly used supervised learning algorithms, including the current state-of-the-art neural networks; (4) some suggested improvements to these models that are currently being explored in the field; and (5) how best to rigorously test the performance of such models before their application in anger.

Data collection and curation

The first task in any machine learning project is not an enviable one. Collecting and cleaning data to train ML algorithms for VS presents its own unique set of difficulties for which no perfect solutions exist—but it is hoped that this section can raise awareness of these and suggest the outlines of a suitable approach.

Before discussion of the fine details of databases and individual fields, it is important to note that is still some controversy in the field about the best philosophy: should a training set for QSAR modeling optimize quality and internal consistency, or size? Unfortunately, there is an unavoidable trade-off here. While it might be hoped that any two measurements of binding affinity between the same compound and protein would be interchangeable, this is not the case in reality. As will be discussed in more detail below, different variants of the same protein, different assay conditions, and different assay techniques with different bioactivity measurement types all create questions over whether different sources of data can be combined into the same training set. Counter to this runs the obvious fact that machine learning algorithms are most powerful when trained on huge datasets.

Historically, there has been an understandable prevailing view that consistency between measurements must be prioritized above all else. In *Pitfalls in QSAR*,[48] Cronin and Shultz argue that only data measured by the same protocol, and perhaps even only that measured by the same laboratory and workers, should be combined. However, more recently, there has been a shift in the other direction: to observe this, we can note the large number of recent publications that combine data from a variety of heterogeneous sources.[49–51] The most explicit and informative recent study of this issue of which the authors are aware also supports this view: Lagunin et al.'s *Rational Use of Heterogeneous Data in Quantitative Structure-Activity Relationship (QSAR) Modeling of Cyclooxygenase/Lipoxygenase Inhibitors.*[52] Here the authors experiment systematically with alternative approaches for combined use of data from different assay types, finding that the most accurate QSAR models were those trained on sets which combined diverse assay types into the same general set, and that scores were improved even by including records where the assay type was not actually known (a surprisingly common scenario!). These general-set models were found to be superior even when compared with models working by combining results from a set of individual models which were each built on data from single assay types in an ensemble.

Two further arguments can be made on the side of training VS algorithms on a combination of heterogeneous datasets. First a virtual screening model intended for general use across a vast chemical space must plainly be trained using a sample from across as much of that space as possible to extend its applicability domain. Second, models now coming into more frequent use, such as deep neural networks (DNNs) and random forests, are widely understood to thrive on the largest datasets and be relatively robust to noisy training sets.

Databases and access

The data used to train a supervised machine learning model for activity prediction is nearly always reused data from past large-scale assays. Access to large internal databases is an advantage for those affiliated with major pharmaceutical companies and their collaborators, but there are several publicly available or pay-to-access databases which offer data on the scale needed for machine learning. Some of the most widely used are as follows:

(1) ChEMBL[53]—A manually curated (and so high-quality) public database of bioactivity data, with data on 1.9 m compounds and 12,000 targets from 1.1 M assays at the time of writing.
(2) GOSTAR (Global Online Structure Activity Relationship Database)[54]—Another manually curated SAR database, larger still, but pay-to-access: check whether your organization is a client. 7.8 m compounds, 9000 targets, 8.7 m assays.
(3) PubChem BioAssay[55]—268 m bioactivity measurements on 18,000 targets from 1.1 m assays. Freely available, but not manually curated—so some caution suggested.

Databases such as these are often best accessed through their purpose-built API, which will provide functions for selecting and filtering their contents, or through a custom web-scraping script which programmatically accesses pages.

The data that must be compiled can be thought of as a table with three fields: the compound, the target, and an observed measure of activity. Each of these is discussed in turn in the following sections. While these are the fields used directly in modeling, practitioners are encouraged to retain all available fields, even if just in raw form, during collection and curation if possible: the importance of this additional data may only be realized at a later stage in the project.

Compounds

The compound is best represented at the stage of collection as a *canonical SMILES string*. SMILES are an unambiguous, compact representation of molecules which is human-readable and available from all notable databases. Several SMILES strings can represent the same compound, so *canonicalization* places the represented atoms and bonds in a standardized order—an important step to ensure the proper integration of data on the same compound.

Targets

While conventional models have been single-task, predicting an activity just for one target, there is increasing interest in *multitask* models which predict for a range of targets simultaneously (see later). A second field representing the target is important if the model will predict not just for one, but for multiple targets—but this can introduce complications when integrating data from multiple sources.

The choice of a suitable common labeling system to represent proteins can present difficulty for several reasons. To begin with, many nomenclatures exist for proteins in the scientific community, and data from different sources often use different identifiers or (nonoverlapping!) selections of identifiers. Some use common gene names, on which there are many variations, others Uniprot or Entrez Gene IDs, and others further systems still. Moreover, proteins in themselves present a range of further complexities: assay results which appear to pertain to the same compound and protein may have tested different conformations of the protein, or just fragments containing the active site, or the protein may have more than one distinct active site. Assay descriptions should be examined carefully to pin down these details if possible.

It is important to maintain awareness of these complexities when curating data and choose a protein labeling system with a level of granularity appropriate to the project at hand—or, if necessary, to create a customized system for the task which separates different forms as required.

Activity measurement

The last critical field is the measure of bioactivity. Measurements of binding affinity between a compound and protein are made using a variety of different assay types and in a variety of different conditions. Some important features of assays to look out for during data collection are as follows.

- **Broad assay types**. Bioactivity assays may be in vivo or in vitro. Notwithstanding the arguments above, these assay types are very fundamentally different, and caution should be taken over whether they should be combined into the same dataset.
- **Specific assay technologies**. There is a large diversity here; as argued earlier, the authors here take the position that data from different assay technologies can be combined, but awareness is still important.
- **Same technology, different conditions**. To give an example, binding affinity of compounds to kinases is usually measured by observing their competitiveness with the natural ATP substrate—but the concentrations of ATP used may vary.
- **Measurement types**. This is one of the thorniest issues. Activity may be measured as a single-shot inhibition percentage, an IC50 or pIC50 from multiple inhibition percentages, a K_d or K_i, a thermal shift value, or even just a simple binary active or inactive flag. If hoping to combine such data into the same set, one approach is to simplify the problem by binarizing all data into active and inactive classes and limiting the project to the construction of a classification model. This requires the choice of an appropriate threshold in each value type: an example of where a set of such thresholds have been chosen can be found in the methods of Giblin et al., whose successful models validate this approach.[56]

Cleaning collected data—Best practices

Curation of the collected data after collection is a necessary process, as any large combined dataset will contain inappropriately large or small compounds, duplicate measures of activity, and other undesired attributes. Fourches et al. have studied this problem in detail and published a set of guidelines, which include the following:

- Removal of compounds with invalid structures (unreadable SMILES, valence violations, etc.).
- Removal of non-druglike (or nonsmall-molecule-like) compounds (inorganics, organometallics, counterions, biologics, etc.).
- Verification of stereochemistry.
- Aggregation of duplicates, with deletion of perfect repeats, abandonment of suspiciously different measurements and combination of relatively consistent ones (e.g., by averaging or taking the median).

Fourches et al. provide a comprehensive discussion which describes many more potential issues and measures that can be taken beyond this, which can be seen in their work.[57–59] It is important to note, however, that in practical work on a machine learning project, the value gained from implementing extra steps in a workflow should always be weighed against time invested. If a given inconsistency or issue affects just 0.1% of a dataset, simply dropping the problematic records and moving on may be a better approach than adding a complex corrective process.

A final remark on the data collection and curation stage is that, here in particular, the proper cooperation of a multidisciplinary team with expertise in machine learning, medicinal chemistry and structural biology is crucial to the production of a useful VS model: there is no substitute for input from contacts with real domain knowledge of the data and its many complexities.

Representing compounds to machine learning algorithms

Morgan and ECFP fingerprints are the most commonly used representation of molecules for machine learning. These representations were developed for QSAR modeling, have been validated for model-building across a substantial body of literature. Various methods are used to map the very large number of possible circular groups of atoms ($\sim 10^{10}$ for fingerprints of radius 3) onto a reasonable fixed number of features for learning (typically $\sim 10^{2-3}$), such as hashing or selection of the groups showing highest variance across the dataset.

Graph convolutions are a new alternative that creates a learned representation of molecules starting right from the molecular graph itself and have now shown considerable promise as a way to improve on fingerprints. However, these are specific to neural networks, and so are fully discussed after the section on this subject below (see *Graph convolutional and message passing neural networks*). What follows describes past work that has been done primarily using fingerprints as features.

Candidate learning algorithms

A large diversity of different supervised learning algorithms have been used to produce successful machine learning models, but the majority of practitioners use one from the following selection: (i) Naive Bayes (NB), (ii) k-nearest neighbors (kNN), (iii) support vector machines (SVMs), (iv) Random

Forests (RF), or (v) artificial neural networks (ANNs). This section briefly reviews the use of each in virtual screening. Most time is spent on ANNs as there is a prevailing view that neural networks are the future of the field.

Naive Bayes

Using NB for VS involves calculating a probability that a compound falls into the active or inactive class given a molecular descriptor by direct application of Bayes' theorem. The overall conditional probability of activity is calculated by multiplying contributions from different descriptors.

Many comparisons find NB to perform poorly in comparison with other candidate models for VS.[60, 61] This is due to the model working on the assumption that all descriptors of a molecule will be conditionally independent, which is far from true, and is acutely vulnerable to undersampling of chemical space.[62] To understand why, suppose that some given span of a particular descriptor is only occupied by one compound in the given dataset, and that this compound is active: the contribution from this descriptor toward the estimated probability of activity will automatically be 100%, despite the model knowing close to nothing about the impact of this descriptor on activity. Even worse, if the compound is inactive, this will set the contribution from the descriptor to zero—and, when multiplied with the contributions from the other descriptors, this will unilaterally set the overall probability to zero, again on the basis of almost no information!

However, it must be said that several studies have found it to have utility, nonetheless. Jang et al. successfully found novel and diverse hits for mGlu1 receptor inhibitors using NB as a part of their workflow,[63] and Wang et al. found it to actually outperform other methods in correctly predicting whether new agents would be active against multiresistant *Staphylococcus aureus*.[64] It has also worked well as a complement to other VS techniques such as pharmacophore modeling,[65] or as a component of ensembles.[66]

k-Nearest neighbors

kNN classification is conceptually very simple, classifying a new unknown point as belonging to the modal class of the nearest k known points. For VS, the points are compounds which may belong to active or inactive classes, or if a regression model is desired, the average activity of the nearest k compounds can be used. The most common distance measures used are the Euclidean and Manhattan distances.[62] The main question when applying kNN to VS is which features should be used to create the space in which the compounds are embedded.

Support vector machines

Support vector machines draw a hyperplane through chemical space to divide active and inactive molecules. This is arguably one way of viewing any classification, but SVMs are distinct from others in the way they choose the optimal position of this plane, which will occupy the halfway point in the largest possible gap that divides the points of each class.

SVMs are one of the most popular model types used in VS today. Where comparisons are made, they are found to be high-performing,[61, 67, 68] and implementations exist in popular libraries such as LIBSVM[69] which handle complexities such as hyperparameter optimization and significantly lower the barriers to application. SVMs have a clear track record of identifying novel actives which hold up in experimental verification: for example, Deshmukh et al. used them to identify novel FEN1 inhibitors,[67] and Chandra et al. found two new PTP1B inhibitors showing significant activity in experiments from five candidates selected using SVMs.[68]

Random forests

Conceptually very different to the other members of this list, RFs use votes from an ensemble of decision trees to pick a class in classification, or a value in regression. Each Decision Tree (DT) can be imagined as a flow chart consisting of a series of simple questions: typically, these may be whether the value of a specific descriptor falls above or below a certain threshold. During training, the questions and thresholds are chosen so as to split the data as cleanly as possible by class, with more such questions being added as extra nodes in the tree until complete separation is achieved.

While individual DTs are usually found to generalize poorly, they are much more interpretable than the other techniques on this list, and large ensembles of such trees, each posing a different set of questions, are found to be powerful models. Indeed, many studies find their performance to be roughly on a par with SVMs and neural networks, such as that by Svetnik et al. who compared these other techniques with RFs for the modeling of a variety of QSAR and other endpoints, including the inhibition of COX2, *P*-glycoprotein transport activity, and blood-brain barrier permeability.[70] Giblin et al. found RFs to outperform both SVMs and generalized linear models in VS for bromodomain proteins.[56]

Artificial neural networks

As has been found across several fields, it has now been borne out by a considerable body of literature that DNNs outperform all other techniques in virtual screening.

In *Deep Learning as an Opportunity in Virtual Screening* (2014), Unterthiner et al. demonstrate the superiority of DNNs over a long list of potential competitors, which included SVMs, Logistic Regression, and kNNs.[71] Each technique was optimized by a hyperparameter search and assessed by its ability to predict active versus inactive compounds over 1230 targets in a cluster-based test set by average AUC. It is worthy of note that Unterthiner and some of the other authors later went on to win the Tox21 Data Challenge, an international competition for the prediction of toxicity endpoints from chemical structures, using the very same techniques.[72] They suggest two main reasons for the high observed performance of neural networks in learning from chemical structures.

First, while this is universal across all learning tasks, neural networks excel because of their ability to learn a complex, abstract internal representation of compounds from combinations of simpler features. To understand this, we can look at how neural networks might learn features relevant to, for example, facial recognition in image data.

Features learned in such a scenario form a hierarchy: the lower layers learn to detect simple patterns such as edges, which are then combined into more complex patterns resembling components of a human face. At the last hidden layer, we can see that these parts have been further combined into recognizable faces—showing how a neural network builds understanding of high-level concepts via a hierarchy of simpler components.

While the learning of networks developed for VS is harder to visualize, Unterthiner et al. suggest that there is similar mechanism at work. Used as features, ECFPs represent the presence (or absence) of particular groups of atoms, usually two or three bonds in diameter. Learned combinations of these components in the first hidden layers might be recognizable as something akin to reactive centers, which are then combined further to create representations of fully fledged pharmacophores, enabling the network to predict pharmacological effects such as the likelihood of binding. It could be argued that human drug designers understand therapeutic compounds in very loosely analogous way—building an intuitive understanding of larger chemical structures from knowledge of a hierarchy of simpler ones, and combining this with experience of their impact on properties.

Multitask deep neural networks

A second reason for good performance is that neural networks naturally allow for the training of models that predict on multiple endpoints simultaneously—or *multitask learning*—by simply adding a distinct node at the output for each. In the context of virtual screening, this usually means the construction of models which predict activity not just for one, but for multiple target proteins.

To understand how this can be an advantage, consider the balance between tasks in bioactivity data available to us today: some proteins, as well understood and popular therapeutic targets, have a wealth of bioactivity data available from assays against many thousands of structurally diverse compounds. However, others may have just a handful—not nearly enough to support the construction of models viable for general use across chemical space.

Multitask learning offers a particular advantage in these cases. By training the network on data from multiple tasks, an informative, general representation of the kind described above can be learned. A task with few data can then exploit these representations, which could never have been learned using its own data alone, to make better predictions.

Beyond the work of Unterthiner et al.[71, 72] can be found several further studies vindicating the state-of-the-art performance of multitask deep neural networks (MTDNNs) in VS. Dahl et al. found an MTDNN outperformed RFs, gradient-boosted decision tree ensembles, and single-task neural nets in active-inactive classification for 14 of 19 assays modeled[73]; Ma et al. found similar results comparing against RFs and single-task nets with an MTDNN instead trained for regression on 15 tasks that included several different pharmacological endpoints (IC50, solubility, clearance, permeability).[74] Some studies have created MTDNNs with output nodes for hundreds of kinases, finding them to consistently outperform single-task alternatives.[75, 76]

In a technique that is used across machine learning in general, to push MTDNNs to their limit, their predictions can be combined with those from other models in ensembles to increase accuracy further still. Unterthiner and colleagues complemented the output from their MTDNNs with those from RFs and SVMs to produce their most competitive models in the Tox21 challenge.[72]

Future directions: Learned descriptors and proteochemometric models

While a plethora of intriguing propositions for improving the performance of VS models are always being made and explored in the QSAR community, here we discuss two extensions to current models which have some following and may be of particular interest to the reader.

Graph convolutional and message passing neural networks

While fingerprints are still the main molecular representation used in VS, an exciting new class of representations has emerged in recent years. Rather than selecting some set of specific attributes of a molecule and encoding these as numerical features, they make use of the complete molecular graph, and rather than being predefined, they learn a representation which is optimal for the task.[77]

Graph convolutional networks (GCNs) were first used on molecular graphs by Duvenaud et al. when they introduced their "neural fingerprint" technique for learning data-driven descriptors of compounds.[78] Since this time, a variety of further work has suggested related approaches.[79, 80] A common framework into which these techniques fit was proposed by researchers from Google in their 2017 paper, *Neural Message Passing for Quantum Chemistry*,[81] which describes each of them as variations of a more general class of *Message Passing Neural Networks* (MPNNs)—so understanding an MPNN is perhaps the best way to understand these different schemes.

MPNNs assign each atom in the molecule a "hidden state"—a vector that carries information about it—and create a representation in two phases. First, in a "message passing" phase, these hidden states are augmented with the addition of information from the surrounding atoms and bonds—so they represent much more of the molecule than just the atom in isolation. This uses two functions:

(1) M_t, which calculates a message vector to carry information to each atom from (i) the current hidden state of the atom, (ii) the current hidden states of the atom's neighbors, and (iii) the current state of the bonds that connect them.

(2) U_t, which calculates a new hidden state of the atom from (i) the incoming message vector and (ii) the current hidden state of the atom.

This process can be repeated many times, allowing information about the atom's whole local region to collect within its hidden state. The subscript t is added to indicate that the framework allows for a different M and U function at each iteration.

In a second "readout" phase, a third and final function R takes the hidden states for all atoms in the molecule and combines them into the final representation of the molecule. This usually involves a summation over the hidden states (or some function of them) to allow an arbitrary number of atoms in the molecule to be combined into a vector of fixed length.

This process is used to generate the representation in the forward pass of the neural net—but each of M_t, U_t, and R can be chosen to contain learnable parameters, meaning that they can be updated in the backward pass like the weights in any neural network. We can now see both how MPNNs use information from the whole molecular graph to create a fixed-length descriptor, and how they learn adaptively to create better descriptors of the data on which they are trained.

There is now considerable evidence to support the idea that these advantages mean that MPNN techniques represent a new frontier in improving QSAR model accuracy: for example, Withnall et al. showed that their variations on the MPNN framework can outperform state-of-the-art models for VS,[77] and Wu et al. found graph-based featurizations to significantly outperform other methods on 11 of 17 datasets studied.[82]

Proteochemometric models

While the discussion thus far has considered machine learning for VS exclusively as a ligand-based technique, frameworks which incorporate target structural information, if it is available, are also possible and are often found to improve performance. These are known as "proteochemometric" models (PCMs).

The main question when incorporating such information is how to represent the protein—which, like the choice of compound descriptor, has several possible answers. Rifaioglu and co-authors discuss a variety of target protein descriptors breaking them into categories such as sequence composition, physicochemical properties, similarity measures, topological properties, geometrical characteristics, and functional sites.[83]

For QSAR modeling, the region of the protein of greatest relevance is undoubtedly the binding site, and one approach is to build a sequence-based descriptor which varies depending on the amino acids present here. This approach is well-suited to creating models to predict across a family of proteins with different but comparable binding sites, such as the kinases, which have significant conservation at their binding site for ATP.

In order for this to be possible, an alignment between the different targets being modeled is required: that is, a mapping which shows which amino acid occupies each spatial position in the binding site of each target. High-quality alignments may already be available from sources such as the Kinase-Ligand Interaction Fingerprints and Structures database (KLIFS).[84] Using this information, a descriptor can be constructed by simply concatenating a short vector representing each amino acid in the binding site. Amino acids themselves can be represented in several ways, but a competitive option is the Z-scales 5 descriptor[85]—these assign each amino acid a series of 5 numbers, condensed via Principal Component Analysis from 26 physicochemical property values of amino acids.[86]

Such descriptors have been used to good effect in previous work. Giblin et al. constructed a variety of models to predict inhibition of bromodomain proteins and carefully compared models with and without such target information, finding that their PCM models hugely outperformed the purely ligand-based alternatives.[56] Work using neural networks in a PCM context is still scarce, but Lenselink et al. observed an increase in the performance of their ANNs even using very crude descriptors based on averaged physicochemical properties across regions of the primary sequence[60]. More complex descriptors using 3D representations of proteins are beginning to be explored also: Subramanian et al. constructed accurate QSAR models for 95 kinases by using physics-based 3D spatial descriptions of their binding sites.[87]

Evaluating virtual screening models
Train-test splits: Random, temporal, or cluster-based?

The standard approach to evaluating the performance of a machine learning model is to *randomly* select a proportion of the dataset as a held-out test set, which remains unseen throughout model development. The remainder may be split further into training and validation or cross-validation folds for experimentation with alternative hyperparameters, but the test set remains totally unseen until a final model evaluation after development is complete, with its purpose being to provide a fair estimate of performance in real use. Unless the intention is merely to build a small local model for use on a specific scaffold, the goal of VS models is to be able to predict usefully across drug-like chemical space—including on new and unseen scaffolds—and this is the future performance we wish to evaluate. Unfortunately, a randomly selected test set is very unlikely to provide a realistic sense of this.

Bioactivity datasets, especially those originating from the pharmaceutical industry, naturally consist of groups of very similar compounds. Among the reasons for this is the fact that drug development projects typically take a lead and systematically optimize its properties by repeated experimentation

Random train-test split   **Cluster-based or temporal split**

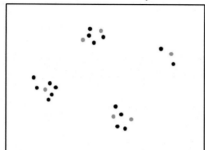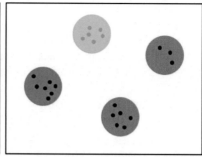

FIG. 1

Illustration of chemical space with groups of similar compounds in a bioactivity dataset. A random split is likely to split groups across the train *(burgundy)* and test *(cyan)* sets, meaning test set compounds have close relations in the train set. Alternative methods such as temporal and cluster-based splits create more realistic measures of future performance.

with variants that share the same scaffold, generating data to match. This being the case, compounds in a conventional randomly selected test set are likely to have close analogs in the train set, creating an inflated measure of performance (Fig. 1).

There are two commonly used approaches to increasing test-set realism. One possibility is to cluster the dataset into groups of similar compounds, allowing the whole of any given group to be allocated either to the train or test set. Two possible clustering algorithms—Butina clustering and single-linkage hierarchical clustering—which have both been used for this purpose are discussed in more detail later in this chapter (see *Clustering in hit discovery*). Both methods have tuneable parameters that give the practitioner some control over cluster size: the larger the clusters created, the greater the degree of difference between the molecules in different sets, and the harsher the test. The problem then becomes knowing how large clusters should be to form a realistic test set. Unfortunately, unless the degree of chemical difference between the compounds in the dataset and the compounds on which the model will be used in practice is known in advance, this is ultimately a subjective issue and up to the practitioner.

Alternatively, in the case that a date is available for each measurement in the compiled dataset, a temporal split can be made by taking all measurements made beyond a given time as the test set. This approach is also widely used and has significant advantages: it is simple to apply, introduces no subjective choice about cluster size and, arguably, provides the most realistic possible measure of future model performance. However, it is only applicable if date information is available for all measurements, which is unlikely to be true in a large, combined heterogeneous set of the kind that is argued for above. Furthermore, since drug projects often run over a period of years, it is likely that some chemical series present in the dataset will stretch over any given cut-off in time, meaning there are still questions about the validity of test metrics as an estimate of performance on new projects.

External validation

Another useful and widely used approach to VS model evaluation is external testing, using data from new sources—publications, benchmark sets from online competitions etc.—which are unrelated to the original dataset and hence likely to contain substantially novel content. Despite this, care should be taken that any molecules that are present in the original set are removed, and it is also recommended that some analysis be made of how similar the remaining molecules are to members of the original set if possible. A good example of this can be seen in the work of Li et al., who used t-SNE to generate 3D visualizations of their dataset alongside each of four external validation sets that they used to evaluate their model, allowing their distributions to be compared in a clear and visual way.[76]

Prospective experimental validation

In light of the complications to each evaluation technique discussed above, it is important to reflect that there is no perfect way to evaluate the performance of a virtual screening model before use. Arguments can be made against any choice of test set and ultimately, any calculated metrics are always in part a function of the train-test similarity and will vary across chemical space.

The gold standard of VS model validation is simply to show that it is useful in practice[88]—a goal that, with the right dataset, the right algorithm and careful work through each of the steps discussed above, is certainly achievable and has ample precedent.[56, 67, 68, 89–92] If a VS model can help drug hunters to find more novel, potent and structurally diverse hits than they could without it, even the meanest skeptics will be hard-pressed to argue against its value!

Clustering in hit discovery

Finding groups of similar compounds within a given set has several applications in hit discovery.

- Prior to any screening experiments, clustering may be used in the construction of a virtual screening model intended to help select likely hits. As discussed above, by creating clusters of similar compounds for assignment to the train or test set, they can help to ensure that the test set gives a realistic measure of future performance on novel scaffolds.
- In the design of screening experiments, clustering can help to select structurally diverse compounds for inclusion. By selecting representatives from different clusters, a wider exploration of chemical space is possible with the same number of compounds.
- In the analysis of results from screens, whether real or virtual, clustering can help to identify any grouping or categories that exist among the identified hits.

Clustering algorithms require a definition of chemical space with two elements: first, a way of describing each compound as a vector, which is typically done using chemical fingerprints such as ECFPs, and second, a way to measure the distance between compounds, of which the most common is the Tanimoto coefficient. Built on these definitions, a large number of clustering algorithms exist, but we restrict our discussion here to three main types which are most often used in hit discovery: the Butina, k-means, and hierarchical clustering algorithms. This section provides a brief discussion of each.

Butina clustering

The Butina algorithm, introduced by Butina[93] for use in cheminformatics specifically, is the crudest method of the three, but retains considerable popularity in practical use due to its simplicity and low memory cost—just $O(N)$ (where N is the number of compounds in the set). It is very similar to the (also popular) Jarvis-Patrick clustering method and works by iterative sphere exclusion, in three steps:

1. **Generation of molecular fingerprints for all compounds in the set**. In the 1999 paper, Butina used Daylight fingerprints for this purpose, but other fingerprint types could be used; the use of some form of binary fingerprints is recommended as it enables more rapid calculation of Tanimoto distances, important in the next step.

2. **Sorting of compounds by number of neighbors**. Prior to this step, a Tanimoto distance cut-off, T_c, must be selected. A "neighbor" of any given compound is defined as another compound at a distance equal to or less than T_c. The compounds are then sorted by the number of neighbors they have in descending order. This is the most computationally expensive step, but there is a memory-saving trick here. While this step does necessitate the calculation of the distance between every possible pair of molecules, and so carries an $O(N^2)$ time cost, these values do not need to be retained after calculation—only a single integer count of the number of neighbors of each molecule needs to be held in memory. Thus, in comparison with the other methods described here, the memory cost is just $O(N)$.

3. **Creation of clusters by iterative sphere exclusion**. Taking the compound with the most neighbors as the first cluster centroid, this compound and all its neighbors as defined above become the first cluster. They are removed from the set, which can be imagined as the exclusion of a sphere of radius T_c from chemical space. Once this is complete, the next compound in the list (with the next-highest number of neighbors) is assigned as the next cluster centroid, and the process is repeated, creating and removing a second cluster, etc., until the whole dataset has been allocated to a cluster.

As an adjustable parameter, T_c can be varied to exercise some control over cluster size.

While this process was developed to allow clustering in a time when memory was much scarcer (Butina's own workstation had just 32 MB of memory!), the low memory requirement and simplicity mean it is still commonly used. In their 2019 paper on high-performing machine learning-based QSAR models, Martin et al. used Butina clustering to create train-test splits for assays with more than 10,000 compounds, where hierarchical clustering became impractical.[94] Guidi et al. used it to select 110 diverse hits from 275 returned by HTS in their work on therapeutic compounds for Schistosomiasis.[95]

This simple approach does, of course, carry significant downsides. The original paper mentions a tendency of the approach to produce a high number of "singletons"—individual compounds which are allocated as the only member of their own cluster. While some of these will be molecules which are simply very different from other members of the set and so justifiably define their own cluster, others arise because they are just outside the cut-off of a nearby sphere. This is shown, along with some further suggested problems, in Fig. 2.

K-means clustering

K-means clustering is a more robust technique which does not suffer from the drawbacks of Butina clustering, with no problem of false singletons and an ability to recognize clusters of more varied shape and size. It is a very general unsupervised machine learning method which is used extensively outside hit discovery and works by finding k cluster centroids which minimize the sum of squared Euclidean distances between the centroids and their cluster members (each point is assigned to the closest centroid). Instead of setting a cluster size, the user instead specifies the number of clusters, k. Of course, this is still a partly subjective choice that must be made depending on the intended application.

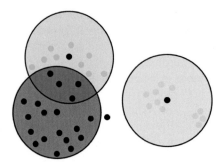

FIG. 2

Problems with Butina clustering. Cluster centroids in *black*, clusters in colored groups. An example "false singleton" can be seen below center, which might more sensibly belong to the *burgundy* cluster, but was just outside the cut-off. Another problem can be seen in the *blue* cluster, which adopts a strange crescent-like shape as many of its neighbors were already allocated to the larger *burgundy* cluster. Finally, it is clear to the human eye that the olive cluster contains two distinct groups, but these are not recognized as all exclusion spheres have the same radius.

Uses of k-means in previous work include that of Gagarin et al., who used it to aid the analysis of noisy HTS data on the inhibition of *Escherichia coli* dihydrofolate reductase. By clustering hits from the HTS, they were able to find groups which together showed statistically significant activity, increasing the number of true positive hits that they were able to identify from the noisy data compared to approaches that did not use clustering.[96]

Hierarchical clustering

Hierarchical clustering produces a complete dendrogram showing the range of levels at which compounds are related. When it works correctly, it can present a more complete picture of the underlying structure of the set than any other method, and it is included as standard in several software packages used by medicinal chemists.

There are several variations of hierarchical clustering methods which fall into two main categories.[97] *Divisive* methods initialize with all examples as members of a single cluster, and split this cluster recursively, while *agglomerative* methods begin instead with each point defining its own cluster, and iteratively link nearby points into larger and larger clusters. Within the more popular agglomerative class, there are further variations based on the rules for linking together points and clusters. For example, the "average-linkage" variant links together at each iteration the two clusters whose means are closest together, while "single-linkage" instead links clusters based on the closest pair of individual points among their members. The popular Wards Clustering method is also an example of agglomerative hierarchical clustering.

While hierarchical methods are widely used and can be highly informative, their core disadvantage is that they are very computationally costly, making them impractical for chemical sets beyond a certain size. General agglomerative clustering methods have a time complexity of $O(N^3)$ and a memory complexity of $O(N^2)$ due to the need to calculate and recalculate full pairwise distance matrices. However, for some specific variants, better solutions exist: for example, the commonly used single-linkage variant has an optimal algorithm with an $O(N^2)$ time complexity.[98]

References

1. Hughes JP, Rees SS, Kalindjian SB, Philpott KL. Principles of early drug discovery. *Br J Pharmacol* 2011;**162**:1239–49.
2. Lipinski C, Hopkins A. Navigating chemical space for biology and medicine. *Nature* 2004;**432**:855–61.
3. Polishchuk PG, Madzhidov TI, Varnek A. Estimation of the size of drug-like chemical space based on GDB-17 data. *J Comput Aided Mol Des* 2013;**27**:675–9.
4. Bohacek RS, McMartin C, Guida WC. The art and practice of structure-based drug design: a molecular modeling perspective. *Med Res Rev* 1996;**16**:3–50.
5. Rosén J, Gottfries J, Muresan S, Backlund A, Oprea TI. Novel chemical space exploration via natural products. *J Med Chem* 2009;**52**:1953–62.
6. Dandapani S, Marcaurelle LA. Grand challenge commentary: accessing new chemical space for 'undruggable' targets. *Nat Chem Biol* 2010;**6**:861–3.
7. Thomas GL, Wyatt EE, Spring DR. Enriching chemical space with diversity-oriented synthesis. *Curr Opin Drug Discov Devel* 2006;**9**:700–12.
8. Clardy J, Walsh C. Lessons from natural molecules. *Nature* 2004;**432**:829–37.
9. Kirkpatrick P, Ellis C. Chemical space. *Nature* 2004;**432**:823.
10. Medina-Franco J, Martinez-Mayorga K, Giulianotti M, Houghten R, Pinilla C. Visualization of the chemical space in drug discovery. *Curr Comput Aided Drug Des* 2008;**4**:322–33.
11. Ashenden SK. Screening library design. In: *Methods in enzymology*. Academic Press; 2018. p. 73–96. https://doi.org/10.1016/bs.mie.2018.09.016.
12. Gordon EM. Libraries of non-polymeric organic molecules. *Curr Opin Biotechnol* 1995;**6**:624–31.
13. Nilakantan R, Nunn DS. A fresh look at pharmaceutical screening library design. *Drug Discov Today* 2003;**8**:668–72.
14. Terrett NK, Gardner M, Gordon DW, Kobylecki RJ, Steele J. Combinatorial synthesis—the design of compound libraries and their application to drug discovery. *Tetrahedron* 1995;**51**:8135–73.
15. Hansch C, Leo A, Hoekman D. *Exploring QSAR. Fundamentals and applications in chemistry and biology.* Washington: American Chemical Society; 1995.
16. Lassalas P, et al. Structure property relationships of carboxylic acid isosteres. *J Med Chem* 2016. https://doi.org/10.1021/acs.jmedchem.5b01963.
17. Roth H-J. There is no such thing as 'diversity'! *Curr Opin Chem Biol* 2005;**9**:295.
18. Ferguson AM, Patterson DE, Garr CD, Underiner TL. Designing chemical libraries for lead discovery. *J Biomol Screen* 1996;**1**:65–73.
19. Ashenden S. *On the dissemination of novel chemistry and the process of optimising compounds in drug discovery projects.* University of Cambridge; 2019. https://doi.org/10.17863/CAM.38232.
20. MacArron R, et al. Impact of high-throughput screening in biomedical research. *Nat Rev Drug Discov* 2011;**10**:188–95.
21. Sliwoski G, Kothiwale S, Meiler J, Lowe EW. Computational methods in drug discovery. *Pharmacol Rev* 2014;**66**:334–95.
22. Macalino SJY, Gosu V, Hong S, Choi S. Role of computer-aided drug design in modern drug discovery. *Arch Pharm Res* 2015;**38**:1686–701.
23. Schneider G, Fechner U. Computer-based de novo design of drug-like molecules. *Nat Rev Drug Discov* 2005;**4**:649–63.
24. Schneider P, Schneider G. De novo design at the edge of chaos. *J Med Chem* 2016;**59**:4077–86.
25. Husby J, Bottegoni G, Kufareva I, Abagyan R, Cavalli A. Structure-based predictions of activity cliffs. *J Chem Inf Model* 2015;**55**:1062–76.
26. Livingstone DJ. Pattern recognition methods in rational drug design. *Methods Enzymol* 1991;**203**:613–38.
27. Hartenfeller M, Schneider G. De novo drug design. *Methods Mol Biol* 2011;**672**:299–323.

28. Schneider G, Funatsu K, Okuno Y, Winkler D. De novo drug design—ye olde scoring problem revisited. *Mol Inf* 2017;**36**.

29. Suryanarayanan V, Panwar U, Chandra I, Singh SK. De novo design of ligands using computational methods. In: *Methods in molecular biology 1762*. Humana Press, Totowa, NJ; 2018. p. 71–86.

30. Kawai K, Nagata N, Takahashi Y. De novo design of drug-like molecules by a fragment-based molecular evolutionary approach. *J Chem Inf Model* 2014;**54**:49–56.

31. Kutchukian PS, Lou D, Shakhnovich EI. FOG: fragment optimized growth algorithm for the de novo generation of molecule: occupying druglike chemical space. *J Chem Inf Model* 2009;**49**:1630–42.

32. Markov Chains. *Brilliant math & science wiki*. Available at: https://brilliant.org/wiki/markov-chains/. [Accessed 8 September 2020].

33. Hartenfeller M, et al. Dogs: reaction-driven de novo design of bioactive compounds. *PLoS Comput Biol* 2012;**8**:e1002380.

34. Ghiandoni GM, et al. Enhancing reaction-based de novo design using a multi-label reaction class recommender. *J Comput Aided Mol Des* 2020;**34**:783–803.

35. Takeda S, Kaneko H, Funatsu K. Chemical-space-based de novo design method to generate drug-like molecules. *J Chem Inf Model* 2016;**56**:1885–93.

36. Blaschke T, et al. Reinvent 2.0: an AI tool for de novo drug design. *J Chem Inf Model* 2020. https://doi.org/10.1021/acs.jcim.0c00915.

37. Miranker A, Karplus M. An automated method for dynamic ligand design. *Proteins Struct Funct Bioinf* 1995;**23**:472–90.

38. Rotstein SH, Murcko MA. GenStar: a method for de novo drug design. *J Comput Aided Mol Des* 1993;**7**:23–43.

39. Nishibata Y, Itai A. Automatic creation of drug candidate structures based on receptor structure. Starting point for artificial lead generation. *Tetrahedron* 1991;**47**:8985–90.

40. Lengauer T, Rarey M. Computational methods for biomolecular docking. *Curr Opin Struct Biol* 1996;**6**:402–6.

41. Hawkins PCD, Skillman AG, Nicholls A. Comparison of shape-matching and docking as virtual screening tools. *J Med Chem* 2007;**50**:74–82.

42. Wermuth CG, Ganellin CR, Lindberg P, Mitscher LA. Glossary of terms used in medicinal chemistry (IUPAC recommendations 1998). *Pure Appl Chem* 1998;**70**:1129–43.

43. Yang SY. Pharmacophore modeling and applications in drug discovery: challenges and recent advances. *Drug Discov Today* 2010;**15**:444–50.

44. Sanders MPA, et al. From the protein's perspective: the benefits and challenges of protein structure-based pharmacophore modeling. *Med Chem Commun* 2012;**3**:28–38.

45. Garcia-Hernandez C, Fernández A, Serratosa F. Ligand-based virtual screening using graph edit distance as molecular similarity measure. *J Chem Inf Model* 2019;**59**:1410–21.

46. Willett P. Similarity-based virtual screening using 2D fingerprints. *Drug Discov Today* 2006;**11**:1046–53.

47. Kubinyi H. Success stories of computer-aided design. In: *Computer applications in pharmaceutical research and development*. Wiley Online Library; 2006. p. 377–424. https://doi.org/10.1002/0470037237.ch16.

48. Cronin MTD, Schultz TW. Pitfalls in QSAR. *J Mol Struct THEOCHEM* 2003;**622**:39–51.

49. Tarasova OA, et al. QSAR modeling using large-scale databases: case study for HIV-1 reverse transcriptase inhibitors. *J Chem Inf Model* 2015;**55**:1388–99.

50. Sagardia I, Roa-Ureta RH, Bald C. A new QSAR model, for angiotensin I-converting enzyme inhibitory oligopeptides. *Food Chem* 2013;**136**:1370–6.

51. Cortes-Ciriano I, Bender A. Improved chemical structure-activity modeling through data augmentation. *J Chem Inf Model* 2015;**55**:2682–92.

52. Lagunin AA, Geronikaki A, Eleftheriou P, Pogodin PV, Zakharov AV. Rational use of heterogeneous data in quantitative structure-activity relationship (QSAR) modeling of cyclooxygenase/lipoxygenase inhibitors. *J Chem Inf Model* 2019;**59**:713–30.

53. Gaulton A, et al. The ChEMBL database in 2017. *Nucleic Acids Res* 2017;**45**:D945–54.

54. GOSTAR (GVK BIO Online Structure Activity Relationship Database), 2010. https://gostardb.com/gostar/.
55. Wang Y, et al. PubChem BioAssay: 2017 update. *Nucleic Acids Res* 2017;**45**:D955–63.
56. Giblin KA, Hughes SJ, Boyd H, Hansson P, Bender A. Prospectively validated proteochemometric models for the prediction of small-molecule binding to bromodomain proteins. *J Chem Inf Model* 2018;**58**:1870–88.
57. Fourches D, Muratov E, Tropsha A. Trust, but verify: on the importance of chemical structure curation in cheminformatics and QSAR modeling research. *J Chem Inf Model* 2010;**50**:1189–204.
58. Fourches D, Muratov E, Tropsha A. Curation of chemogenomics data. *Nat Chem Biol* 2015;**11**:535.
59. Fourches D, Muratov E, Tropsha A. Trust, but verify II: a practical guide to chemogenomics data curation. *J Chem Inf Model* 2016;**56**:1243–52.
60. Lenselink EB, et al. Beyond the hype: deep neural networks outperform established methods using a ChEMBL bioactivity benchmark set. *J Cheminformatics* 2017;**9**:45.
61. Li Y, et al. Predicting selective liver X receptor β agonists using multiple machine learning methods. *Mol BioSyst* 2015;**11**:1241–50.
62. Carpenter KA, Huang X. Machine learning-based virtual screening and its applications to Alzheimer's drug discovery: a review. *Curr Pharm Des* 2018;**24**:3347–58.
63. Jang JW, et al. Novel scaffold identification of mGlu1 receptor negative allosteric modulators using a hierarchical virtual screening approach. *Chem Biol Drug Des* 2016;**87**:239–56.
64. Wang L, et al. Discovering new agents active against methicillin-resistant Staphylococcus aureus with ligand-based approaches. *J Chem Inf Model* 2014;**54**:3186–97.
65. Yu M, Gu Q, Xu J. Discovering new PI3Kα inhibitors with a strategy of combining ligand-based and structure-based virtual screening. *J Comput Aided Mol Des* 2018;**32**:347–61.
66. Lian W, et al. Discovery of Influenza A virus neuraminidase inhibitors using support vector machine and Naïve Bayesian models. *Mol Divers* 2016;**20**:439–51.
67. Deshmukh AL, Chandra S, Singh DK, Siddiqi MI, Banerjee D. Identification of human flap endonuclease 1 (FEN1) inhibitors using a machine learning based consensus virtual screening. *Mol BioSyst* 2017;**13**:1630–9.
68. Chandra S, Pandey J, Tamrakar AK, Siddiqi MI. Multiple machine learning based descriptive and predictive workflow for the identification of potential PTP1B inhibitors. *J Mol Graph Model* 2017;**71**:242–56.
69. Chang CC, Lin CJ. LIBSVM: a library for support vector machines. *ACM Trans Intell Syst Technol* 2011;**2**:1–27.
70. Svetnik V, et al. Random forest: a classification and regression tool for compound classification and QSAR modeling. *J Chem Inf Comput Sci* 2003;**43**:1947–58.
71. Unterthiner T, et al. Deep learning as an opportunity in virtual screening. *Adv Neural Inf Process Syst* **27**, 2014, 1–9.
72. Mayr A, Klambauer G, Unterthiner T, Hochreiter S. DeepTox: toxicity prediction using deep learning. *Front Environ Sci* 2016, 80;**3**.
73. Dahl GE, Jaitly N, Salakhutdinov R. *Multi-task neural networks for QSAR predictions.* arXiv Prepr; 2014.
74. Ma J, Sheridan RP, Liaw A, Dahl GE, Svetnik V. Deep neural nets as a method for quantitative structure-activity relationships. *J Chem Inf Model* 2015;**55**:263–74.
75. Allen BK, Ayad NG, Schurer SC. Data-driven interrogation of kinase inhibitor classification using multitask neural networks. *J Chem Inf Model.* *https://scholar.harvard.edu/bryce_allen/publications/data-driven-interrogation-kinase-inhibitor-classification-using-multitask.*.
76. Li X, et al. Deep learning enhancing kinome-wide polypharmacology profiling: model construction and experiment validation. *J Med Chem* 2020;**63**:8723–37.
77. Withnall M, Lindelöf E, Engkvist O, Chen H. Building attention and edge message passing neural networks for bioactivity and physical-chemical property prediction. *J Cheminformatics* 2020, 1–18;**12**.
78. Duvenaud D, et al. Convolutional networks on graphs for learning molecular fingerprints. In: Cortes C, Lawrence ND, Lee DD, Sugiyama M, Garnett R, editors. *Advances in neural information processing systems.* vol. 7. Curran Associates; 2015. p. 2224–32.

79. Kearnes S, McCloskey K, Berndl M, Pande V, Riley P. Molecular graph convolutions: moving beyond fingerprints. *J Comput Aided Mol Des* 2016;**30**:595–608.

80. Coley CW, Barzilay R, Green WH, Jaakkola TS, Jensen KF. Convolutional embedding of attributed molecular graphs for physical property prediction. *J Chem Inf Model* 2017;**57**:1757–72.

81. Gilmer J, Schoenholz SS, Riley PF, Vinyals O, Dahl GE. *Neural message passing for quantum chemistry.* arXiv (JMLR. org); 2017. p. 1263–72.

82. Wu Z, et al. MoleculeNet: a benchmark for molecular machine learning. *Chem Sci* 2018;**9**:513–30.

83. Rifaioglu AS, et al. Recent applications of deep learning and machine intelligence on in silico drug discovery: methods, tools and databases. *Brief Bioinform* 2019;**20**:1878–912.

84. Kooistra AJ, et al. KLIFS: a structural kinase-ligand interaction database. *Nucleic Acids Res* 2016;**44**:D365–71.

85. Van Westen GJP, et al. Benchmarking of protein descriptor sets in proteochemometric modeling (part 2): modeling performance of 13 amino acid descriptor sets. *J Cheminformatics* 2013, 1–20;**5**.

86. Sandberg M, Eriksson L, Jonsson J, Sjöström M, Wold S. New chemical descriptors relevant for the design of biologically active peptides. A multivariate characterization of 87 amino acids. *J Med Chem* 1998;**41**:2481–91.

87. Subramanian V, Prusis P, Xhaard H, Wohlfahrt G. Predictive proteochemometric models for kinases derived from 3D protein field-based descriptors. *Med Chem Commun* 2016;**7**:1007–15.

88. Ripphausen P, Nisius B, Peltason L, Bajorath J. Quo vadis, virtual screening? A comprehensive survey of prospective applications. *J Med Chem* 2010;**53**:8461–7.

89. Luo M, Wang XS, Roth BL, Golbraikh A, Tropsha A. Application of quantitative structure-activity relationship models of 5-HT1A receptor binding to virtual screening identifies novel and potent 5-HT1A ligands. *J Chem Inf Model* 2014;**54**:634–47.

90. Allen BK, et al. Large-scale computational screening identifies first in class multitarget inhibitor of EGFR kinase and BRD4. *Sci Rep* 2015, 1–16;**5**.

91. Neves BJ, et al. Discovery of new anti-schistosomal hits by integration of QSAR-based virtual screening and high content screening. *J Med Chem* 2016;**59**:7075–88.

92. Li H, Visco DP, Leipzig ND. Confirmation of predicted activity for factor XIa inhibitors from a virtual screening approach. *AIChE J* 2014;**60**:2741–6.

93. Butina D. Unsupervised data base clustering based on daylight's fingerprint and Tanimoto similarity: a fast and automated way to cluster small and large data sets. *J Chem Inf Comput Sci* 1999;**39**:747–50.

94. Martin EJ, et al. All-assay-Max2 pQSAR: activity predictions as accurate as four-concentration IC50s for 8558 novartis assays. *J Chem Inf Model* 2019;**59**:4450–9.

95. Guidi A, et al. Discovery by organism based high-throughput screening of new multi-stage compounds affecting Schistosoma mansoni viability, egg formation and production. *PLoS Negl Trop Dis* 2017;**11**:e0005994.

96. Gagarin A, Makarenkov V, Zentilli P. Using clustering techniques to improve hit selection in high-throughput screening. *J Biomol Screen* 2006;**11**:903–14.

97. Rokach L, Maimon O. Clustering methods. In: *Data mining and knowledge discovery handbook.* Springer; 1980. p. 321–52. https://doi.org/10.1007/0-387-25465-X_15.

98. Sibson R. SLINK: an optimally efficient algorithm for the single-link cluster method. *Comput J* 1973;**16**:30–4.

Lead optimization

Stephanie Kay Ashenden

Data Sciences and Quantitative Biology, Discovery Sciences, R&D, AstraZeneca, Cambridge, United Kingdom

What is lead optimization

A lead compound has been identified and now must be optimized to become a candidate drug.[1] Lead optimization refers to the process of designing and improving a preidentified lead compound, and involves manipulation of multiple parameters of the compound,[2] relying on chemical modifications to the compound. During this process, synthetic changes are made to optimize the compounds ADMET properties (absorption, distribution, metabolism, excretion, and toxicity) as well as the compound's activity, potency, and selectivity.[2] In addition, the compound needs to remain novel to allow for it to be patented.[2] These ADMET properties relate to ability of the compound to be taken up into the blood stream (absorption), how the compound is moved around the body to its desired site (distribution), how well the compounds are broken down once in the body (metabolism), how the compounds and any metabolites are removed from the body (excretion) and any negative effect the compound will have on the body (toxicity). However, these properties are not only considered during the lead optimization phase and are monitored throughout earlier drug development stages.[3] The use of drug metabolism and pharmacokinetics (DMPK) parameters in lead optimization (both in vitro and in vivo) is a focus of research organizations to aid in producing compounds that fall with an acceptable range in terms of these properties[4] by supplying the tools and methodologies.[5] These effects and improvements on a compounds properties are made in what is known as the design, make, test, and analyze (DMTA) cycle (Fig. 1).[6]

These properties are explored in the following sub sections and are also summarized in Table 1. These properties are assessed through various assays discussed later, which allow for changes in the property values to be recorded as the compound is modified.

Applications of machine learning in lead optimization

Trotter and co-author explained that a key area for machine learning in the lead optimization phase of drug discovery is the prioritization of particular compounds for development.[13] Scoring functions can be used to aid in prioritizing molecules in lead optimization as summarized by Li and co-authors.[14] As in the hit identification steps, docking can be applied to predict the binding capabilities of a compound to a target and the authors explain that the predicted potency of the compounds can be used to rank the molecules. 3D-convolutional neural networks have also been used to rank compounds based on their potency against a particular target. As explained by Yamashita and co-authors,[15] convolutional neural networks have a grid pattern and typically contain three types of layers of which the fully connected

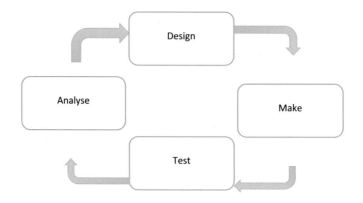

FIG. 1

The design, make, test, and analyze (DMTA) cycle.

Table 1 Examples of important assays used to test suitability of compounds.			
Test	**Applications**	**Limitations**	**Advantages**
Caco-2[7]	Assess the permeability and absorption and model the intestinal barrier	Does not tolerate organic solvents Sensitive to excipients Sensitive to different environments and cultures Takes 21 days for the cells to differentiate	Reduces need for animal studies Aid in understanding of transport mechanism and drug pathway
LogD[8]	Measure of lipophilicity of the compound specifically, the dissociation of weak acids and bases	Mathematical and experimental derived values cannot be easily compared	Takes into consideration the pH dependence of the molecule in an aqueous solution
Solubility[9]	Assess the saturation concentration of a solute in a solvent	Depends on the presence of other species in the solvent	Can be characterized into levels of solubility, regardless of solvent used
Microsomal metabolism[10]	Used to investigate compound metabolism and clearance	Microsomes do not exist in healthy, human cells Expression and activity of CYP enzymes is variable depending on factors such as genetics and environmental aspects	Can observe CYP (cytochrome P) enzymes
Hepatocyte Metabolism[11]	Used to investigate compound metabolism in hepatocytes	Limited by donor availability	The liver is the primary place that drug metabolism takes place
hERG IC_{50}	Used to assess cardiac toxicity by inhibition of hERG[12]	The binding assay cannot identify when a compound only binds to one state or other sites of the channel Manual methods are technically difficult	The binding assay is low cost and can be used in high-throughput Can use automated or manual methods depending on individual requirements/aims

layer (mapping the output), the pooling layer and the convolutional layer (both feature extraction). The approach published by Yasuo and co-authors aimed to develop hit compounds into optimized compounds focusing on the exploration and evaluation of candidates (by predicting the drug-likeness).[16] To do this the authors used learning to rank models of which they describe as models that instead of predicting a class or a value, predict the order of data set.[16] The authors go on to explain that learning to rank models can be split into three different types, pointwise (assignment of a value to a data point and then ordered), pairwise (takes data point pairs and orders them before reordering into a final order), and listwise (predicted order of the dataset).[16]

Optimizing the molecules themselves or the ADMET properties is another area of interest in machine learning. The tool admetSAR contains a number of models that can be used for the prediction of ADMET properties and a module (ADMETopt) is specifically designed for lead optimization.[17–19] The optimization method uses scaffold hopping and property (such as ADMET) screening.[19] Scaffold hopping, as described by Böhm and Stahl, takes a known active compounds central core and modifies it and resulting in a novel and structurally different compound.[20]

Deep learning has been used to aid in the optimization of molecules. One such example is published by Zhou and co-authors.[21] The authors describe MolDQN (Molecule Deep Q-Networks), which they describe as using reinforcement learning (decision-making to achieve best reward[22]) combined with domain knowledge. In their method, the authors explain several rules and design choices that are applied in regard to the modifications that are made (which can be additions as well as removals).[21] Additionally, the authors explain that they control the diversity of the compounds by limiting how many changes from the starting molecule can be made and ensure that such changes are valid.[21]

The prediction of DMPK properties has been undertaken using available time series data where the authors used data generated early in the project as training sets to predict properties.[23] Property prediction is something of great interest in the pharmaceutical industry. If successful, it could allow for faster separation of compounds with desirable and undesirable properties. However, complications in using machine learning approaches often arise from the lack of high quality data.[24] Despite this, there have been a number of different approaches to attempting the prediction of compound properties.

Assessing ADMET and biological activities properties

Hits that are intended to become leads are assessed for the chemical, synthetic, and functional behaviors by using understandings of structure-activity relationships (SAR) as well as their physicochemical and potential toxicology profiles by analyzing the compounds ADMET properties. These profiles should be determined as early as possible to prevent failure later on which amounts to increased costs.[25] Experimentally, these properties are tested and measured in various assays as seen in Table 1.[6]

A variety of different assays are performed by chemists to assess how a compound will fair within the body, particularly in areas such as permeability and solubility. Generally, in vitro assays on their own do not represent a whole human system; therefore, the predictions of adverse effects must be carefully considered. To assess permeability and absorption, chemists analyze Caco-2 cell monolayer in and out of the cell, allowing them to understand and model the intestinal barrier (most notable the human small intestinal mucosa). Caco-2 cells originate from human colorectal adenocarcinoma and growth in monolayer epithelial cells.[6]

Chris Lipinski discussed[26] screening doses on Caco-2 permeability assays and explained that if the dose is too low, the importance of efflux transporters can be overestimated but if the drug concentration is too high, the transporters in which the compound translocate via can become saturated.[26] Because of the heterogeneous nature of the cells, a variety of transports (absorptive and efflux) may be present and results will vary from laboratory and laboratory.[6, 26]

Guangli and Yiyu demonstrated that support vector machines can be used to predict Caco-2 permeability and were shown to be an effective method.[27] The authors noted that the number of hydrogen bond donors as well as properties relating to the molecular surface area were determinates of membrane permeability.[27] In another study, structure-property relationships (QSPR) were used to predict caco-2 permeability and Boosting (method for improving learning algorithm performance as described by Freund and Schapire[28]) was found to have good predicting ability[29] alongside the use of a genetic algorithm for descriptor selection.

LogD (distribution coefficient) is the log of the partition of a compound between the lipid and aqueous phases and represents a compound's lipophilic nature and is used to predict the in vivo permeability.[30]Frequently, $\log P$ is used in place of $\log D$; however, $\log P$ only considers neutral compounds rather than ionizable compounds.[30] $\log D$ accounts for a molecules pH dependence in aqueous solution.[30] This is an important factor as the human body does not maintain a single pH, instead it changes throughout the body as well as whether the individual has fasted or been fed.[6, 31]

Solubility is another important property that is measured. It is the property of a solute to dissolve in a solvent. It has been suggested that over 40% of new chemical entities are insoluble.[9] Solubility is measured as the saturation concentration so when adding more solute, the concentration in the solution does not increase.[32] There are several techniques that formulation scientists can use to improve the solubility of the compounds. These methods can include changing the pH and using buffers or even the use of novel excipients.[9] Solubility can be influenced by the species in the solvent, however, the USP (United States Pharmacopeia)[33] and BP (British Pharmacopoeia)[34] have classified solubility, regardless of the solvent used. It is based on part of solvent that is required per part of solute. The lower the part of solvent required per part solute, the more soluble it is considered.[6]

Because of the number of new molecules that are found to be poorly soluble the potential for prediction of intestinal solubility to aid the lead optimization phase is of high importance.[35] Both qualitative and quantitative approaches have been used in the prediction of solubility.[36] Many computational models for predicting solubility have been published and are based on well-known algorithms including neural networks, multiple linear regression, and group contributions[37] (a relatively simple approach that uses the occurrences of fragments in a molecule as descriptors and then calculates the solubility using Eq. 1).[37] Of course problems with this method can arise when trying to predict the solubility of a compound containing an unknown fragment.[38]

$$\log S = \sum i a_i n_i + a_0 \qquad (1)$$

Eq. (1) represents group contribution method for predicting solubility of a molecule.

The combination of neural networks and molecular topological descriptors have been shown to yield reasonable predictive capabilities.[39] The authors did note that there are limitations to using topological indices including the lack of three-dimensional effects being represented, as well as conformational. However, topological descriptors can be calculated rapidly.[39] As more and more complicated methods for the prediction of solubility are used, authors have asked whether human experts can actually predict solubility better than computational methods.[40] The authors provided experts with drug like compounds and asked the experts to predict the aqueous solubility. In comparison they applied a variety of machine learning algorithms on the same dataset. The best algorithm was a multilayer perceptron. Multi-layer perceptron's are a type of neural network that is also known as a type of feedforward network works by having the input pass through the hidden layers, applying any weights and biases and calculates the output of the model. A loss function is applied to compare the predicted output and the true output. Backpropagation is then applied by updating the model weights. The authors of the study found that a human predicted just as well as the computational method, however, generally the computational methods performed better when compared with how well each human performed. The authors concluded that combining both computational efforts and human decision-making was the most beneficial.[40]

Recursive neural networks have also been applied to attempt the prediction of physical-chemical properties as demonstrated by Bernazzani and co-authors.[41] Chinea published a paper explaining recursive neural networks and described them as being able to learn complex structured information and are nonlinear and adaptive (update with new information[42]) in nature.[43]

Microsomes are not present in living, healthy cells, but are reformed from parts of the endoplasmic reticulum during laboratory procedures to break up cells.[44] Microsomes are used to assess the metabolism of compounds and they can express key enzymes in the drug metabolizing process, notably cytochrome P enzymes.[45] It is estimated that 60% of marketed drugs are substrates for CYP enzymes.[10] However, it is important to note that the expression and activity of these enzymes varies greatly from individual to individual as they are affected by a wide range of genetic and environmental factors.[6, 46]

As well as microsomal metabolism, hepatocyte metabolism is also assessed in vitro. Hepatocytes, make up approximately 80% of the livers mass.[47] However, this analysis is limited by hepatocyte donors as well as the ability of adult differentiated hepatocytes to proliferate in culture.[6, 48]

Prediction of excretion and renal clearance is difficult for a number of biological reasons,[49] however, attempts have been made from compound structure. Renal-ex Predictor is an online available tool that can be used to predict excretion and clearance.[50, 51] In the paper, the authors explain their model construction process consisted of randomly splitting the datasets into training and test sets, removing highly correlated values and columns with zero-variance as well as feature selection approaches.[51] They explored a variety of different models to find the best accuracy.[51] The prediction method for clearance in humans was split into two steps where first it operates as a classification task to predict the excreted type (reabsorption, intermediate, or secretion) and then based on that value, a regression model is selected to predict the clearance value.[50, 51] Predicting which drug elimination route is also of interest and attempts have been made to do so, such as when the major route of drug elimination is via biliary excretion where published attempts showed good predictive accuracy.[52]

Finally, functional scientists analyze hERG (human Ether-à-go-go-Related Gene), which contributes to the hearts electrical activity and is used in assays to assess cardiac toxicity. Because of this contribution, it is a common reason that drugs fail preclinical testing when the compounds interact with this target, as interruptions to the hearts electrical activity can be fatal. Therefore the risk is weighed up against the disease the compound is being used to treat. For example, a compound, that is being used to treat a non-life-threatening illness, that interacts with hERG that could potentially induce torsade de pointes arrhythmia. Cisapride is a drug that was removed from the market for this reason and it was shown that it only induced torsade de pointes arrhythmia in approximately 1 of 120,000 patients.[53]

The hERG binding assay, is not a functional test and therefore does not highlight the agonistic or antagonistic effects of the compound and it does not identify which state the compound binds to on the channel, nor which sites it binds to.[54] Functional assays can also be performed.[6]

Evaluating safety and toxicity is discussed in the following chapter, however, there are several machine learning approaches to predicting hERG specifically that we mention here. Ogura and co-authors note that information related to hERG has increased in the public domain which has allowed for machine learning to be used to successfully predict it.[55] In their paper, the authors develop a classification model using support vector machines using structural fingerprints alongside a descriptor selection algorithm [NSGA-II (Non-dominated Sorting Genetic Algorithm II)].[55] NSGA-II is a multi-objective genetic algorithm that has been used for selecting descriptors to ensure the best predictive performance for a minimal amount of required descriptors.[55, 56] Deep learning has also been applied to the task, such as in the case of deephERG of which predicts hERG blockers using a multitask deep neural network algorithm.[57] Multitask deep learning share information among multiple learning tasks to improve the performance.[58] Their use in the pharmaceutical industry has been discussed and open source implementations have been made available, such as in DeepChem.[59]

There are a number of ways biological activity of a compound may be assessed. Medvedev and co-authors demonstrated that it can be inferred by assessing transcription factor activity[60] as an example. For activity prediction of a compound, attempts have been made to predict the ligand activity using 3D models which were either pocket-based or ligand property-based. This was performed for G-Protein coupled receptors and nuclear hormone receptors. For over half the models, that were pocket-based, the model was able to discriminate the activities from decoys with acceptable and above accuracy. Although the property-based model performed better with over one-half of their models showing high accuracy suggesting that these 3D models can be used for compound activity prediction.[61] Another approach in predicting activity from compound structure used conformal prediction.[62] The method used by the authors is called Inductive Mondrian Conformal Predictor and allowed for identification of compounds most likely considered active for a specific target.

In a tutorial on conformal prediction, Shafer and Vovk explain conformal prediction to learn from previous experiences to determine the levels of confidence in a new prediction.[63] Applying this during the lead optimization phase will allow a library of "similar" compounds to the lead compound with slight changes in them [such as matched molecular pairs (MMPs) of the lead compound] to be assess for the highest predicted biological activity compared with the lead compound.[64] Additionally, this method is also of interest during the previous hit identification phase as it can be hugely beneficial to predict with a level of confidence the activity of a compound to decide whether it should be taken forward for further investigation. The combination of structural and bioactivity-based fingerprints has been shown to improve predictivity as well as the ability to scaffold hop (searching for compounds with similar activity but different core structures).[65]

The use of multiple instance learning via joint instance and feature selection has also been explored and discussed for predicting activity.[66] Multiple instance learning has training instances arranged into sets (known as bags) and a label is provided for the entire bag.[67] The label is the result of one or more instances in the bag. In terms of applying this to compound activity, the bag would be the molecule and the bags labels would be the activity. However, we do not know what conformer (molecule isomer differing by the rotation in a bond) is bioactive.[66] This type of learning is suitable when it is not clear which conformer of a molecule results in the observed activity whilst also recognizing the most representative molecular descriptors.[66]

Tools for activity prediction have been developed such as the PASS (prediction of activity spectra for substance) tool.[68, 69] It predicts the biological activity spectrum of a compound using the 2D structure of a compound. The activity spectrum estimation is based on Bayesian statistics and predicts the active to inactive ratio at various prediction thresholds. Leave-one-out cross validation estimation was used to determine the accuracy.[69] The tool is also able to predict a variety of pharmacological effects such as various types of toxicity and the tool can also display the predicted activity of a molecule at different threshold levels.[69]

There are not only tools for activity prediction, but many tools used to predict ADMET properties of which have been summarized in many publications. The website VLS3D.com[70] describes itself as a directory of tools and databases for computational approaches in the drug discovery pipeline. It contains a very comprehensive list of available tools of approximately over 200 tools at time of writing, for ADMET and physicochemical properties predictions.[71] Peach and coauthors previously published a paper on the tools and resources available for predictions related to metabolism.[72] In their paper, they provide a table of 22 different software that are publicly available (with different levels of availability, such as commercial or open source) for the use of predicting metabolism. In another study, by Czodrowski and co-authors,[73] the authors also review methods to predict metabolism. In the review the authors summarized the different computational approaches to achieve such a prediction.[73] The authors highlight the uses of rule-based, ligand-based (such as quantum mechanical methods which consider the electronic nature of the ligands, descriptor methods of which are derived from the compound structure and pharmacophore modeling which considers the 3D structure information), and protein-based methods.[73]

Despite many tools and prediction methods available, chemical intuition is still important.[74]

The discussion of chemical intuition from a human versus computational methodologies has been discussed with the overall consensus being that the best approach is to utilize both to combine and complement. Chemical intuition plays such an important part in any development, there is great value in ensuring that biases and downfalls of the methods need to be accurately communicated.[74] In addressing the influence of medicinal chemist's intuition, one report discussed how chemists simplify problems, the amount of agreement between chemists on the criteria used, and the accuracy of reporting the relevant criteria.[75] To do this, chemists were asked to select chemical fragments from a set of approximately 4000. The findings showed that chemists greatly simplify the problem by using few criteria and generally, although there was agreement on what parameters should be used, there was no strong agreement between chemists on how the parameter preferences were determined and thus what constituted undesirable parameters. Overall, the study highlighted that there is a low consensus between chemists.[75] In another study,[76] the authors assessed how consistent the medicinal chemists' opinion is and investigated this along with a compound acquisition program that was conducted by Pharmacia. This report also showed a lack of consistency between chemists and highlighted the danger of declaring a compound as undesirable as it is then excluded from further assessment (as well as structurally similar compounds). What was identified is a conflict between chemists' designation of the undesirability of a compound, which inevitably can influence the related computational models. Therefore the use of data-driven MMP analysis help mitigate against the variation in opinions between medicinal chemist intuition and provide further evidence for decision-making.[6]

During the lead optimization phase the aim is to improve the compound toward a more desirable structure that yields suitable properties. To assess how changes to the chemical structure will affect these properties—single small changes in the compound can be made to assess and determine the change in properties. These changes are performed in the same location on the compound and are known as MMPs. These have been used extensively in computational chemistry and are discussed in more detail later.

Matched molecular pairs

MMPs can be described as two compounds that are identical with exception of a molecular fragment that differs in the same position.[77] The term was first used in the book Chemoinformatics in Drug Discovery written by Kenny and Sadowski.[77] MMPs are useful for observing step by step how the changes in the compound's chemistry affect properties.[6]

MMP algorithms can be split into two categories, namely, supervised and nonsupervised methods. The difference between the two categories is that in supervised methods the transformations are pre-defined (the transformations that make the MMP) whereas in unsupervised methods, an algorithm is used to identify all the potential pairs.[6, 78]

Unsupervised methods frequently use maximum common substructure (MCS) algorithms, whereas supervised methods such as fragment-based methods rely on known transformations. Therefore, for all algorithms that are available, the advantages and disadvantages can be described by these two methods. For example, unsupervised methods can identify new MMPs whereas supervised methods have precise control of what the MMP is.[6]

The algorithms involved have all been summarized and discussed[78] very recently and shows that a range of methods are available, although some are proprietary. Table 2 highlights some of the key MMP algorithms that have been developed.[6]

Table 2 Some key MMP algorithms available.

Algorithm	Based on	Method type	Advantages	Disadvantages
Huassain-Rea[79]	Hussain-Rea fragmentation	Nonsupervised	Computationally efficient	Does not allow for considerations of the environment
Hajduk et al.[80]	Pairwise comparison of compound using findsubs routine (Daylight)— uses specified transformations	Supervised	Uses specified substructure for a more targeted identification process	Limited to terminal or side group changes only
ThricePairs[81]	Specified transformations and SMARTS	Supervised	Yields good Tanimoto scores of which suggests chemical diversity	Yields a low number of transformations
WizePairs[82]	Maximum common substructure and SMIRKS	Nonsupervised	Can capture the local single site environment	Can be applied to larger datasets, however, the authors example is very small

The Hussain-Rea[79] methodology, as described in the authors paper, takes all molecules that are inputted and enumerates all acyclic single cuts (in the single cut example that the authors describe) and then indexes each fragment into the start and ending fragments. This therefore allows identification of the transformation. However, this algorithm also allows for double cuts to be made and the methodology follows the same as single cuts. The Hussain-Rea fragmentation method forms the basis of several applications of MMPs by such as that of Matsy,[83] a knowledge-based methodology to predict R groups that are likely to improve biological activity.[6]

Another example of an unsupervised method is the WizePair algorithm.[82] This method is based on the MCS approach to identify the potential MMPs of which are then verified to ensure that the MMPs are located at a single site and encoded in SMIRKS reaction notation and is able to capture the local single site environment.[82] The method can be applied to large datasets. The authors use this method on some set 11-histone deacetylase inhibitors where the system could be used to apply medicinal chemistry knowledge from one project to the next. In addition, the method allows for common bioisosteres identification.[6, 82]

An example of a supervised algorithm is that of Hajduk and Sauer[80] of whom used pairwise comparisons utilized from Daylight software[84] and retained SMILES pairs, along with the potency and substituent data, where the difference between the two SMILES strings was a sole substituent.[80] Furthermore, the authors explain that SMIRKS were also used for specific instances.[80]

ThricePairs which has defined transformations and SMARTS.[81] The ThricePairs method is an in-house proprietary software and due to the defined transforms, yields a low number of matched pairs, but those that did had a desirable mean Tanimoto score, suggesting chemical diversity. The authors use their method on a large dataset to assay in vitro human liver microsomal turnover assay results.[6, 81]

The MCS is used to detect the largest identical substructures between two compounds and thanks to its unsupervised approach can lead to the identification of novel matched pairs.[85] Raymond and Willett discussed algorithms for maximum common subgraph isomorphisms.[86] The authors explain that a 2D structure is represented as atoms (vertices) and bonds (the edges) whereas a 3D structure again represents the atoms as vertices, but the edges as represent the geometric distance between the vertices.[86] Furthermore the authors explain that it can be shown that 2D graphs, due to their sparseness, have approximately equal number of edges and vertices whereas for 3D chemical graphs, there is an edge between each pair of vertices and so the number of vertices is approximately the square of the number of edges (Eqs. 2, 3).[6, 86]

$$|E(G) \approx O|V(G)| \tag{2}$$

Eq. (2) shows the number of edges in proportion to the number of vertices in 2D representations.

$$|E(G) \approx O|V(G)|^2 \tag{3}$$

Eq. (3) shows the number of edges in proportion to the number of vertices in 3D representations.

Raymond and Willet explain that MCS can donate two types of graph subtypes, the maximum common induced subgraph (MCIS) and the maximum common edge subgraph (MCES).[86] These represent a graph with the largest number of vertices (with edges in between) and edges, respectively, common to the two graphs being compared.[86] The MCS can also be split into whether it is connected (there is only one subgraph—each pair of vertices forms the endpoints of a path) or disconnected (multiple subgraphs).[6]

Improving how MMPs are computed have been discussed where the study highlights that MCS (compares two molecules and identifies the largest possible substructure that is identical between the two) and the fragment and index (cleaves the acyclic single bonds and compares all possible fragments between the two molecules) methods are the most prevalent methods at finding rules. It is suggested that combining the two methods increases the effectiveness of finding rules.[87] Additionally, it has been shown that using an MCS based similarity measure is more effective than atom-pair based methods when it comes to searching chemical databases and compliments the atom-pair based methods well. The authors suggest using a hybrid of the two methods for prediction models, namely, bioactivity prediction models.[6, 85]

Deep learning approaches have been applied (graph neural network) to detect the maximum common subgraph, such as that published by Bai and co-authors which is based on a reinforcement learning framework.[88] The approach uses Deep Q-Learning.

MMPA can be further extended to a match molecular series where instead of just comparing two compounds you extend this to a series of compounds.[83] Again, each of these compounds in the series is identical with exception of a single molecular fragment in the same location. This type of analysis allows for medicinal chemists to analyze trends in activity over a project.[83] It has also been proposed that matched molecular series analysis can be used in generation and prioritization of novel molecules.[6, 89]

The limitations of many methods is their time-consuming nature relating to computational efficiency and is often a problem with calculating MMPs.[79] Whilst MMPs have great use in understanding activity cliffs[90] and differences between compounds in terms of similarity, the methodology is not without its limitations.[78] Furthermore, it is advised to include contextual information when using MMPs as it has been shown that in cases of predicting hERG inhibition, solubility, and lipophilicity, the prediction ability is enhanced.[91] The algorithms that derive the MMPs can themselves be a source of limitations. Those using a set of predefined molecular transformations of which are used as a starting point are limited by the fact that transformations that differ from those in the starting source will not be identified.[91] This limitation can be avoided by using Most Common Molecular fragment algorithms, however, despite being shown that they can work well on large datasets,[92] they can be computationally exhaustive.[6, 91]

Machine learning with matched molecular pairs

Machine learning has been used in conjunction with MMPs. Such an example is the work of Turk and coauthors of which used the combination for virtual compound optimization.[93] In their study, the authors use a fragment-based MMP method alongside different machine learning algorithms (including deep learning) and designed compound optimization scenarios to predict the effect of changes in the structure on the compounds activity.[93]

Potency changes between MMPs have also been predicted using methods such as support vector regression approaches.[94] The MMPs were considered in a direction-dependent manner (included the recorded potency difference) and were calculated using an in-house method that is based on the Huassain-Rea[79] algorithm.[94] The authors also developed six new kernels (used to map data into a higher dimensional space) where each of the new kernels differed by the type of fingerprint representation used but all based on Tanimoto similarity) and found that the accuracy of the prediction was influenced by the kernel used.[94]

In another study, MMPs were used alongside descriptor importance analysis to understand relationships between structure and $\log D$ where the authors explained that they could be used to aid the property predictions.[95] In another study, predicted values were used in addition to experimental measurements to identify significant transformations (for particular endpoints) which the authors explained reduce limitations of only using experimental measurements [such as data limitation and interpretability (through the model not from the data)].[96]

Computational approaches alongside chemist intuition will be able to aid the discovery of novel molecules with desirable properties especially as chemical space is so large. Duros and coauthors noted that automated processes including decision-making algorithms can fall victim to poor data.[97] The authors compared the collaborative capabilities of humans and machine learning approaches in exploring chemical space and found that working together generated the best results.[97]

References

1. Lead optimization—Latest research and news | Nature. Available at: https://www.nature.com/subjects/lead-optimization [Accessed 19 June 2019].
2. Lead optimisation—Drug discovery | Sygnature discovery. Available at: https://www.sygnaturediscovery.com/drug-discovery/integrated-drug-discovery/lead-optimisation/ [Accessed 3 August 2018].
3. Eddershaw P, Beresford A, Bayliss M. ADME/PK as part of a rational approach to drug discovery. *Drug Discov Today* 2000;**5**:409–14.
4. Cheng K-C, Korfmacher WA, White RE, Njoroge FG. Lead optimization in discovery drug metabolism and pharmacokinetics/case study: the Hepatitis C Virus (HCV) protease inhibitor SCH 503034. *Perspect Med Chem* 2007;**1**:1–9.
5. Cheng K, Korfmacher W, White R, Njoroge F. Lead optimization in discovery drug metabolism and pharmacokinetics/case study: the Hepatitis C Virus (HCV) protease inhibitor SCH 503034. In: *Dyes and drugs 1*. Apple Academic Press; 2011. p. 196–209.
6. Ashenden S. *On the dissemination of novel chemistry and the process of optimising compounds in drug discovery projects*. University of Cambridge; 2019. https://doi.org/10.17863/CAM.38232.
7. Transport Across Caco-2 monolayer: biological, pharmaceutical and analytical considerations. Available at: http://pharmaquest.weebly.com/uploads/9/9/4/2/9942916/caco2.pdf (Accessed 21 June 2018).
8. Gao Y, Gesenberg C, Zheng W. Oral formulations for preclinical studies: principle, design, and development considerations. In: *Developing solid oral dosage forms: pharmaceutical theory and practice*. 2nd ed. Elsevier; 2016. p. 455–95. https://doi.org/10.1016/B978-0-12-802447-8.00017-0.
9. Savjani KT, Gajjar AK, Savjani JK. Drug solubility: importance and enhancement techniques. *ISRN Pharm* 2012;**2012**.
10. Knights KM, Stresser DM, Miners JO, Crespi CL. In vitro drug metabolism using liver microsomes. *Curr Protoc Pharmacol* 2016;**74**:7.8.1–7.8.24.
11. Sahi J, Grepper S, Smith C. Hepatocytes as a tool in drug metabolism, transport and safety evaluations in drug discovery. *Curr Drug Discov Technol* 2010;**7**:188–98.
12. hERG safety assay. Available at: https://www.cyprotex.com/toxicology/cardiotoxicity/hergsafety (Accessed 15 September 2020).
13. Trotter MWB, Buxton BF, Holden SB. Support vector machines in combinatorial chemistry. *Meas Control* 2001. https://doi.org/10.1177/002029400103400803.
14. Li H, Sze KH, Lu G, Ballester PJ. Machine-learning scoring functions for structure-based drug lead optimization. *Wiley Interdiscip Rev Comput Mol Sci* 2020;**10**.

15. Yamashita R, Nishio M, Do RKG, Togashi K. Convolutional neural networks: an overview and application in radiology. *Insights Imaging* 2018;**9**:611–29.
16. Yasuo N, Watanabe K, Hara H, Rikimaru K, Sekijima M. Predicting strategies for lead optimization via learning to rank. *IPSJ Trans Bioinform* 2018;**11**:41–7.
17. Cheng F, et al. AdmetSAR: a comprehensive source and free tool for assessment of chemical ADMET properties. *J Chem Inf Model* 2012;**52**:3099–105.
18. Yang H, et al. AdmetSAR 2.0: Web-service for prediction and optimization of chemical ADMET properties. *Bioinformatics* 2019;**35**:1067–9.
19. Yang H, et al. ADMETopt: a web server for ADMET optimization in drug design via scaffold hopping. *J Chem Inf Model* 2018;**58**:2051–6.
20. Böhm HJ, Flohr A, Stahl M. Scaffold hopping. *Drug Discov Today Technol* 2004;**1**:217–24.
21. Zhou Z, Kearnes S, Li L, Zare RN, Riley P. Optimization of molecules via deep reinforcement learning. *Nat Sci Rep* 2019;**9**.
22. Sutton R, Barto A. *Reinforcement learning: an introduction. Adaptive computation and machine learning series.* MIT Press; 2018.
23. von Korff M, et al. Predictive power of time-series based machine learning models for DMPK measurements in drug discovery. In: *Lecture notes in computer science (including subseries lecture notes in artificial intelligence and lecture notes in bioinformatics)*; 2019. p. 741–6. https://doi.org/10.1007/978-3-030-30493-5_67.
24. Vamathevan J, et al. Applications of machine learning in drug discovery and development. *Nat Rev Drug Discov* 2019;**18**:463–77.
25. Bleicher KH, Böhm HJ, Müller K, Alanine AI. Hit and lead generation: beyond high-throughput screening. *Nat Rev Drug Discov* 2003;**2**:369–78.
26. Lipinski CA. Avoiding investment in doomer drugs, is poor solubility an industry wide problem. *Curr Drug Discov* 2001;17–9.
27. Guangli M, Yiyu C. Predicting Caco-2 permeability using support vector machine and chemistry development kit. *J Pharm Pharm Sci* 2006;**9**:210–21.
28. Freund Y, Schapire RE. Experiments with a new boosting algorithm. In: *Proceedings of the 13th international conference on machine learning*; 1996. doi:10.1.1.133.1040.
29. Wang NN, et al. ADME properties evaluation in drug discovery: prediction of Caco-2 cell permeability using a combination of NSGA-II and boosting. *J Chem Inf Model* 2016;**56**:763–73.
30. Bhal SK, Kassam K, Peirson IG, Pearl GM. The rule of five revisited: applying log D in place of log P in drug-likeness filters. *Mol Pharm* 2007;**4**:556–60.
31. Dressman JB, Amidon GL, Reppas C, Shah VP. Dissolution testing as a prognostic tool for oral drug absorption: immediate release dosage forms. *Pharm Res* 1998;**15**:11–22.
32. Solubility—an overview | ScienceDirect Topics. Available at: https://www.sciencedirect.com/topics/biochemistry-genetics-and-molecular-biology/solubility (Accessed 15 September 2020).
33. United States Pharmacopeial Convention. *The United States Pharmacopeia 37—The national formulary 32.* United States Pharmacopeial Convention; 2014.
34. Stationery Office (Great Britain). British pharmacopoeia 2009. In: *Stationery Office*; 2008.
35. Augustijns P, et al. A review of drug solubility in human intestinal fluids: implications for the prediction of oral absorption. *Eur J Pharm Sci* 2014;**57**:322–32.
36. Bergström CAS, Larsson P. Computational prediction of drug solubility in water-based systems: qualitative and quantitative approaches used in the current drug discovery and development setting. *Int J Pharm* 2018;**540**:185–93.
37. Jorgensen WL, Duffy EM. Prediction of drug solubility from structure. *Adv Drug Deliv Rev* 2002;**54**:355–66.
38. Klopman G, Zhu H. Estimation of the aqueous solubility of organic molecules by the group contribution approach. *J Chem Inf Comput Sci* 2001;**41**:439–45.

39. Huuskonen J, Salo M, Taskinen J. Aqueous solubility prediction of drugs based on molecular topology and neural network modeling. *J Chem Inf Comput Sci* 1998;**38**:450–6.

40. Boobier S, Osbourn A, Mitchell JBO. Can human experts predict solubility better than computers? *J Cheminform* 2017;**9**.

41. Bernazzani L, et al. Predicting physical-chemical properties of compounds from molecular structures by recursive neural networks. *J Chem Inf Model* 2006;**46**:2030–42.

42. Kridel D, Dolk D, Castillo D. Adaptive modeling for real time analytics: the case of 'big data' in mobile advertising. In: *Proceedings of the annual Hawaii international conference on system sciences*; 2015. https://doi.org/10.1109/HICSS.2015.111.

43. Chinea A. Understanding the principles of recursive neural networks: a generative approach to tackle model complexity. In: *Lecture notes in computer science (including subseries lecture notes in artificial intelligence and lecture notes in bioinformatics)*; 2009. p. 952–63. https://doi.org/10.1007/978-3-642-04274-4_98.

44. Vrbanac J, Slauter R. ADME in drug discovery. In: *A comprehensive guide to toxicology in preclinical drug development*. Elsevier; 2013. p. 39–67. https://doi.org/10.1016/B978-0-12-803620-4.00003-7.

45. Parmentier Y, Bossant MJ, Bertrand M, Walther B. In vitro studies of drug metabolism. In: *Comprehensive medicinal chemistry II*. vol. 5. Elsevier; 2006. p. 231–57.

46. Zanger UM, Schwab M. Cytochrome P450 enzymes in drug metabolism: regulation of gene expression, enzyme activities, and impact of genetic variation. *Pharmacol Ther* 2013;**138**:103–41.

47. Bowen R. *Hepatic histology: hepatocytes*. Colorado State University; 1998. Available at: http://www.vivo.colostate.edu/hbooks/pathphys/digestion/liver/histo_hcytes.html. (Accessed 25 June 2018).

48. Castell JV, Jover R, Martínez-Jiménez C, Gmez-Lechn MJ. Hepatocyte cell lines: their use, scope and limitations in drug metabolism studies. *Expert Opin Drug Metab Toxicol* 2006;**2**:183–212.

49. Fagerholm U. Prediction of human pharmacokinetics—renal metabolic and excretion clearance. *J Pharm Pharmacol* 2007;**59**:1463–71.

50. Renal-ex predictor. Available at: https://adme.nibiohn.go.jp/renal_ex (Accessed 11 September 2020).

51. Watanabe R, et al. Development of an in silico prediction system of human renal excretion and clearance from chemical structure information incorporating fraction unbound in plasma as a descriptor. *Sci Rep* 2019;**9**.

52. Hosey CM, Broccatelli F, Benet LZ. Predicting when biliary excretion of parent drug is a major route of elimination in humans. *AAPS J* 2014. https://doi.org/10.1208/s12248-014-9636-1.

53. Sanguinetti MC, Tristani-Firouzi M. hERG potassium channels and cardiac arrhythmia. *Nature* 2006;**440**:463–9.

54. Priest BT, Bell IM, Garcia ML. Role of hERG potassium channel assays in drug development. *Channels (Austin)* 2008;**2**:87–93.

55. Ogura K, Sato T, Yuki H, Honma T. Support Vector Machine model for hERG inhibitory activities based on the integrated hERG database using descriptor selection by NSGA-II. *Sci Rep* 2019;**9**.

56. Deb K, Pratap A, Agarwal S, Meyarivan T. A fast and elitist multiobjective genetic algorithm: NSGA-II. *IEEE Trans Evol Comput* 2002;**6**:182–97.

57. Cai C, et al. Deep learning-based prediction of drug-induced cardiotoxicity. *J Chem Inf Model* 2019;**59**:1073–84.

58. Zhang Y, Yang Q. An overview of multi-task learning. *Natl Sci Rev* 2018;**5**:30–43.

59. Ramsundar B, et al. Is multitask deep learning practical for pharma? *J Chem Inf Model* 2017;**57**:2068–76.

60. Medvedev A, et al. Evaluating biological activity of compounds by transcription factor activity profiling. *Sci Adv* 2018;**4**. eaar4666.

61. Kufareva I, Chen Y-C, Ilatovskiy A, Abagyan R. Compound activity prediction using models of binding pockets or ligand properties in 3D. *Curr Top Med Chem* 2012;**12**:1869–82.

62. Toccaceli P, Nouretdinov I, Gammerman A. Conformal prediction of biological activity of chemical compounds. *Ann Math Artif Intell* 2017;**81**:105–23.

63. Shafer G, Vovk V. A tutorial on conformal prediction. *J Mach Learn Res* 2008;**9**:371–421.

64. Toccaceli P, Gammerman A. Combination of inductive mondrian conformal predictors. *Mach Learn* 2019;**108**:489–510.
65. Laufkötter O, Sturm N, Bajorath J, Chen H, Engkvist O. Combining structural and bioactivity-based fingerprints improves prediction performance and scaffold hopping capability. *J Cheminform* 2019;**11**:54.
66. Zhao Z, et al. Drug activity prediction using multiple-instance learning via joint instance and feature selection. *BMC Bioinform* 2013. https://doi.org/10.1186/1471-2105-14-S14-S16.
67. Carbonneau MA, Cheplygina V, Granger E, Gagnon G. Multiple instance learning: a survey of problem characteristics and applications. *Pattern Recogn* 2018;**77**:329–53.
68. Lagunin A, Stepanchikova A, Filimonov D, Poroikov V. PASS: prediction of activity spectra for biologically active substances. *Bioinformatics* 2000;**16**:747–8.
69. Parasuraman S. Prediction of activity spectra for substances. *J Pharmacol Pharmacother* 2011;**2**:52–3.
70. VLS3D.com. Available at: http://www.vls3d.com/index.php [Accessed 10 September 2020].
71. VLS3D.com. ADMET and physchem predictions and related tools. Available at: http://www.vls3d.com/index.php/links/chemoinformatics/admet/admet-and-physchem-predictions-and-related-tools.
72. Peach ML, et al. Computational tools and resources for metabolism-related property predictions. 1. Overview of publicly available (free and commercial) databases and software. *Future Med Chem* 2012;**4**:1907–32.
73. Czodrowski P, Kriegl JM, Scheuerer S, Fox T. Computational approaches to predict drug metabolism. *Expert Opin Drug Metab Toxicol* 2009;**5**:15–27.
74. Gomez L. Decision making in medicinal chemistry: the power of our intuition. *ACS Med Chem Lett* 2018;**9**:956–8.
75. Kutchukian PS, et al. Inside the mind of a medicinal chemist: the role of human bias in compound prioritization during drug discovery. *PLoS One* 2012;**7**, e48476.
76. Lajiness MS, Maggiora GM, Shanmugasundaram V. Assessment of the consistency of medicinal chemists in reviewing sets of compounds. *J Med Chem* 2004;**47**:4891–6.
77. Kenny PW, Sadowski J. Structure modification in chemical databases. In: *Chemoinformatics in drug discovery*. Wiley-VCH Verlag GmbH & Co. KGaA; 2005. p. 271–85. https://doi.org/10.1002/3527603743.ch11.
78. Tyrchan C, Evertsson E. Matched molecular pair analysis in short: algorithms, applications and limitations. *Comput Struct Biotechnol J* 2017;**15**:86–90.
79. Hussain J, Rea C. Computationally efficient algorithm to identify matched molecular pairs (MMPs) in large data sets. *J Chem Inf Model* 2010;**50**:339–48.
80. Hajduk PJ, Sauer DR. Statistical analysis of the effects of common chemical substituents on ligand potency. *J Med Chem* 2008;**51**:553–64.
81. Dossetter AG. A statistical analysis of in vitro human microsomal metabolic stability of small phenyl group substituents, leading to improved design sets for parallel SAR exploration of a chemical series. *Bioorg Med Chem* 2010;**18**:4405–14.
82. Warner DJ, Griffen EJ, St-Gallay SA. WizePairZ: a novel algorithm to identify, encode, and exploit matched molecular pairs with unspecified cores in medicinal chemistry. *J Chem Inf Model* 2010;**50**:1350–7.
83. O'Boyle NM, Boström J, Sayle RA, Gill A. Using matched molecular series as a predictive tool to optimize biological activity. *J Med Chem* 2014;**57**:2704–13.
84. Daylight. Available at: https://www.daylight.com/ [Accessed 16 September 2020].
85. Cao Y, Jiang T, Girke T. A maximum common substructure-based algorithm for searching and predicting drug-like compounds. *Bioinformatics* 2008;**24**:i366–74.
86. Raymond JW, Willett P. Maximum common subgraph isomorphism algorithms for the matching of chemical structures. *J Comput Aided Mol Des* 2002;**16**:521–33.
87. Lukac I, et al. Turbocharging matched molecular pair analysis: optimizing the identification and analysis of pairs. *J Chem Inf Model* 2017;**57**:2424–36.
88. Bai Y, et al. *Fast detection of maximum common subgraph via deep Q-learning*; 2020.

89. Awale M, Riniker S, Kramer C. Matched molecular series analysis for ADME property prediction. *J Chem Inf Model* 2020;**60**:2903–14.

90. Hu X, Hu Y, Vogt M, Stumpfe D, Bajorath J. MMP-cliffs: systematic identification of activity cliffs on the basis of matched molecular pairs. *J Chem Inf Model* 2012;**52**:1138–45.

91. Papadatos G, et al. Lead optimization using matched molecular pairs: inclusion of contextual information for enhanced prediction of hERG inhibition, solubility, and lipophilicity. *J Chem Inf Model* 2010;**50**:1872–86.

92. Raymond JW, Watson IA, Mahoui A. Rationalizing lead optimization by associating quantitative relevance with molecular structure modification. *J Chem Inf Model* 2009;**49**:1952–62.

93. Turk S, Merget B, Rippmann F, Fulle S. Coupling matched molecular pairs with machine learning for virtual compound optimization. *J Chem Inf Model* 2017;**57**:3079–85.

94. De La Vega De León A, Bajorath J. Prediction of compound potency changes in matched molecular pairs using support vector regression. *J Chem Inf Model* 2014;**54**:2654–63.

95. Fu L, et al. Systematic modeling of log D7.4 based on ensemble machine learning, group contribution, and matched molecular pair analysis. *J Chem Inf Model* 2020;**60**:63–76.

96. Sushko Y, et al. Prediction-driven matched molecular pairs to interpret QSARs and aid the molecular optimization process. *J Cheminform* 2014;**6**.

97. Duros V, et al. Intuition-enabled machine learning beats the competition when joint human-robot teams perform inorganic chemical experiments. *J Chem Inf Model* 2019;**59**:2664–71.

Evaluating safety and toxicity

7

Aleksandra Bartosik[a] and Hannes Whittingham[b]

[a]*Clinical Data and Insights, Biopharmaceuticals R&D, AstraZeneca, Warsaw, Poland,* [b]*Data Sciences and Quantitative Biology, Discovery Sciences, R&D, AstraZeneca, Cambridge, United Kingdom*

Introduction to computational approaches for evaluating safety and toxicity

Drug-induced toxicity poses a significant challenge for late-stage attrition of drugs. Potential drug candidate safety profile needs to be monitored closely to avoid adverse drug reaction (ADRs). As of 2021, pharmaceutical companies and regulators in many countries rely on in vitro and in vivo testing. Although this approach provides researchers with necessary data, it has certain disadvantages. Certain types of toxicity remain undetected in the nonclinical phase due to physiological and metabolic variability between humans and other species[1] and due to the presence of synergistic effects. Sitaxentan is an example of a drug that is not hepatotoxic in animal trials[2] but causes significant liver injury in humans.[3]

Thus, toxicity issues are often recognized only as late as during clinical trials or even as a result of postmarket surveillance. The risk of late toxicity detection increases in cases of idiosyncratic side effects, drug candidates tested in clinical trials with small patient population size, and little diversity of genotypes and phenotypes.

Once a drug has already reached the market and if serious adverse events occur at a prescribed dose, they can cause harm to many patients, cause the addition of a black box warning to the label, or even a drug market withdrawal. Cardiotoxicity and hepatotoxicity are the two most challenging safety issues leading to phase III failure or postapproval withdrawal due to safety liabilities.[4] In OECD countries, ADRs are one of the most common causes of hospitalization, primary care, and long-term treatments. They result in approximately 15% of total hospital activity and expenditure, the most common being venous thromboembolism and pressure ulcers.[5] Furthermore, ADRs often lead to situations where even more medicines are prescribed to treat the toxic effects of other medicines.

In 2016, the Frank R. Lautenberg Chemical Safety for 21st Century Act amended and updated the Toxic Substances Control Act (TSCA) that was in force since 1976. The new regulation emphasizes the need for the development of new computational methods and predictive models for safety and toxicity assessment.[6] This statement follows recent developments in high throughput screening, combinatorial medicinal chemistry, big data available thanks to industry cooperation, and large publicly available databases.

Furthermore, nonclinical toxicology studies still require animal testing, which is not only resource-heavy but also raises numerous ethical concerns. Each year, over 2000 new chemical molecules are tested on animals and some of them require 2-year-long chronic studies.[7] Development of explainable

The Era of Artificial Intelligence, Machine Learning, and Data Science in the Pharmaceutical Industry. https://doi.org/10.1016/B978-0-12-820045-2.00008-8

and interpretable drug toxicity prediction models is a game-changing opportunity for the pharmaceutical industry. During the past decade, rapid development in this field is possible thanks to an increase in the number of publicly available scientific, medical, and regulatory databases. Unprecedented growth in data science (DS) and artificial intelligence (AI) methods is also caused by an increase in computational power and falling costs of computations.

The use of computational models can help in decreasing the amount of necessary empirical data, reduce animal testing, allow to personalize therapies with the use of biomarkers, and to connect "siloed" domains. An applicability domain of a model is a subset of a domain of classification function, for which it is applicable to make predictions. Notably, any valid training set belongs to applicability domain. This means that a model can only generate valid predictions for molecules structurally similar to those used to train and build the model.[8]

There are multiple DS and AI models with an extended applicability domain that support the integration of a variety of complex data sets. The datasets come from a variety of sources, for instance, protein targets, protein-protein interaction networks, multiomics data, chemical databases, medical records, and pharmacovigilance reporting data.

Different kinds of studies usually involve different data sets. Premarketing surveillance focuses on data from preclinical screening (e.g., drug targets, chemical structure, bioassay data) and phases I to III clinical trials. Postmarketing surveillance focuses on data accumulated once a drug is already on the market (spontaneous reports submitted to national surveillance systems and their validation in additional sources).

DS- and AI-based solutions are increasingly becoming accepted as cost-effective, time-saving, and accurate methods of estimating potential toxicity endpoints of new drug candidates and support data-driven informed decision-making. Biomarker-based approaches have simplified diagnostics of tuberculosis,[9] breast cancer metastasis,[10] and diabetic-driven retinal changes.[11] Data samples collected with use of these approaches enrich biochemical databases that can serve as training datasets for safety modeling. In this chapter, we review the latest advances and challenges in the field of computational approaches to safety evaluation during preclinical development and pharmacovigilance.

In silico nonclinical drug safety

The purpose of preclinical safety evaluation of a drug candidate is to define any potential toxicity endpoints that could put patient's health and well-being into danger during first-in-human trials. Safety profiling is very different from establishing drug's efficacy because it is not possible to a priori define all mechanisms of toxicity. Adverse reactions can occur, for example, due to a mechanism-related side effect, an off-target activity such as hERG channel inhibition, drug-drug interaction, or variability in the metabolomic response. Some adverse reactions are dose-related or idiosyncratic, which means that they are not related to the mechanism, and only some patients develop them. In the case of antiepileptic drugs, idiosyncratic drug reactions accounts for up to 10% of all adverse reactions.[12] In silico safety profiling of potentially toxic compounds from screening tests involves quantitative structure-activity relationships (QSAR), toxicophore mapping, knowledge-based approaches, data mining, classification, regression, class probability estimation, similarity matching, co-occurrence grouping, causal modeling, pattern recognition, and clustering methods. In this section, we will present an overview (Fig. 1) of DS, machine learning (ML), and AI methods applied to most common nonclinical activities and useful chemical and biological data sources.

What is the challenge?

drug safety = right dose to the right patient at a right time

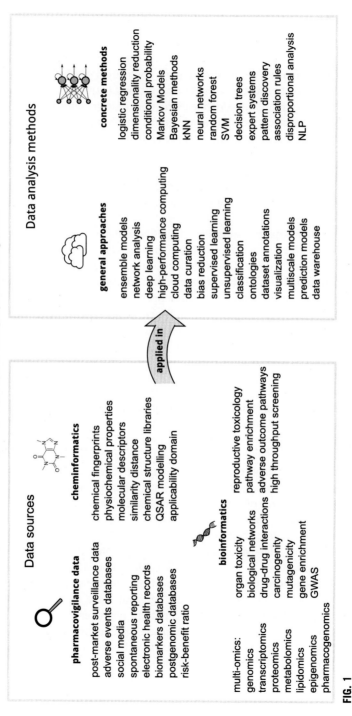

Data sources

pharmacovigilance data

post-market surveillance data
adverse events databases
social media
spontaneous reporting
electronic health records
biomarkers databases
postgenomic databases
risk-benefit ratio

cheminformatics

chemical fingerprints
physiochemical properties
molecular descriptors
similarity distance
chemical structure libraries
QSAR modelling
applicability domain

bioinformatics

multi-omics:
genomics
transcriptomics
proteomics
metabolomics
lipidomics
epigenomics
pharmacogenomics

organ toxicity
biological networks
drug-drug interactions
carcinogenity
mutagenicity
gene enrichment
GWAS

reproductive toxicology
pathway enrichment
adverse outcome pathways
high throughput screening

applied in

Data analysis methods

general approaches

ensemble models
network analysis
deep learning
high-performance computing
cloud computing
data curation
bias reduction
supervised learning
unsupervised learning
classification
ontologies
dataset annotations
visualization
multiscale models
prediction models
data warehouse

concrete methods

logistic regression
dimensionality reduction
conditional probability
Markov Models
Bayesian methods
kNN
neural networks
random forest
SVM
decision trees
expert systems
pattern discovery
association rules
disproportional analysis
NLP

FIG. 1

Overview of in silico drug safety approaches, methods, and data sources used in safety and toxicity evaluation.

Machine learning approaches to toxicity prediction

In this section, we present the most common nonlinear ML models used in toxicity modeling. Quantitative structure-activity relationships (QSAR) prediction method is currently the most common approach to toxicity prediction and the most prominent application of ML in this field. QSAR modeling has been applied to predict ecotoxicity, cardiotoxicity, oral toxicity, carcinogenicity, hepatotoxicity, respiratory toxicity, skin irritation, endocrine disruption, mutagenicity, gastrointestinal absorption, drug metabolism, and pharmacokinetics.[13] ICH M7 guideline was the first international regulation that considers QSAR Ames mutagenicity test in silico predictive models equal to toxicological studies for human safety evaluation.[14]

The main algorithms used in QSAR classification and regression problems include k-nearest neighbors (kNN), logistic regression with regularization, decision trees (DTs), random forest (RF), support vector machines (SVM), genetic algorithms (GA), Naïve Bayesian classifiers, and artificial neural networks (ANN).

In comparison with traditional preclinical safety and toxicity evaluation approaches, ML is a relatively low-cost, rapid, and efficient technique. Thus, it has recently been gaining popularity in pharmacokinetic and pharmacodynamic (PK/PD) predictions, dose and time response modeling, QT interval prolongation testing, hERG, and CYP isoforms binding.[15] In this review, we present several QSAR ML applications in genotoxicity, mutagenicity, carcinogenicity, and reproductivity effect assessments. However, effective models have not been found yet for all of these areas.[16] Below we review the basic concepts and some representative work in this field.

k-nearest neighbors

kNN is a nonparametric supervised learning method used for pattern classification (the output is a class membership) and regression (the output is feature value for a particular specimen) when there is no prior knowledge on the distribution of data. Interpretability is the main advantage of kNN as the k neighbors are easily retrieved and understood by humans unless there are too many dimensions (features) for a human to comprehend the data. In 1967, Cover and Hart[17] proposed a kNN method to classify points based on their spatial features. In toxicity prediction, kNN estimates the toxicity score of a molecule basing on toxicity scores of the closest k neighboring compounds.[18] Neighbor molecules are those which have a high degree of structural resemblance of structural representations to the target molecule. kNN classifier performance depends strongly on the chosen metric that defines the distance between points and therefore their similarity, for example, Euclidean, Minkowski, Manhattan, Chebyshev, Jaccard-Tanimoto, Hassanat, Lance-Williams, or Mahalanobis.[19]

kNN has multiple applications in various QSAR models. For example, in QSAR ocular toxicity prediction,[20] kNN was combined with RF and trained on Dragon and MOE (Molecular Operating Environment) molecular descriptors. Solimeo et al.[20] generated the descriptors for the library of 75 compounds compiled by the National Toxicology Program Interagency Center for the Evaluation of Alternative Toxicological Methods. They have validated their models on an external validation set containing 34 compounds from other sources. Correct classification rate ranged from 72% to 87% for individual models and reached 93% for a consensus model.

In another study aiming to create a QSAR model for acute toxicity prediction,[21] kNN model learned on experimental LD50 data and from TTC database[22] release to the public by EFSA. Similarly to Solimeo et al., Chavan et al. used Dragon molecular descriptors, however, they also preselected features with genetic algorithms. In this case, compounds were correctly classified in 66% for the internal test set and only in 57% for the external validation set, which might have resulted from large number and complexity of mechanisms driving acute toxicity. Correct classification was also a challenge in hERG K + channel study,[23] where kNN model was built on eight PaDEL software-generated fingerprints[24]: CDK, Estate, Extended CDK, CDK Graph, MACCS, PubChem, substructures presence, and substructure count. The training set was balanced and consisted of 93 hERG active compounds and 79 hERG inactive compounds. The test set contained 221 and 1574 compounds, respectively. Although internal CV (cross-validation) showed that the best substructure count model did not make an error in 68%, the external validation set did not exceed 58%.

In the case of kNN, the challenging part is to find the optimal k-value. In this study, the authors performed cross-validation to address this problem. Consensus models are based on three, five, and seven different fingerprint-based models to increase classification specificity and sensitivity. The best consensus model based on Extended-CDK, PubChem, and Substructure count fingerprint-based demonstrated 63% sensitivity and 54% specificity for the external validation dataset from PubChem.

Methods to select optimal subsets of features for the space—*variable selection*—have received considerable attention from those using kNN for QSAR modeling. Variable selection (in any statistical modeling) involves a systematic approach in which models are fitted using different subsets of the available variables to find a combination which produces best performance on held-out data. In the case of kNN models for QSAR, performance is most often measured using a q^2 value (comparable with an r^2 value for a typical regression fit, but instead assessing the fit for unseen data). Zheng and Tropsha[25] introduced a method for variable selection in kNN for QSAR using simulated annealing, a method loosely based on thermodynamics. Beginning with a random subset of variables, a model is trained and a q^2 value calculated. Then, in a repeated process, a fraction of the current variables are exchanged for other randomly selected variables, and the q^2 assessed again. If the new q^2 is higher, the new set of variables is automatically accepted; if it is lower, the new set may still be accepted—but with a probability that is scaled by the q^2 of the new model, and which diminishes with the number of repeats. This "lowering of temperature" allows for exploration of substantially different combinations early on but is then more likely to accept only the best solutions later.

Using kNN with simulated annealing, successful VS models have been built to predict activity in G-Protein Coupled Receptors[26] and estrogen receptor-mediated endocrine disruptors.[27] However, comparative studies do not find kNN to be the most competitive option when cast against other ML techniques,[28–30] and most practitioners make use of SVMS, RFs, or ANNs.

For chemical datasets with a relatively small number of data points and a relatively high number of features (dimensions), kNN requires dimensionality reduction. In general, the dataset should be dense and all similar data points must be close in every dimension. The more features there are, the harder it is to achieve it. A good solution is to increase the number of data points, however, in case of some biological experiments, this might be a significant limitation. Additionally, classification is sensitive to noisy features. Thus, in the case of every kNN model, applicability domain should be clearly defined.[31]

Logistic regression

Logistic regression is another fundamental method initially formulated by David Cox in 1958[32] that builds a logistic model (also known as the logit model). Its most significant advantage is that it can be used both for classification and class probability estimation, because it is tied with logistic data distribution. It takes a linear combination of features and applies to them a nonlinear sigmoidal function. In the basic version of logistic regression, the output variable is binary, however, it can be extended into multiple classes (then it is called multinomial logistic regression). The binary logistic model classifies specimen into two classes, whereas the multinomial logistic model extends this to an arbitrary number of classes without ordering them.

The mathematics of logistic regression rely on the concept of the "odds" of the event, which is a probability of an event occurring divided by the probability of an event not occurring. Just as in linear regression, logistic regression has weights associated with dimensions of input data. In contrary to linear regression, the relationship between the weights and the output of the model (the "odds") is exponential, not linear.

One can say that the interpretability of logistic regression is not as easy as the interpretation of kNN or linear regression, but still much easier than more "black-box" models such as Neural Networks. The main obstacle is the multiplicative nature of the "odds."

In QSAR toxicity modeling, logistic regression has found multiple applications. For example, Li et al. applied logistic regression to predict skin sensitization with use of data from murine Local Lymph Node Assay studies and similarity 4D-fingerprint descriptors.[33] The training set consisted of 196 compounds, and the test set contained 22 compounds divided into four sensitizers classes: weak, moderate, strong, and extreme. The paper has shown that test set accuracy can vary strongly depending on the initial classification of the training set (whether we have separate classes for different strengths of sensitization, or we have a binary "weak or nonweak" classification). In this case, the approach with a smaller number of classes provided better accuracy. Logistic regression is also prone to restrictive expressiveness and complete separation. In cases when a feature correctly separating two classes in the first iteration is found, data are not separated further. This issue can be, however, solved by penalization of the weights or defining a prior probability distribution of weights. Ren et al. have compared logistic regression with another interpretable model, linear discriminant analysis (LDA) model in aquatic toxicity prediction. They used experimental data and chemical structure-based descriptors calculated by the CODESSA and DRAGON software packages.[34] Both models were validated internally and externally and LR has outperformed LDA for compounds that exhibit excess toxicity versus nonpolar narcotic compounds and for more reactive compounds versus less reactive compounds. Logistic regression was also used in consensus models with other techniques to increase prediction capabilities. In a study on QSAR nanotoxicity of nanoparticles[35] in PaCa2 (pancreatic cancer cells), a consensus model consisting of logistic regression, Naïve Bayes, kNN, and SVM was developed with PADEL-Descriptor version 2.8 generated 1D and 2D chemical descriptors basing on SMILES, lipophilicity, and hydrogen bonding.

SVM

SVM is a nonprobabilistic linear classifier applied to classification problems, regression analysis, and in pattern recognition. In an N-dimensional space, SVM tries to find an $(N-1)$-dimensional hyperplane (so-called decision surface) that separates two classes of instances and maximizes the margin between the decision surface and each of the classes. If two classes are linearly separable, then it is a "hard-border" case and the location of the hyperplane is defined only by the data points of each class that are closest to the hyperplane (so-called support vectors). In real life, however, this is rarely the case that data are perfectly linearly separable, so "soft-border" approach is applied, which aims at minimizing an arbitrary loss function, which is an attempt to minimize the classification error.

Modifications can be made to enable the technique to cope with such situations where the datasets cannot be separated linearly by a single plane, with the two most important being (1) transformations of the space so that the points become separable and (2) the introduction of slack variables—an additional term in the loss function which allows, while penalizing, a few points which fall on the wrong side of the boundary. With some further changes, SVMs can also be adapted for use in regression.

SVM belongs to a class of methods called kernel methods, which allow for using the "kernel trick." For data that have nonlinear frontier between classes, it can be transformed using a kernel function so that data in new coordinates have a linear border (this border can still be fuzzy, though). Examples of kernel functions are polynomial kernel, which extends the original features $x_1, \dots x_n$, with combinations of those (e.g., x_1x_2, x_2x_3) and Gaussian kernel (also known as radial basis function kernel, RBF) which replaces original features with features that describe pairwise similarity between each original feature.

The primary advantage of SVMs is that they perform well in high dimensional spaces and do not require dimensionality reduction and the second one is that they are memory efficient. Thus, they are suitable methods for classification of high dimensional biological data. SVM major drawback is limited interpretability, tendency to overfit if a number of features significantly exceed the number of data points, lack of probability estimation, and dimensionality "blow-up" when applying kernel functions.

Kotsampasakou et al.[36] applied SVM to drug-induced liver toxicity studies. The model was trained on 2D and 3D chemical descriptors, preselected with a Genetic Algorithm (GA). Tan et al. developed an SVM-based carcinogenicity prediction-based physicochemical properties, constitutional, topological, and geometrical descriptors. In this case, the model reached a high overall accuracy of 88.1%.[37] SVM models were also used to predict vascular endothelial growth factor receptor (VEGFR)2 inhibition by aminopyrimidine-5-carbaldehyde oxime derivatives[38] and in a classification model for neuraminidase inhibitors of influenza A virus (H1N1).[39] SVM output is also used in consensus models together with Naïve Bayes, kNN, and RF algorithms; for example, in a recent study oral rat acute toxicity on a dataset coming from the industrial context.[40] It is worth noting that the authors managed to increase the fraction of industrial compounds with the model applicability domain from 58% (NICEATM model which was state of the art) to 94% for the new model. Additionally, balanced accuracies increased from 0.69 for already existing NICEATM model to 0.71 for the new consensus model.

Decision tree

A DT is a nonparametric supervised learning method that was first used in QSAR in the late 1970s in drug potency pattern analysis.[41] DTs can be built based on either discrete (regression trees) or continuous input values (classification trees). Leaves of a tree represent class labels, nonleaf nodes represent logical conditions, and root-to-leaf paths represent conjunctions of the conditions on its way.

The most substantial advantage of DTs is direct interpretability and explainability since this white-box model reflects the human decision-making process. The model works well for massive datasets with diverse data types and has an easy-to-use mutually excluding feature selection embedded. Thus, DTs are useful in exploratory analysis and hypothesis generation based on chemical databases queries.[42,43] For instance, DT-based algorithm has been used by Su et al. to classify chemical structure features associated with cytochrome P450 (CYP) enzymes inhibition with use of an input dataset of 10,000 chemical compounds.[44]

In some conditions, DTs are more prone to overfitting and biased prediction resulting from class imbalance. The model strongly depends on the input data and even a slight change in training dataset may result in a significant change in prediction.[45] Currently, its application is limited because there exist other models with better prediction capabilities. Nevertheless, DTs are a staple of ML, and this algorithm is embedded as voting agents into more sophisticated approaches such as RF or Gradient Boosting Classifier.

Random forest and other ensemble methods

To create robust and highly predictive models, multiple classifiers of the same ML decision model can be combined into an ensemble of classifiers.[46] For instance, an RF algorithm is an ensemble algorithm combining single DTs. It selects DTs randomly and averages their predictions or chooses the class pointed by DTs voting. Each DT has binary leaves, and thus RF reciprocates so well the presence or absence of chosen molecular descriptor or molecular fingerprints.[47]

Combination of multiple classifiers and decision models for a single classification problem reduces variance, decreases the risk of overfitting and class imbalance, which are inherent to any ML method. The central assumption behind Ensemble Methods is that combined diverse and independent predictions of many single approaches results in better performance because generalization error is reduced. Additionally the potential error is decreased with various ensemble techniques of counting votes of single models. Max voting, averaging, and weighted averaging are basic examples such ensemble techniques used for calculating probabilities in classification problems.

More advanced ensemble approaches applied in QSAR modeling include stacking, blending, bagging (bootstrap aggregating), and boosting.[45] At first, the dataset is divided into training, test, and validation datasets. Stacking approach uses output predictions from one model as input features to another model. The model makes predictions on the test set. In the blending approach, predictions are made only on the validation set. The validation set predictions are subsequently used to build the final model. From the perspective of QSAR toxicity prediction, the two last ensemble approaches are most relevant.

The first one being Bootstrapping, which is a sampling technique that generalizes the results of multiple models into one result. It is applied in algorithms such as Bagging metaestimator[48] and RF,[49] in which training set is subsetted with replacement.[50] The second technique called Boosting is a sequential process combining several weak learners to form a strong learner. AdaBoost Decision Tree (ABDT),[51,52] GBM,[53] XGBoost,[54] Light GBM,[55] and CatBoost[56] are examples of ML models using this technique.

Naïve Bayes classifier

Naïve Bayes classifier (also known as just Naïve Bayes) is a set of supervised learning classifiers based on the Bayes theorem of conditional probabilities. While Bayes' theorem dates to 1700s, researchers use Naïve Bayes since 1960s. For each feature, the model estimates class membership probability. For each feature class, the model calculates membership probability independently, relying on the naive assumption of feature independence. The assumption that any correlation is irrelevant distinguishes Naïve Bayes classifier from other ML models. Another advantage is that the method requires a relatively small training dataset to perform classification. Thus, it suits well for the problems such as reproductive toxicity prediction, where animal testing is expensive, both in economy and ethics terms and human testing is impossible. Following REACH (Registration, Evaluation, and Authorisation of

Chemicals; EC 1907/2006) legislation in Europe and TSCA in the United States, Marzo and Benfenati[57] developed a chemicals classification method based on Naïve Bayes that is applicable in reproductive toxicity prediction. The model was trained on ECFP2 molecular descriptors and Leadscope databases and Procter and Gamble data (1172 compounds in total data curation). According to the authors, the Matthews Correlation Coefficient (which allows to assess quality of binary classification when classes are of various sizes) value for the model was high as it reached ≥ 0.4 in validation. In another study, a Naïve Bayes classifier model was trained using molecular descriptors such as AlogP, Molecular weight (MW), number of H. The model was validated by the internal fivefold cross-validation and external test set and reached accuracy of $90.0\% \pm 0.8\%$ for the training set and $68.0\% \pm 1.9\%$ for the external test set.[58]

Clustering and primary component analysis

The ML methods mentioned earlier in this chapter were largely supervised learning methods. There are, however, several important methods that belong to the unsupervised learning category, the most important being Primary Component Analysis (PCA) and Cluster Analysis (also known as clustering).[59]

These methods are useful in the primary phase of the data mining effort. PCA allows for dimensionality reduction and assessment which factors contribute the most information, while clustering allows finding potential classes in unclassified data. Chavan et al.[21] applied PCA to clustering studies on Munro chemical database using no no-observed-adverse-effect level (NOEL) values as response variable.

Deep learning

QSAR modeling took significant advantage of development in deep learning in recent years.[60] Deep learning is ML with use of Deep Neural Networks, a case of Artificial Neural Networks (ANN). ANN was first applied in drug design in 1973 to classify molecules into active and inactive classes.[61] In 1990, Aoyama et al.[62] suggested ANN as a more advanced alternative to linear regression performance in prediction on drug-drug interactions. In 1991, Andrea and Kalayeh[63] applied neural networks to QSAR model of dihydrofolate reductase inhibitors and in 1993, Wikel and Dow[64] used ANN for variable selection in QSAR. In 2002, ANN ensemble was used in toxicity prediction to minimize uncertainty accompanying individual models.[65]

In 2012, Kaggle Merck Molecular Activity Challenge was won by a team which applied Deep Neural Networks in drug target prediction.[66] Subsequently, in 2014, another challenge named Tox21 Data Challenge organized by NIH, FDA, and NCATS were also won by a project applying deep learning to off-target drug and environmental chemicals toxicity detection.[67] DeepTox model normalized representations of chemical compounds and subsequently generated DL-specific chemical descriptors.

The best performing DL descriptors were later applied to toxicity prediction in an ensemble model, which has outperformed other ML approaches such as Naïve Bayes, SVM, and RF. Computation of DL-dedicated chemical molecules representation is necessary as classical molecular fingerprints[47] may sometimes describe the chemical structure in too many details.

Another type of ANN, convolutional neural networks (CNN) were trained directly on graphs to develop new molecular fingerprints, data-driven, and interpretable.[68] Fully Connected Recurrent Neural Network (FCRNN) and their ensembles were fed with DL descriptors: vector representations of SMILES strings, numerical feature values, molecular images, and IGC50 (50% inhibition growth concentration of the test agent) served as toxicity metrics.[69] In another study, authors demonstrated that the so-called "one-shot learning" approach could significantly decrease the amount of data necessary to train a DL model, which is preferably a unique feature in this field.[70] In this study, Karim et al. performed better in terms of accuracy (84%–88%) than the state-of-the-art toxicity prediction methods, that is, TopTox, AdmetSAR, and Hybrid2D.

QSAR Neural Networks have, unfortunately, certain limitations. One of them is that redundant and autocorrelated features decrease model performance and that unknown descriptors influence model predictions in unpredictable ways. Thus, feature preselection is advised, similarly as in case of shallow architectures. Second, once the best descriptors and parameters are found on the training dataset, the risk of overfitting increases. NN are also strongly data-dependent, however, in this case, regularization techniques seem to be a reasonable solution, for example, dropout as mentioned earlier or ReLU. Finally, low interpretability issues and ANNs being "black-box models" are the major obstacle for regulatory safety applications of DL. More interpretable and explainable methods may be more appropriate to facilitate responsible decisions related to human safety and efficacy. Some promising methods include generative adversarial networks (GAN), Latent GAN long short-term memory (LSTM), and variational autoencoders (VA).[46]

Pharmacovigilance and drug safety

In recent years, DS and ML methods have facilitated information processing both in computational toxicology and molecular science and the postmarketing ADR surveillance. An increasing amount of data in the form of individual case safety report forms (ICSRs) from a number of sources provide invaluable support for drug safety monitoring outside the well-controlled clinical trial environment. These sources include spontaneous reports from healthcare professionals (HCPs) and data mining of medical reports and social media.

The real-world patient population is more numerous and diverse in terms of phenotypes than patient cohorts in clinical trials. Also, the postmarketing data include information on patients with numerous comorbid diseases that might not have been considered in a clinical trial cohort because they had not met strict inclusion and exclusion criteria. Therefore, analyzing postmarketing data gives a more holistic picture. Successful postmarketing data acquisition, assessment, and interpretation depend strongly on the ability to automate the manual work and the ability to integrate multiple sources and types of information about potential signals, both structured and unstructured.

This section provides a general overview of DS and AI applications in monitoring postmarketing drug safety and outlines future perspectives and challenges.

Data sources

Excellent quality of data is substantial to the efficacy of statistical and ML models. In the case of pharmacovigilance data, this task is similar to finding needles from a haystack of information sources. The more safety data is available, the more accurate the ML models can be developed. According to ICH Topic E 2 D Post Approval Safety Data Management guideline, Individual Case Safety Reports on ADRs may come from two types of sources: solicited and unsolicited (Fig. 2).

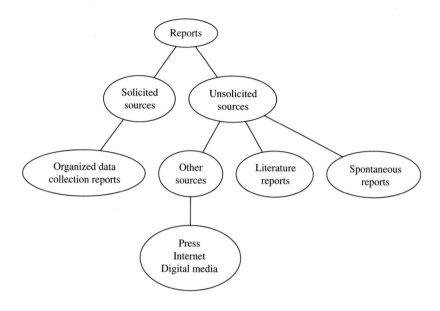

FIG. 2

Overview of adverse drug reaction (ADR) information sources according to ICH Topic E 2 D Post Approval Safety Data Management guideline.

Solicited sources are structured data collection systems, for instance, clinical trials reports, noninterventional studies, organ toxicity databases, postapproval named patient programs and other patient support, disease management program data, surveys of patients or healthcare service providers, as well as data sources related to patient compliance, compassionate, or named patient use.

Unsolicited sources include spontaneous ADR reporting schemes, scientific and medical literature (e.g., Medline database), nonmedical resources (e.g., lay press, media, patient forums, and social media logs) as well as Market Authorization Holder-related media (e.g., sponsor's website), electronic medical records (EMR), licensed partners, payers claims data, and internet search logs.

ADRs are reported by HCPs (Healthcare Professionals), members of public, or by a pharmaceutical company to a national drug safety monitoring system. The UK Yellow Card system, established in 1964, was the first adverse event reporting system (AERS).[71] Since 1968, the global largest SRS (Spontaneous Reporting System) database, WHO VigiBase, has aggregated over 15 million reports related to ADRs concerning medical products from 100 countries.[72] According to ICH-E2D GVP Annex IV, SRS submission to a national pharmacovigilance center is voluntary for HCPs and mandatory for pharmaceutical companies. Using standardized terminology that aligns with MedDRA medical thesaurus[73] and VeDDRA[74] facilitates data analysis and scientific understanding and helps to avoid naming ambiguities. The structure and content may vary, however, each ICSR should contain: name of medicine suspected to cause ADR, demographic details of the patient who experienced ADR, their symptoms, reporting person's name, and contact details.[75]

Disproportionality analysis

Spontaneous reporting is the largest source of postmarketing surveillance data. To assess the impact of reported ADRs on patients' safety and identify Signals of Disproportionate Reporting (SDRs), NCAs perform disproportionality analysis (DA).[76] SDRs were reported in over 90% of studies.[77] In simple terms, DA compares the rate of AEs co-occurrence with a drug of interest and rate of AEs occurrence without the drug.[78] Noteworthy, DA applies to hypothesis generation and not to hypothesis testing, which is emphasized by all regulators. FDA OpenVigil is an example of an online tool for basic DA.[79] NCAs around the globe use multiple approaches to perform DA, however, three of them, frequentist, Bayesian, and questionnaire-based, are most common. Frequentist approaches rely on occurrence frequency comparison between what is expected and what is observed. It is used by the UK Medicines Control Agency (MCA, now MHRA).[80] Calculation of relative reporting ratio (RRR), proportional reporting ratio (PRR), and reporting Odds Ratio are frequentist techniques.

Bayesian methods combine probability function inference and elements of frequentist approaches. Uppsala Monitoring Center, among others, applies Bayesian Confidence Propagation Neural Network (BCPNN) (Medical Applications of Artificial Intelligence, 2017). Other Bayesian methods are Gamma Poisson Shrinker (GPS), Multiitem Gamma Poisson Shrinker (MGPS) used by FDA, and empirical Bayesian geometric means (EBGMs). Another method called Change Point Analysis allows estimating changes observed in ordered data aggregated over time, for example, cases where drug exposure is counted in years and subsequent ADRs appear over time.[81]

Causality assessment[82] may include questionnaire-based methods, such as Naranjo questionnaire,[83] Vanulet questionnaire,[84] WHO-UMC system for standardized case causality assessment, multiitem association analysis,[85] and Gallagher et al. algorithm. One can find more causality assessment algorithms and their analysis in the Varallo review.[86]

Mining medical records

Peer-reviewed biomedical literature can be an additional source of high-quality toxicological data supporting the analysis of drug-ADR associations generated from drug safety databases. In recent years, researchers have been witnessing information and literature overload.[87] NLM Medline, one of the biomedical publication aggregation databases, has over 25 million records and growing.

Information retrieval, text mining, and natural language processing techniques are applied to information extraction from unstructured data with complex sentences and punctuation: books, monographs, review articles, clinical trials reports, case reports, observational studies documentation, and scientific publications. Development of such techniques requires interdisciplinary collaboration between experts in multiple domains, for example, HCPs, molecular scientists, biologists, regulators, pharmacologists, computational linguists, software engineers, and user experience designers.[88]

Electronic health records

Data from Electronic Health Records (EHRs) are primarily used to validate and confirm spontaneous reporting signals. For example, this approach was applied by Wang et al. to pharmacovigilance signal detection related to Rheumatoid Arthritis treatment with disease-modifying antirheumatic drugs.[74] EHRs are resources of higher quality than SRS since HCPs maintain them. They contain data related to the entire medical history of a patient and present longitudinal record about their routine outside clinical trial environment.[72]

Additionally, EHRs are aggregated for a relatively large patient population in comparison with clinical trials, and contrary to SRS, they do not suffer from underreporting and survivor bias. Thus, they facilitate detecting signals with high background incidence rate and research off label use. EHRs volume is impressive and is usually analyzed with the use of NLP methods for unstructured data.

Social media signal detection

Social media is another valuable complementary data source for pharmaceutical companies enhancing postmarketing surveillance activities.[89] Social media continue to be plentiful sources of information related to the user's mood and well-being that can be extracted and annotated with use of NLP.[90] ADR-related signals come equally from explicit status updates as well as patient support groups posts and implicit data such as likes on Facebook and retweets on Twitter,[91] the structure of one's social network and the temporal patterns of updates.

Mith et al. have compared information derived from social media and from the official US FDA Adverse Event Reporting System as well as with drug information databases and systematic reviews. Authors reported that there had been a moderate agreement between ADRs in social media and traditional, however, social media can reveal some additional features not detected with traditional methods, for example, anxiety, that was not available in traditional sources. Doshi-Velez and Kim study shows that a simple rule-based algorithm analyzing the presence of a particular wording in social media status updates can lead to the classification of the Individual Case Safety Report (ICSR).[92] When applying text mining on a person's status updates and messages on Facebook, it is possible to predict a depressive disorder.[93] Another study proposes a method to detect ADRs basing on Yahoo Health discussion forums.[94] The authors suggest that in order to guarantee the highest quality ML predictions should be revised by humans.

As every information source, social media have their limitations too. Most of ML text mining apply to the post of English-speaking individuals and annotation is required.[95] Quality of data is low; datasets are often sparse and unbalanced similarity of spontaneous reports. Finally, ethical issues and GDPR bring additional challenges to be addressed before developing new applications in this area.

Knowledge-based systems, association rules, and pattern recognition

Well-annotated and manually created by human experts, high-quality databases are used as a knowledge base to create knowledge graphs (also referred to as ontologies in information science). Pattern recognition methods utilize knowledge stored as ontologies, thesauri, dictionaries, and grammars to clarify polysemous phrases.[87]

An example of a knowledge graph is PEGASE, which is used for searching and exploring pharmacovigilance data.[96] PEGASE utilizes OntoADR ontology and MedDRA terminology and includes FAERS database patients' data. Bean et al. proposed a knowledge graph designed to predict unknown ADRs with use of electronic health records (EHR), such as patient's medical history, diagnostic test results, allergies, imaging data, diagnosis, treatment, and medications. Their knowledge graph consists of five types of nodes: ADRs, indications, protein targets, gene enrichment, and drugs.[97] The model reached AUC equal to 92%.

Knowledge graphs are applicable in DILI prediction,[98] in gathering ADR-related information in social media[99] and prediction of drug-drug interactions.[100] Generalized association rules and co-occurrence and term lookup are two other strategies used in knowledge-based systems and pattern recognition. Association rules are human-interpretable, and they support decision processes by finding relationships among large sets of data items at a confidence threshold. LimTox is a web tool for applied text mining of adverse event and toxicity basing on associations rules of compounds, drugs, and genes.[101] LimTox preprocesses and standardizes .pdf files containing literature sources related to organ level toxicity. Subsequently, the model creates a classification text ranking, and pattern detection occurs, rule matching and SVM models perform classification. Results are validated. Association rules and co-occurrence grouping play a significant role in drug-drug interactions detection (DDI). Qin et al.[102] and Harpaz et al.[85] found several DDIs undiscovered during clinical trials but detected in real-world patients' data with the use of association rules mining. Nevertheless, generated rules must always be validated, and DDI cause-and-effect relationships often need to be established by a human expert.

Conclusions

In early 2020, a molecule DSP-1181 entered Phase I clinical trials for the treatment of obsessive-compulsive disorder (OCD). A message hit the headlines: the long-acting, potent serotonin 5-HT1A receptor agonist, was declared to be developed by AI. The chemical structure is not published yet.[103] The two companies Sumitomo and Exscientia, which developed DSP-1181, based on their drug discovery processes on previous knowledge and wealth of experience. Thus, ML is primarily supporting drug discovery and development process. This example shows that although there are still many challenges ahead, the direction of pharmaceutical research is clear and hopes are high. In the recent years, computational approaches have facilitated reduction, replacement, and refinement of animal models (3R) with the use of massive datasets.[7] The DS, ML, and AI field continues to develop at an impressive pace. To realize its potential, balanced and annotated databases, semantic harmonization, and more precise ontologies are indispensable. High-quality massive datasets allow to reduce the number of false-positives and to develop DS and ML approaches with higher predictive abilities.

Furthermore, more partnership and collaboration initiatives like the NIH Big Data to Knowledge (BD2K) initiative are necessary across the globe.[104] The field is also moved forward by experience and ideas exchanges during challenge related to DS, ML, and AI and massive dataset analysis challenges, for example, Critical Assessment of Structure Prediction protein-folding competition,[105] Merck Kaggle molecular activity challenge,[106] CAMDA challenge (http://www.camda.info/), Genomics TrackBioNLP competition BioCreative and Data mining (KDD) Cup (https://www.kdd.org/kdd2019/kdd-cup), and NCATS Challenge (https://ncats.nih.gov/aspire/challenges/challenge1).

From the regulatory and patient's safety perspective, there is a need for more interpretable AI and ML models from which features can be extracted in a human-readable format. Regulatory authorities and members of the public are (and should be) reluctant to replace relatively well-understood traditional animal models with unclear mechanism "black box" models of feature selection.[92] In April 2019, the FDA published a discussion paper "Proposed Regulatory Framework for Modifications to Artificial Intelligence/Machine Learning (AI/ML)-Based Software as a Medical Device (SaMD)—Discussion Paper and Request for Feedback" that describes the FDA's opinion on potential approach to premarket review for AI- and ML-driven software modifications. Hopefully, this will also be possible for medicinal products in the future and it will help to deliver the right dose to the right patient at the right time.

References

1. Seok J, et al. Genomic responses in mouse models poorly mimic human inflammatory diseases. *Proc Natl Acad Sci* 2013;**110**:3507–12.
2. Owen K, Cross DM, Derzi M, Horsley E, Stavros FL. An overview of the preclinical toxicity and potential carcinogenicity of sitaxentan (Thelin®), a potent endothelin receptor antagonist developed for pulmonary arterial hypertension. *Regul Toxicol Pharmacol* 2012;**64**:95–103.
3. Galiè N, Hoeper MM, Gibbs R, Simonneau G. Liver toxicity of sitaxentan in pulmonary arterial hypertension. *Eur Respir J* 2011;**37**:475–6.
4. Laverty H, et al. How can we improve our understanding of cardiovascular safety liabilities to develop safer medicines?: cardiovascular toxicity of medicines. *Br J Pharmacol* 2011;**163**:675–93.
5. Slawomirski L, Auraaen A, Klazinga N. The economics of patient safety: strengthening a value-based approach to reducing patient harm at national level. *OECD Health Working Paper No. 96.* Paris: OECD; 2017, https://doi.org/10.1787/5a9858cd-en.
6. Ciallella HL, Zhu H. Advancing computational toxicology in the big data era by artificial intelligence: data-driven and mechanism-driven modeling for chemical toxicity. *Chem Res Toxicol* 2019;**32**:536–47.
7. Stokes W. Animals and the 3Rs in toxicology research and testing: the way forward. *Hum Exp Toxicol* 2015;**34**:1297–303.
8. Sahigara F, et al. Comparison of different approaches to define the applicability domain of QSAR models. *Molecules* 2012;**17**:4791–810.
9. Rajpurkar P, et al. Deep learning for chest radiograph diagnosis: a retrospective comparison of the CheXNeXt algorithm to practicing radiologists. *PLoS Med* 2018;**15**:e1002686.
10. Ehteshami Bejnordi B, et al. Diagnostic assessment of deep learning algorithms for detection of lymph node metastases in women with breast cancer. *JAMA* 2017;**318**:2199–210.
11. Sayres R, et al. Using a deep learning algorithm and integrated gradients explanation to assist grading for diabetic retinopathy. *Ophthalmology* 2019;**126**:552–64.
12. Zaccara G, Franciotta D, Perucca E. Idiosyncratic adverse reactions to antiepileptic drugs. *Epilepsia* 2007;**48**:1223–44.
13. Yang H, Sun L, Li W, Liu G, Tang Y. Identification of nontoxic substructures: a new strategy to avoid potential toxicity risk. *Toxicol Sci* 2018;**165**:396–407.
14. Honma M, et al. Improvement of quantitative structure–activity relationship (QSAR) tools for predicting Ames mutagenicity: outcomes of the Ames/QSAR international challenge project. *Mutagenesis* 2019;**34**:3–16.
15. Klon AE. Machine learning algorithms for the prediction of hERG and CYP450 binding in drug development. *Expert Opin Drug Metab Toxicol* 2010;**6**:821–33.
16. Yang H, et al. Evaluation of different methods for identification of structural alerts using chemical Ames mutagenicity data set as a benchmark. *Chem Res Toxicol* 2017;**30**:1355–64.
17. Cover T, Hart P. Nearest neighbor pattern classification. *IEEE Trans Inf Theory* 1967;**13**:21–7.
18. Willett P, et al. Chemical similarity searching. *J Chem Inf Comput Sci* 1998;**38**(6):983–96. https://doi.org/10.1021/ci9800211.
19. Prasath VBS, et al. Distance and similarity measures effect on the performance of K-nearest neighbor classifier—a review. *Big Data* 2019. https://doi.org/10.1089/big.2018.0175.

20. Solimeo R, Zhang J, Kim M, Sedykh A, Zhu H. Predicting chemical ocular toxicity using a combinatorial QSAR approach. *Chem Res Toxicol* 2012;**25**:2763–9.
21. Chavan S, et al. Towards global QSAR model building for acute toxicity: munro database case study. *Int J Mol Sci* 2014;**15**:18162–74.
22. Munro IC, Ford RA, Kennepohl E, Sprenger JG. Correlation of structural class with no-observed-effect levels: a proposal for establishing a threshold of concern. *Food Chem Toxicol* 1996;**34**:829–67.
23. Chavan S, Abdelaziz A, Wiklander JG, Nicholls IA. A k-nearest neighbor classification of hERG K+ channel blockers. *J Comput Aided Mol Des* 2016;**30**:229–36.
24. Yap CW. PaDEL-descriptor: an open source software to calculate molecular descriptors and fingerprints. *J Comput Chem* 2011;**32**:1466–74.
25. Zheng W, Tropsha A. Novel variable selection quantitative structure-property relationship approach based on the k-nearest-neighbor principle. *J Chem Inf Comput Sci* 2000;**40**:185–94.
26. Luo M, Wang XS, Tropsha A. Comparative analysis of QSAR-based vs. chemical similarity based predictors of GPCRs binding affinity. *Mol Inf* 2016;**35**:36–41.
27. Zhang L, et al. Identification of putative estrogen receptor-mediated endocrine disrupting chemicals using QSAR- and structure-based virtual screening approaches. *Toxicol Appl Pharmacol* 2013;**272**:67–76.
28. Chandra S, Pandey J, Tamrakar AK, Siddiqi MI. Multiple machine learning based descriptive and predictive workflow for the identification of potential PTP1B inhibitors. *J Mol Graph Model* 2017;**71**:242–56.
29. Li Y, et al. Predicting selective liver X receptor β agonists using multiple machine learning methods. *Mol BioSyst* 2015;**11**:1241–50.
30. Wang L, et al. Discovering new agents active against methicillin-resistant *Staphylococcus aureus* with ligand-based approaches. *J Chem Inf Model* 2014;**54**:3186–97.
31. Sahigara F, Ballabio D, Todeschini R, Consonni V. Defining a novel k-nearest neighbours approach to assess the applicability domain of a QSAR model for reliable predictions. *BMC J Cheminformatics* 2013;**5**(27). https://doi.org/10.1186/1758-2946-5-27.
32. Cramer JS. The origins of logistic regression. *Tinbergen Institute Working Paper No 2002-119/4* 2002. https://doi.org/10.2139/ssrn.360300.
33. Li Y, et al. Categorical QSAR models for skin sensitization based upon local lymph node assay classification measures part 2: 4D-fingerprint three-state and two-2-state logistic regression models. *Toxicol Sci* 2007;**99**:532–44.
34. Ren YY, et al. Predicting the aquatic toxicity mode of action using logistic regression and linear discriminant analysis. *SAR QSAR Environ Res* 2016;**27**:721–46.
35. Chau YT, Yap CW. Quantitative nanostructure–activity relationship modelling of nanoparticles. *RSC Adv* 2012;**2**:8489–96.
36. Kotsampasakou E, Montanari F, Ecker GF. Predicting drug-induced liver injury: the importance of data curation. *Toxicology* 2017;**389**:139–45.
37. Tan NX, Rao HB, Li ZR, Li XY. Prediction of chemical carcinogenicity by machine learning approaches. *SAR QSAR Environ Res* 2009;**20**:27–75.
38. Nekoei M, Mohammadhosseini M, Pourbasheer E. QSAR study of VEGFR-2 inhibitors by using genetic algorithm-multiple linear regressions (GA-MLR) and genetic algorithm-support vector machine (GA-SVM): a comparative approach. *Med Chem Res* 2015;**24**:3037–46.
39. Algamal ZY, Qasim MK, Ali HTM. A QSAR classification model for neuraminidase inhibitors of influenza A viruses (H1N1) based on weighted penalized support vector machine. *SAR QSAR Environ Res* 2017;**28**:415–26.
40. Lunghini F, et al. Consensus models to predict oral rat acute toxicity and validation on a dataset coming from the industrial context. *SAR QSAR Environ Res* 2019;**30**:879–97.
41. Topliss JG. A manual method for applying the Hansch approach to drug design. *J Med Chem* 1977;**20**:463–9.

42. Schöning V, Hammann F. How far have decision tree models come for data mining in drug discovery? *Expert Opin Drug Discovery* 2018;**13**:1067–9.

43. Hammann F, Drewe J. Decision tree models for data mining in hit discovery. *Expert Opin Drug Discovery* 2012;**7**:341–52.

44. Su B-H, et al. Rule-based prediction models of cytochrome P450 inhibition. *J Chem Inf Model* 2015;**55**:1426–34.

45. Giordanetto F, Mannhold R, VCH-Verlagsgesellschaft. *Early drug development: bringing a preclinical candidate to the clinic.* vol. 2. Wiley-VCH; 2018.

46. Basile AO, Yahi A, Tatonetti NP. Artificial intelligence for drug toxicity and safety. *Trends Pharmacol Sci* 2019;**40**:624–35.

47. Wu Y, Wang G. Machine learning based toxicity prediction: from chemical structural description to transcriptome analysis. *Int J Mol Sci* 2018;**19**:2358.

48. Singh KP, Gupta S, Kumar A, Mohan D. Multispecies QSAR modeling for predicting the aquatic toxicity of diverse organic chemicals for regulatory toxicology. *Chem Res Toxicol* 2014;**27**:741–53.

49. Polishchuk PG, et al. Application of random forest approach to QSAR prediction of aquatic toxicity. *J Chem Inf Model* 2009;**49**:2481–8.

50. Breiman L. Random forests. *Mach Learn* 2001;**45**:5–32.

51. Ekins S, Kostal J, et al. Accessible machine learning approaches for toxicology, quantum mechanics approaches in computational toxicology. *Computational toxicology: risk assessment for chemicals.* 1st. Hoboken, New Jersey: Wiley; 2018. p. 18–44.

52. Svetnik V, et al. Boosting: an ensemble learning tool for compound classification and QSAR modeling. *J Chem Inf Model* 2005;**45**:786–99.

53. Kwon S, Bae H, Jo J, Yoon S. Comprehensive ensemble in QSAR prediction for drug discovery. *BMC Bioinformatics* 2019;**20**. https://doi.org/10.1186/s12859-019-3135-4, 521.

54. Zhang L, et al. Applications of machine learning methods in drug toxicity prediction. *Curr Top Med Chem* 2018;**18**:987–97.

55. Zhang J, Mucs D, Norinder U, Svensson F. LightGBM: an effective and scalable algorithm for prediction of chemical toxicity—application to the Tox21 and mutagenicity data sets. *J Chem Inf Model* 2019;**59**:4150–8.

56. Wang Y, Xiao Q, Chen P, Wang B. In silico prediction of drug-induced liver injury based on ensemble classifier method. *Int J Mol Sci* 2019;**20**:4106.

57. Marzo M, Benfenati E. Classification of a Naïve Bayesian Fingerprint model to predict reproductive toxicity. *SAR QSAR Environ Res* 2018;**29**(8):631–45. https://doi.org/10.1080/1062936X.2018.1499125.

58. Zhang H, Cao Z-X, Li M, Li Y-Z, Peng C. Novel naïve Bayes classification models for predicting the carcinogenicity of chemicals. *Food Chem Toxicol* 2016;**97**:141–9.

59. Ferreira MMC. Multivariate QSAR. *J Braz Chem Soc* 2002;**13**:742–53.

60. Ghasemi F, Mehridehnavi A, Pérez-Garrido A, Pérez-Sánchez H. Neural network and deep-learning algorithms used in QSAR studies: merits and drawbacks. *Drug Discov Today* 2018;**23**:1784–90.

61. Hiller SA, Golender VE, Rosenblit AB, Rastrigin LA, Glaz AB. Cybernetic methods of drug design. I. Statement of the problem—the perceptron approach. *Comput Biomed Res* 1973;**6**:411–21.

62. Aoyama T, Ichikawa H. Neural networks applied to pharmaceutical problems. VI reconstruction of weight matrices in neural networks a method of correlating outputs with inputs. *Chem Pharm Bull (Tokyo)* 1991;**39**:1222–8.

63. Andrea TA, Kalayeh H. Applications of neural networks in quantitative structure-activity relationships of dihydrofolate reductase inhibitors. *J Med Chem* 1991;**34**:2824–36.

64. Wikel JH, Dow ER. The use of neural networks for variable selection in QSAR. *Bioorg Med Chem Lett* 1993;**3**:645–51.

65. Agrafiotis DK, Cedeño W, Lobanov VS. On the use of neural network ensembles in QSAR and QSPR. *J Chem Inf Comput Sci* 2002;**42**:903–11.

66. Dahl GE, Jaitly N, Salakhutdinov R. *Multi-task neural networks for QSAR predictions* [arXiv Prepr.]; 2014.

67. Mayr A, Klambauer G, Unterthiner T, Hochreiter S. DeepTox: toxicity prediction using deep learning. *Front Environ Sci* 2016;**3**. https://doi.org/10.3389/fenvs.2015.00080.

68. Duvenaud D, et al. Convolutional networks on graphs for learning molecular fingerprints. In: Cortes C, Lawrence ND, Lee DD, Sugiyama M, Garnett R, editors. *Advances in neural information processing systems.* vol. 7. Red Hook, New York: Curran Associates; 2015. p. 2224–32.

69. Karim A, et al. *Toxicity prediction by multimodal deep learning.* arXiv:1907.08333 [physics, stat]; 2019.

70. Altae-Tran H, Ramsundar B, Pappu AS, Pande V. Low data drug discovery with one-shot learning. *ACS Cent Sci* 2017;**3**:283–93.

71. Edwards IR. Spontaneous reporting-of what? Clinical concerns about drugs: spontaneous reporting of what? *Br J Clin Pharmacol* 2001;**48**:138–41.

72. Ventola C, Big L. Data and pharmacovigilance: data mining for adverse drug events and interactions. *P T* 2018;**43**:340–51.

73. Gunnar D, Cédric B, Marie-Christine J. Automatic generation of MedDRA terms groupings using an ontology. *Stud Health Technol Inform* 2012;**180**:73–7.

74. Xu X, et al. Making sense of pharmacovigilance and drug adverse event reporting: comparative similarity association analysis using AI machine learning algorithms in dogs and cats. *Top Companion Anim Med* 2019;**37**:100366.

75. Florian J, Zhuang A. *Benefit-risk assessment methods in medical product development: bridging qualitative and quantitative assessments.* vol. 6. Alexandria, Virginia: Chapman and Hall; 2016.

76. Colilla S, et al. Validation of new signal detection methods for web query log data compared to signal detection algorithms used with FAERS. *Drug Saf* 2017;**40**:399–408.

77. Dias P, Penedones A, Alves C, Ribeiro C, Marques F. The role of disproportionality analysis of pharmacovigilance databases in safety regulatory actions: a systematic review. *Curr Drug Saf* 2015;**10**:234–50.

78. Zink RC, Huang Q, Zhang L-Y, Bao W-J. Statistical and graphical approaches for disproportionality analysis of spontaneously-reported adverse events in pharmacovigilance. *Chin J Nat Med* 2013;**11**:314–20.

79. Böhm R, et al. OpenVigil FDA–Inspection of U.S. American adverse drug events pharmacovigilance data and novel clinical applications. *PLoS One* 2016;**11**:e0157753.

80. Chen M, Zhu L, Chiruvolu P, Jiang Q. Evaluation of statistical methods for safety signal detection: a simulation study. *Pharm Stat* 2015;**14**:11–9. https://doi.org/10.1002/pst.1652.

81. Xu Z, Kass-Hout T, Anderson-Smits C, Gray G. Signal detection using change point analysis in postmarket surveillance: change point analysis. *Pharmacoepidemiol Drug Saf* 2015;**24**:663–8.

82. Agbabiaka TB, Savović J, Ernst E. Methods for causality assessment of adverse drug reactions: a systematic review. *Drug Saf* 2008;**31**:21–37.

83. Naranjo CA, et al. A method for estimating the probability of adverse drug reactions. *Clin Pharmacol Ther* 1981;**30**:239–45.

84. Venulet J, Ciucci AG, Berneker GC. Updating of a method for causality assessment of adverse drug reactions. *Int J Clin Pharmacol Ther Toxicol* 1986;**24**:559–68.

85. Harpaz R, Chase HS, Friedman C. Mining multi-item drug adverse effect associations in spontaneous reporting systems. *BMC Bioinf* 2010;**11**:S7.

86. Varallo FR, Planeta CS, Herdeiro MT, Mastroianni PDC. Imputation of adverse drug reactions: causality assessment in hospitals. *PLoS One* 2017;**12**:e0171470.

87. Hunter L, Cohen KB. Biomedical language processing: what's beyond PubMed? *Mol Cell* 2006;**21**:589–94.

88. Jensen LJ, Saric J, Bork P. Literature mining for the biologist: from information retrieval to biological discovery. *Nat Rev Genet* 2006;**7**:119–29.

89. Sarker A, et al. Utilizing social media data for pharmacovigilance: a review. *J Biomed Inform* 2015;**54**:202–12.

90. Chee B, Berlin R, Schatz B. Measuring population health using personal health messages. *AMIA Annu Symp Proc* 2009;**2009**:92–6.

91. Smith K, et al. Methods to compare adverse events in twitter to FAERS, drug information databases, and systematic reviews: proof of concept with adalimumab. *Drug Saf* 2018;**41**:1397–410.

92. Doshi-Velez F, Kim B. *Towards a rigorous science of interpretable machine learning.* arXiv1702.08608 [cs, stat]; 2017.

93. Eichstaedt JC, et al. Facebook language predicts depression in medical records. *Proc Natl Acad Sci* 2018;**115**:11203–8.

94. Chee B, Karahalios K, Schatz B. Social visualization of health messages in system sciences. In: *42nd Hawaii international conference on system sciences*; 2009. https://doi.org/10.1109/HICSS.2009.399.

95. Pappa D, Stergioulas LK. Harnessing social media data for pharmacovigilance: a review of current state of the art, challenges and future directions. *Int J Data Sci Anal* 2019;**8**:113–35.

96. Bobed C, Douze L, Ferré S, Marcilly R. PEGASE: a knowledge graph for search and exploration in pharmacovigilance data? In: *International conference on semantic web applications and tools for life sciences (SWAT4HCLS)*; 2018.

97. Bean DM, et al. Author correction: knowledge graph prediction of unknown adverse drug reactions and validation in electronic health records. *Sci Rep* 2018;**8**:4284.

98. Overby CL, et al. Combining multiple knowledge sources: a case study of drug induced liver injury. In: Ashish N, Ambite J-L, editors. *Data integration in the life sciences.* vol. 9162. Springer International Publishing; 2015. p. 3–12.

99. Stanovsky G, Gruhl D, Mendes P. Recognizing mentions of adverse drug reaction in social media using knowledge infused recurrent models. In: *Proceedings of the 15th conference of the European chapter of the association for computational linguistics*; 2017. p. 142–51. https://doi.org/10.18653/v1/E17-1014.

100. Abdelaziz I, Fokoue A, Hassanzadeh O, Zhang P, Sadoghi M. Large-scale structural and textual similarity-based mining of knowledge graph to predict drug–drug interactions. *J Web Semant* 2017;**44**:104–17.

101. Cañada A, et al. LimTox: a web tool for applied text mining of adverse event and toxicity associations of compounds, drugs and genes. *Nucleic Acids Res* 2017;**45**:W484–9.

102. Qin X, Kakar T, Wunnava S, Rundensteiner EA, Cao L. MARAS: signaling multi-drug adverse reactions. In: *The 23rd ACM SIGKDD international conference.* Washington, DC: ACM Press; 2017. p. 1615–23. https://doi.org/10.1145/3097983.3097986.

103. Lowe Derek. Another AI-generated drug? *Sci Transl Med* 2019;(February). https://blogs.sciencemag.org/pipeline/archives/2020/01/31/another-ai-generated-drug.

104. Bourne PE, et al. The NIH big data to knowledge (BD2K) initiative. *J Am Med Inform Assoc* 2015;**22**:1114.

105. Moult J, Fidelis K, Rost B, Hubbard T, Tramontano A. Critical assessment of methods of protein structure prediction (CASP)—round 6. *Proteins Struct Funct Bioinf* 2005;**61**:3–7.

106. Jing Y, Bian Y, Hu Z, Wang L, Xie X-QS. Deep learning for drug design: an artificial intelligence paradigm for drug discovery in the big data era. *AAPS J* 2018;**20**:58.

Precision medicine

Sumit Deswal[a], Krishna C. Bulusu[b], Paul-Michael Agapow[c], and Faisal M. Khan[d]

[a]Genome Engineering, Discovery Sciences, BioPharmaceuticals R&D, AstraZeneca, Gothenburg, Sweden, [b]Bioinformatics and Data Science, Translational Medicine, Oncology R&D, AstraZeneca, Cambridge, United Kingdom, [c]Oncology R&D Real World Evidence, AstraZeneca, Cambridge, United Kingdom, [d]AI and Analytics, Data Science and Artificial Intelligence, Biopharma R&D, AstraZeneca, Gaithersburg, MD, United States

It has been shown that several diseases have genetic basis and bias. These changes such as mutations, copy number alteration, epigenetic changes, and RNA expression cause disease progression, adverse events, and drug resistance among other physiological and pharmacological interferences to the patient. A key example here is tumor growth and metastasis to other organs. While medicine has always considered the individual patient's situation and there is a long history of segmenting patient populations by data,[1] the recent explosion of rich biomedical data and informatic tools has galvanized the idea of "precision medicine." By identifying fundamental patient types, we can deliver "the right treatment for the right patient at the right time."[2] Similar ideas can be found in the "5 R" framework for effective drug development, which emphasizes that a drug must reach the right target, in right tissue, with the right safety profile in the right patient.[3] Treatments can be selected that are most likely to help patients based on a molecular understanding of their disease, rather than a possibly deceptive clinical presentation. This is especially keen for the increasing number of high cost therapies with great variation in patient-to-patient outcomes, for example, checkpoint inhibitors.[4] If we can correctly partition a patient population, we get more homogenous cohorts with shared disease states for mechanistic investigation. If we can subtype a trial population to one where a candidate drug will be effective, we can reduce the trial size. If smaller and rarer disease populations—often unrecognized or neglected in therapy development—can be detected within larger populations, they can be investigated and treated appropriately.

However, genomics and transcriptomics (and other relevant omics) are the molecular signature of an individual, not necessarily a disease profile in themselves. Hence a precise and in-depth view of a patient's molecular profile guiding the prescription of a treatment is requisite for precision medicine. While advances in molecular technologies have made it easy to augment the clinical presentation of a patient with mutations and other genetic/epigenetic changes, the challenge is to know which drug or drugs to prescribe for each of these features (or groups of features). The vast number of differences between any two individuals, all possibly interacting and modifying each other, hinder easy discovery of the critical minority.

The answer lies in Big Data and Machine Learning. Vast amount of genomics, proteomics, and drug/gene sensitivity data is now available for the individual cell lines as well as patient samples. These data are also predicted to increase exponentially as the sequencing cost has dropped dramatically and many national projects have been initiated for the whole genome sequencing of large populations. Large

The Era of Artificial Intelligence, Machine Learning, and Data Science in the Pharmaceutical Industry. https://doi.org/10.1016/B978-0-12-820045-2.00009-X

functional genomics and drug screens combined with artificial intelligence and machine learning has enabled us in identifying some of the dependencies that rely on an underlying genetic change/genomic signature. In this chapter, we will discuss recent progress made to identify such context dependencies and the opportunities for data scientists in this field due to availability of large genomics, proteomics, epigenomics, and metabolomics datasets and improvements in computational methods.

Cancer-targeted therapy and precision oncology

Oncology is probably the most widely known application of precision medicine, with a major impact on patient response and quality of life.[5] Cancer is the disease of aberrant cell growth. Some of the early drugs used (and still in use) are the ones which cause a general block in cell proliferation. Such treatment is generally termed as "chemotherapy." However, chemotherapy results in unwanted side effects due to killing of normal cells in the body, particularly the fast-growing cells, for example, cells in the intestinal lining.

As early as 1985, it was realized that if we can inhibit the protein responsible for the aberrant growth phenotype, we will be able to suppress cancer growth with no or little effect on the normal cells (Fig. 1A).[6]

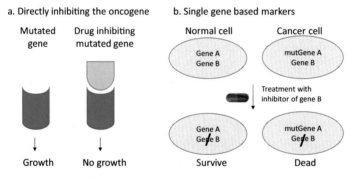

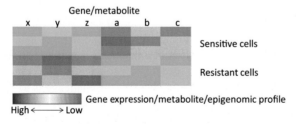

FIG. 1

Targeted therapy and need of machine learning for multigenic biomarker/feature selection. (A) Several approved targeted therapy agents inhibit the oncogene directly. (B) Several single gene-based markers of drug sensitivity or resistance have been identified. (C) Sometime single gene feature is not enough to completely predict the sensitivity or resistance to a targeted drug; hence multiple features need to be combined to generate a signature.

Amplification or overexpression of Human epidermal growth factor receptor 2 (HER2) was shown to play an important role in the development and progression of certain aggressive types of breast cancer. HER2 is the target of the monoclonal antibody trastuzumab (marketed as Herceptin). Trastuzumab is effective only in cancers where HER2 is overexpressed. Another, very successful molecular targeted therapeutic is Imatinib (marketed as Gleevec by Novartis), which is an inhibitor of the oncogenic fusion protein BCR-ABL which is a strong driver of tumorigenesis in chronic myelogenous leukemia (CML). These examples (and others comprehensively archived at OncoKB database[7]) showcase the strength of targeted therapy. Hence currently one of the major goals in oncology research is to find genes that are essential for cancer cell survival in presence of specific mutations, a concept termed synthetic lethality (Fig. 1B). Since normal cells do not have those mutations, they can tolerate the drug without much side effect. Some of such examples are PARP1 inhibition in breast cancer in presence of BRCA1 mutation. Since only cancer cells have BRCA1 mutation, only cancer cells are sensitive to PARP1 inhibition. Recent advances in functional genomics technologies such as CRISPR[8] have been significant in scaling up such efforts to understand "essentiality" of a specific gene for cell survival.

Many functional genomics or drug screens in genetically defined cancer models are currently ongoing in research labs to identify such dependencies.[9] However, such specific dependencies may not rely on a single mutation/aberration. In some cases, a multigenic signature is identified to predict the drug sensitivity. Two successful examples of such signature are MammaPrint and Oncotype Dx diagnostic tests for the chemotherapy sensitivity prediction in breast cancer. Both of these tests are based on the expression profile of genes in cancer (70 and 16 genes in MammaPrint and Oncotype Dx test, respectively). Prediction of such signatures is more data intensive and requires complex data analysis. Machine learning approaches are particularly suitable for this type of multigenic signature/biomarker identification (Fig. 1C).

Personalized medicine and patient stratification

One of the most common uses of machine learning in drug development has been to aid in the creation of sophisticated personalized medicine approaches, usually to stratify patients in terms or risk or response, often through a companion diagnostic. While most common in the oncology setting,[10–13] there are advances in other settings such as nephrology and cardiovascular medicine. These algorithms analyze a variety of different information modalities including demographic, genomic, proteomic, imaging based and more to personally characterize a patient's condition and disease progression.

Scientists leverage statistical and machine learning techniques for survival analysis in these endeavors.[14, 15] Although at first glance survival analysis appears as a regression problem, censored observations in the data complicate the time-to-event prediction.

Healthcare data for prognostic modeling is usually obtained by tracking patients over the course of time in a well-designed study, perhaps lasting years. Often a predefined event such as the relapse of a disease or death due to disease is the focus of the study. The major difference between survival analysis and other regression problems is that the event of interest is frequently not observed in many of the subjects. Patients who did not experience the endpoint during the study or were lost to follow-up for any cause (i.e., the patient moved during a multiyear study) are considered as censored. All that is known about them is that they were disease-free up to a certain point, but what occurred subsequently is unknown. They may have actually experienced the endpoint of interest at a later time. Conversely, patients who have experienced the endpoint of interest are considered as *noncensored* samples or *events*. In many medical prognosis problems, the vast majority of instances (as high as 96%) can be censored. The incomplete nature of the outcome thus challenges traditional regression techniques. Methods which can correctly account for censored observations are essential.

If T_i denotes the actual target time, C_i is the censored time, and U_i is the observed time for all patients, then for measured events $U_i = T_i$ and for censored cases $U_i = C_i < T_i$. The survival outcomes for n patients is then represented by pairs of the random variables (U_i, δ_i) for $i = 1, \ldots, n$. The δ_i indicates whether the observed survival time U_i corresponds to an event ($\delta_i = 1$) or is censored ($\delta_i = 0$). Given a d-dimensional vector $x_i \in \mathbf{R}^d$, the data D for medical prognosis can be represented as:

$$D = \{U_i, x_i, \delta_i\}_{i=1}^n \tag{1}$$

Eq. (1) represents data for medical prognosis.

$$U_i = \min(T_i, C_i) \tag{2}$$

Eq. (2) represents observed survival time for all patients.

$$\delta_i = I(T_i \leq C_i) = \begin{cases} 0, & \text{for censored observation} \\ 1, & \text{for exact observation} \end{cases} \tag{3}$$

Eq. (3) indicates the observed survival time corresponds to an event or is censored.

Methods for survival analysis

The field of prognostic survival analysis has primarily been primarily of interest to biostatisticians. The Cox proportional hazards model is the de facto standard approach[14, 16]; it estimates the log hazard for a patient as a linear combination of the patient's features plus a baseline hazard. A patient's individual predicted hazard function predicts their survival time. A hazard function is the instantaneous rate of decline in survival at a point in time. The Cox model makes the assumption that the hazard functions for any two individuals are proportional; their ratio is constant over time. This assumption is reflected in the formula for the approach:

$$h_i(t) = \exp\left(\sum_{j=1}^p b_j X_{ij}\right) h_0(t) \tag{4}$$

Eq. (4) represents Cox model.

While time-to-event modeling has traditionally been a focus of statisticians, particularly biostatisticians, various machine learning approaches have been applied to the field. The use of decision trees adapted for censored data represent some of the earliest work in the field.[17, 18] Other techniques such as linear programming[19] have also been explored.

Various forms of artificial neural networks have also been applied to explore this space,[20–22] with varying results. Some advantages due to a neural network's ability to model nonlinearities have been observed. However, these algorithms often pose challenges such as requiring a large number of training samples, involving complex tuning of model parameters, treating the time-to-event problem in a classification context rather than regression, struggling with high dimensional data, and/or achieving equivalent or minimally improved performance over the Cox model.

Widespread adoption of SVMs in various machine learning domains has led to recent applications in survival analysis.[23, 24] One example is SVRc: support vector regression for censored data.[14]

SVRc adapts normal support vector regression through an asymmetric loss-penalty function depending on whether a patient's record is a censored observation or an event. SVRc optimizes the following function:

$$\min_{W,b} \frac{1}{2}W^2 + \sum_{i=1}^{n}\left(C_i\xi_i + C_i^*\xi_i^*\right) \tag{5}$$

Eq. (5) represents support vector regression for censored data.
Given the constraints:

$$
\begin{aligned}
y_i - (W \bullet \Phi(x_i) + b) &\le \varepsilon_i + \xi_i \\
(W \bullet \Phi(x_i) + b) - y_i &\le \varepsilon_i^* + \xi_i^* \\
\xi_i^{(*)} &\ge 0, \quad i = 1,\ldots,n
\end{aligned}
\tag{6}
$$

Eq. (6) represents constraints for support vector regression for censored data.
where $s_i = 1$ if censored, $s_i = 0$ if event

$$
\begin{aligned}
C_i^{(*)} &= s_i C_c^{(*)} + \left(1 - s_i\right)C_n^{(*)} \\
\varepsilon_i^{(*)} &= s_i \varepsilon_c^{(*)} + \left(1 - s_i\right)\varepsilon_n^{(*)}
\end{aligned}
\tag{7}
$$

Eq. (7) represents additional factors that the constraints for support vector regression for censored data abide by.

These algorithms are leveraged to develop personalized phenotypes of patient risk, which are then used in variety of applications in clinical drug development, including the development of inclusion/exclusion criteria, patient risk stratification, companion diagnostics, and more.

Finding the "right patient": Data-driven identification of disease subtypes
Subtypes are the currency of precision medicine

Behind the aim of precision medicine is the issue of patient heterogeneity. Many patients with apparently similar clinical presentations possess vastly different pathophysiologies involving different molecular mechanisms.[25] This is especially true for complex diseases, where a single umbrella term may encompass a complex mélange of environment and genetic drivers.[26] For example, asthma has long been understood to be heterogenous, featuring a wide variety of symptoms, clinical values, physiological mechanisms, and responses to treatment.[27] Similarly, optimal intervention for an initial schizophrenic episode results in a full response and recovery for a quarter of the patients, unfavorable or no response for another quarter, and response but later recurrence for the remainder.[28] Even in "noncomplex" conditions, heterogeneity is frequently seen. For example, chemotherapy is often characterized by wide variation in response and adverse reactions.[29, 30]

This obviously makes investigation and treatment less effective. Preclinical research of any apparently homogenous "disease" population may be obfuscated by a cohort that actually possesses a variety of disease mechanisms. A patient may shift from diagnosis to diagnosis across the course of their illness as different symptoms wax and wane, as different tests are used, as different practitioners are involved. They may be given drugs targeting an inappropriate disease mechanism, incurring the burden of a treatment and its possible side effects that provides no relief. Forecasting population health needs is complicated by an overly diverse and ill-defined patient population.

While there is much enthusiasm about and examples of such data-driven subtyping, it has yet to become routine in biomedicine. The reality or usefulness of a proposed subtype may be unclear or may conflict with other subtype proposals. Here I will focus on the pragmatic discovery and validation of patient subtypes using machine learning, outlining the challenges, pointing out possible solutions and moving toward best practices.

As in much of machine learning and data science, there is a thicket of overlapping terms, many of which are ill-defined. Differences in meaning are minimal and in every case they seek to distinguish classes of patients that can be treated differently.[31–33] Likewise, one can speak of features, variables or clinical values; clusters, categories or groupings; patient types, endotypes or phenotypes. For simplicity and consistency, here I will use "biomarker" for any measurable attribute of a patient that can be used for typing, "cluster" to refer to an algorithmically proposed group of patients, "subtype" to refer to an actual group of patients with a distinct pathophysiology that can potentially be targeted for treatment.

The nature of clusters and clustering

The simplicity of the underlying question—what classes of patients there are?—belies the complexity of answering it. Some recent examples are illustrative. It was proposed that adult-onset diabetes could be broken down into five subtypes based on six clinical measurements, with each subtype displaying different clinical outcomes.[34] However, it subsequently was pointed out that such clustering could have simply resulted from dependency between variables.[35] Several lightly correlated variables could produce a similar effect, with algorithms creating plausible yet entirely artefactual clusters from a continuous band of data.[36, 37] Similarly, while another study was able to convincingly cluster rheumatoid patients,[38] critics were able to demonstrate comparable results from purely random data.[39]

In the face of this uncertainty, how can we subtype patients? It is best to regard a proposed cluster as a hypothesis that needs to be validated and tested against other data. As per von Luxburg,[40] "clustering should not be treated as an application-independent mathematical problem, but should always be studied in the context of its end-use." A cluster is only an initial step, a heuristic for suggesting subtypes.

Selection and preparation of data

What data to use for clustering patients is a keen and difficult question. Early attempts at algorithmically subtyping patients used a handful of clinical variables. More recently the rise of high throughput technologies potentially allows subtyping from tens of thousands of different biomarkers of many different modalities. Investigators can use metabolite levels, lifestyle measurements, genomic alterations, clinical assessments of disease stage, radiological images, and many more (for examples of this transition in asthma, see Refs. 41, 42).

How to choose which biomarkers to use? On the one hand, using an overly restrictive set of data may lead to a failure to capture important population differences, resulting in poor or deceptive clustering. Alternatively, it is tempting to use all available data. Disease is increasingly seen as a complex multilevel phenomena,[43, 44] with a network of interactions encompassing genomes, metabolomes, transcriptomes, types of cells, and different body compartments. Disease may manifest differently at these different levels and it makes sense therefore to use different modalities of data. If an investigator is unsure about the value of a particular data modality, it seems better to include it if it might be informative and let the algorithm decide.

This expansive use of data does not come without cost. First, it enlarges the search space for any algorithm, possibly incurring impractical computational needs. It may also create a situation where there are more features than samples and the data are effectively sparse (the so-called Curse of Dimensionality), creating problems for any algorithm searching it. Second, it is in effect doing multiple hypotheses testing, increasing the chance of any variable accidentally aligning to create a pattern. Third, by including lightly correlated but irrelevant biomarkers, interpretation is made much more difficult. For example, neighboring SNPs that are genetically linked could obscure the causal SNP.[45] Even with substantial data, the predictive power of models is observed to rise and then ebb with the number of features, the so-called Hughes paradox.[46]

While we do not want to presuppose or bias the results of subtyping with our choice of data, we must be economical with what is included. It may be necessary, especially with high dimensional data such as omics and imaging, to reduce the number of biomarkers. There are two broad approaches for this: feature selection and feature creation.[47] In the first, biomarkers will be selected for inclusion or exclusion. This selection might be done by consulting domain experts as to which are likely or unlikely to be implicated. For example, uninformative and possibly obfuscating data, e.g., neighboring SNPs linked to an important locus, should be pruned.[48] Alternatively, there is a rich body of work on algorithmic feature selection, although it has a general problem in striking a balance between shrinking the number of features and losing too much information.[49, 50] In contrast, in feature generation, data are reduced by replacing multiple features with new composite features. This approach is most commonly utilized with Principle Component Analysis or Multidimensional Scaling, at the risk of losing some interpretability of the final results. In summary, the data used must be sufficient for clustering without being excessive. There is no panacea for this, judgment must be exercised.

Most methods for clustering need to calculate a distance between records, and thus require features represented as fixed-length numerical vectors. Although many types of data can be easily translated into suitable forms (e.g., presence-absence features as a binary, categoricals via one hot encoding), some heterogenous data are not so easily handled. For example, genomic sequences of variable lengths (e.g., promoter sequences or transcribed regions) present difficulties. These will severely restrict the possible methods, and it may be advisable to encode or transform this data into a suitable form (e.g., aligned sequences of the same length, distances from a prototype sequence).[51, 52]

Approaches to clustering and classification

There are several lengthy reviews of clustering in biomedicine elsewhere.[53–57] I will not attempt to exhaustively catalog the algorithmic possibilities here, instead looking at broad trends, possible issues and interesting developments. Clustering algorithms typically use several parameters, implicit and explicit, whether clusters are based on similarity or difference, how such distance is measured, assumptions about the distribution of the underlying data, treatment of boundaries between clusters, and so on. Unsurprisingly, performance can vary widely across datasets and is impossible to globally recommend any single approach for a given problem. While several of the above reviews benchmark and compare different approaches, it is most ideal to select and tune algorithms on the dataset to be analyzed or a similar one.

Unsupervised and supervised partitional classification

Supervised classification might seem irrelevant to the discovery of patient subtypes as it assumes the outcome cluster labels as a parameter. However, it can still be informative whether we wish to understand how a cluster is composed and by using it to cluster unlabeled individuals. For example, common approaches include using histological or molecular markers such as cell surface receptors as a ground truth label to form clusters against. Performance and assessment of supervised classification is generally much better than the unsupervised, due to this ground truth. However, skepticism needs be exercised about the validity of any such ground truth. For example, the PAM50 approach to classifying breast cancer uses the expression of 50 marker genes to divide breast cancer patients into four subtypes. Yet abundant heterogeneity remains within these subtypes.[58, 59] Something similar can be seen in colorectal cancer, where different molecular typing schemes show only mediocre agreement.[60] Perhaps the best answer is to use supervised classification where the ground truth is inarguable, like treatment outcome or an adverse reaction. Note that sometimes supervised is used in a weaker sense where the algorithm is provided with the number of clusters but not the identity of those clusters.

Unsupervised classification is the most obviously relevant to the case of subtyping patients and discovering subtypes. At the same time, it is the more demanding as it has no restrictions on how the clusters be calculated, potentially makes the most assumptions and is the most difficult to assess because there is no clear ground truth. For these reasons, unsupervised classification has been described as "more art than science."[40] Nonetheless, it is the most common approach used in subtyping efforts. In practice, there may not be such a huge difference between unsupervised and the weaker sense of supervised. A supervised algorithm can be run with the expected number of clusters varying, with the optimal result chosen by relative validity (discussed later).

There is a third class, semisupervised classification which is often used where insufficient labeled data is available or too difficult to obtain. The classification model is trained with both labeled and unlabeled data, using heuristics to infer the class of unlabeled data and thus boost the effective volume of training data. It is infrequently used in subtyping patients.

Hierarchical classification

A common alternative to strict assignment to clusters is to instead build a hierarchy, where more records appear on the tips of the tree, close to similar records. This has the advantage of showing the structure of any grouping, while avoiding the issue of resolution or how many clusters to find in a dataset. It can use any sort of distance measure and been reported as working robustly even on small sample size, high dimensional datasets.[61] Conversely, the lack of explicit clusters often leads to the use of ad hoc cut-offs to identify subtrees as clusters. In addition, the algorithm assembles from the tips up based on local similarity. This leads to good computational performance but can potentially result in suboptimal solutions.

Biclustering

Also known as co-clustering or subspace clustering, this technique simultaneously clusters on rows and columns (in our use case, patients and biomarkers), generating sets of rows that behave similarly across columns and vice versa. It has become popular when using gene expression data to both subtype patients and to reveal modules of co-regulated genes, a situation in which a set of genes might only be a cluster for a set of patients. There is a bewildering array of biclustering variants with different performances and an ongoing debate about interpretation and best practices (reviewed in Refs. 62–64).

Nonetheless, reported performance is good and biclustering should be considered where omics data cannot be reduced in size and the curse of dimensionality would ordinarily defeat analysis.

Clustering trajectories and time series

In more recent years, several new approaches have been developed that attempt to classify and cluster patients based on a temporal sequence of measurements or events. The reasoning behind this is that many diseases are progressive or characterized by periods of exacerbations and remissions and may not be described well by biomarker measurements at a single point in time. Furthermore, by examining the time course of a patient there is a direct path forward to outcomes. Such approaches are currently challenging due to the only partially developed methodology and the computational demands of algorithms used. Broadly, they may be broken down into two categories: those based on trajectories and time series.

A patient trajectory or history is a series of events or diagnoses experienced during the course of a disease. For example, a single patient's trajectory might consist of dated diagnoses of malignant hyperplasia, anemia, obstructive uropathy, and death. Of course, such histories are complex, noisy records. The dominant methodology for identifying common, shared trajectories works by identifying enriched sequential pairs of events and then stepwise assembling these into longer trajectories by searching the population. This approach was prominently used in a nationwide study to identify common paths and comorbidities in over 2 million patients[65] and subsequently used on a smaller scale to predict death from sepsis[66] and breast cancer.[67]

This method is very attractive in that it can use real world data, patient medical records, and discrete diagnoses to build a picture of the evolution of disease. However, it is computationally demanding and assembling trajectories is essentially a heuristic. Statistical support for the significance of any trajectories found is light and interpretation of an intersecting mesh of trajectories is largely ad hoc. Nonetheless, the approach is promising. Possible improvements could be made by using different methods of identifying common trajectories (e.g., this is a long history in purchase history analysis) or by finding ways to cluster different but related trajectories (e.g., Dynamic Time Warping[68]).

While there is a large body of work in time course analysis, matching patients based on time series of measurements is less well-developed. This is largely due to biomedical time series being much shorter and carrying less information than the dense and long sequences used by other studies. Furthermore, much of the biomedical work uses only a single biomarker, such as the expression of a single gene over time. Nonetheless, there is some interesting emerging work. Zhang et al.[69] provide a survey of methodology for clustering gene expression data, finding several methods with good performance. de Jong et al.[70] report a deep learning system capable of handling multivariate short time series with many missing values that was used to successfully subtype Alzheimer's patients.

Integrative analysis

Recent years have seen the rise of integrative or multimodal analysis or multiomics, using different data modalities in a single analysis to reveal patterns and clustering. This approach is appealing as disease is systemic and will leave different signatures at different omic levels. A full resolution of patient subtypes may require examination of the genome, epigenome, and transcriptome. Integrative methods appear to outperform separate analyses of levels that are then reconciled.[71] There is a wide variety of approaches (reviewed in Refs. 71–73), but generally those based on matrix factorization have superior performance. Note that all integrative approaches tend to be computationally demanding.

Deep approaches

Although artificial neural networks and deep learning are commonly used as a form of supervised learning, there are some interesting applications that allow for the discovery of subtypes. For example, autoencoders are a class of neural network that learns efficient encoding of data in an unsupervised manner. Thus, it performs dimension reduction on the data and has been used on electronic health records with semisupervised classifiers to assemble patient subtypes, allowing for noisy or missing data.[74] However, it is unclear how this system could be used in a completely unsupervised context.

Validation and interpretation

Given that the very idea of a cluster is slippery, that the assumptions of a clustering algorithm may not be met, that the distribution of patient biomarkers may be mysterious, and that the data used for clustering may be insufficient or excessive, it is clear that any proposed clustering is simply a hypothesis. How can such hypotheses be strengthened? For patient subtypes, we can look to validate the clustering, essentially looking for evidence that the observed clusters are meaningful and reflect an objective grouping in the world.

Direct validation

There are broadly two types of validation. In the first, direct validation, we seek evidence of the quality of the clustering result from examining the results themselves. Approaches to direct validation fall into three classes[75, 76]:

- Internal validation evaluates the goodness of a proposed clustering by assessing the goodness of the clusters shown without any external information. There are many metrics for this,[77, 78] that typically measure the compactness of individual clusters and their separation from neighboring clusters. However, performance of these metrics falls when clusters are noisy or overlap, unfortunately too common in biomedical data. Note also that these metrics have an implicit assumption of what constitutes a "good cluster" which may not match behavior or distribution of the actual subtypes.
- External validation consists of comparing the result to an external "gold standard" labeling that is held to be true and measuring how correct the clustering is. As with supervised classification, if there is a gold standard for an understood subtyping, there is little need to simply rediscover it. However, such an approach could be used to select the right clustering algorithm for a specific data set, to be used on patients for which there is no labeling.
- In relative validation, many of the same metrics used in internal and external validation are used while varying parameters for an algorithm. It can thus be used to optimize the fit of an algorithm for a given dataset, by selecting the parameter values which maximize metrics. This includes determining the optimal number of clusters for a supervised algorithm, by treating the desired number of clusters as a parameter.

An important aspect of cluster validity is stability, that is, that clusters should not disappear easily after a minor change in the underlying data. Put another way, a data set drawn from the same underlying distribution should give rise to more or less the same clustering. There are a variety of techniques for testing this stability.[79] Most common in subtyping is testing by subsampling or bootstrapping, that is, comparing the results of given by subsets of the original data.[80] More advanced approaches includes adding noise to the model, potentially an important test as some algorithms are fragile to any noise.[47]

A final check of direct validity is for simple reproduction. Some algorithms (notable t-SNE[81]) are nondeterministic. That is, they are not guaranteed to return the same answer if run again. While the result can always be replicated by using the same random seed, some idea of robustness can obtained by running the algorithm multiple times. However, perhaps it is best if possible, to avoid algorithms with this quality, as it introduces extra barriers in reproducing and communicating any analysis.

Indirect validation

While direct validation is essentially a quality check on our data and methodology, it is still possible to generate clusters that are methodologically sound but uninteresting (e.g., the above example of separating patients by eye color) or artifacts of the data. More broad assurance can be gained by looking for evidence outside the data used. The simplest method is to attempt to replicate the clustering in another, independent, dataset. Similar clustering occurring across datasets using the same methodology provides some assurance that the pattern is not peculiar to the original dataset. Note that clustering of two different datasets does not provide a mapping between the two partitions, for example, cluster 1 in dataset A may be cluster 2 in dataset 2. To compare and link clusters across datasets, it is necessary to characterize them (discussed later). However, care has to be taken that the datasets are comparable. If datasets differ epidemiologically or demographically (e.g., COPD cases seen in elderly smokers, young nonsmokers, and industrial workers exposed to pollution), they may consist of different patients' subtypes.

Another common approach to indirect validation is to use several different clustering approaches and compare the results. Again, there is no mapping between the two partitions produced by two different algorithms over the same data. Fortunately, there are a host of metrics for concordance across partitions that do not rely on a ground truth or shared labels (e.g., Rand Index, Jaquard Index, and Normalized Mutual Information; see Ref. 82). In this way, researchers often produce a table of which and how many methods were able to produce a given cluster.

This is obviously an ad hoc approach but given that we may have little idea as to whether the assumptions of any clustering algorithm are met by a dataset, you can argue that it is more robust to use a variety of algorithms. Note that algorithms should be substantially different (e.g., using related variants) or those with shared assumptions would be counter-productive. Also, it may be wise to use concordance metrics that allow comparison of nested clusters (e.g., homogeneity). That is, if one algorithm resolves the data more than another, splitting clusters into smaller clusters, a strict comparison will make the data look more different than it actually is.

Not every cluster found in a given partitioning is equally valuable. Within a single partitioning, some clusters may be more robust than others. Some reflect a real subtype, while others may be artefactual. For example, a common phenomenon seen, although rarely written about, is that of the trash-can or garbage clusters. These are an artifact of clustering algorithms, where divergent records that do not fit with other clusters have been forced together to form a group of outliers. Such clusters can be minimized by elimination of outliers before clustering, and post hoc identified by attention to individual cluster stability and internal validation.[83] Thus a proposed clustering may be useful without every cluster within being meaningful and priority should be given to those with greater support.

Conversely, while points that fall between clusters are generally regarded as nuisance or noise, it is possible they may represent a real biological phenomenon. If a patient is moving between disease states or the diseased cells or tissues of their body exists in a mixture of states (e.g., as in tumor heterogeneity), the patient may genuinely resolve as being between two subtypes. There are families of clustering algorithms that allow for multiple cluster membership, perhaps assigning data to multiple classes in a fuzzy manner. This has been little explored but remains an interesting avenue for the future.[84]

Characterization

Perhaps the ultimate form of indirect validation is characterization of individual clusters, to measure and compare the patients within each, showing that they are different. For example, one could look at the typical outcomes, patient phenotype and biomarkers distribution. The above validation approaches focus largely on ensuring a cluster is well-formed. By looking at the contents, we move toward understanding whether the cluster is meaningful, testing whether a cluster may be an actual patient subtype.

Note that the biomarkers used for this characterization must be chosen carefully. If the same biomarkers are used for clustering and characterization, this is of little value: the clusters were determined on the basis of those features, so they will necessarily differ across clusters. Further, one must watch for biomarkers that are correlated in real but uninteresting ways. For example, clusters based on height will often show a difference in sex, simply due to men and women having different distributions of height.

More informative characterization can be performed using biomarkers independent of the clustering analysis. For example, in a clustering of multiple sclerosis patients based on genomic features, it was demonstrated that each cluster was enriched in particular gene pathways.[45] Similarly, clusters of type 2 diabetes patients differed in their response to treatment.[85] More detailed investigation could also be done by comparisons within individual clusters. For example, the central and peripheral patients in a cluster—those closer and further from its center—could be compared and their centrality measured against outcomes.

Given all the above steps and choices, subtyping can be a long iterative process. A sound result requires good decisions and methodology at every step. Thus there is much scope for error, especially for nonexperts. Fortunately, there is an increasing number of reproducibility tools for making analysis pipelines and workflows that can be used for subtyping analysis. Indeed, there is even a growing number of software packages dedicated just to clustering that can automate algorithm choice, tuning, and validation.[75, 76, 80] While this assumes some risk in the software being used as a black box, it brings the benefit of reproducibility and making it easy to carry out best practices.

Key advances in healthcare AI driving precision medicine

Since the Human Genome Project was completed in 2003,[86] the avalanche of biological data that had subsequently been generated kickstarted the need for "smart" statistical knowledge to objectively move from the theoretical domain to real world applications. With multiomics experiments supporting the field of genomics, more comprehensive datasets giving deeper insights into human/disease biology were generated. There is a plethora of examples where machine learning contributed toward every stage of the drug discovery pipeline—from target identification and evaluation to patient stratification. Random Forest has probably been the most implemented of these methods with applications from predicting therapeutic response in preclinical disease models[87] to prognostic markers/signature identification in patient cohorts[88, 89] to assessing Tumor Purity.[90]

Recent advances in ML/AI techniques tailored toward Healthcare applications have resulted in an impressive array of libraries for the scientific community to implement toward specific preclinical and clinical datasets. While "traditional" ML approaches such as Random Forest, Gradient Boosting Machines, and SVM have shown a lot of value in optimizing molecular characterization of tissue samples, recent state-of-the-art methods have made it possible to ingest huge datasets across chemical and biological domains (multiomics, chemical structure, images, etc.). Deep Learning probably had the most profound effect of these methods on healthcare AI, delivering a step-change in ML/AI applications to patient-benefit. The ability of Deep Learning models (from Autoencoders to Deep Neural Networks) to decipher signals within images resulted in a digitally driven Precision Histology field.[91] Availability of public datasets and whole-slide images helped models to be trained and optimized on real-world data resulting in not only accurate classification of tumors but more significantly being able to predict specific genomic alterations delivering precision histology.[92, 93] Google's AI microscope was one such awe-inspiring technological developments in this space. Radiology is another example where Deep Learning approaches are making a major impact in both the speed and accuracy of disease assessment.[94]

As we, the scientific community, shed more light on the complexities of disease biology, integrative analytics emerge as the only solution to identify/predict the nonlinear relationships that exist across multiomic datasets. Biological Knowledge Graphs evolved as an unbiased integration platform of multilayered healthcare data toward building a "recommendation system" for clinical decision-making.[95] The value of such a platform is in being able to handle sparse data and directed/causal relationships, which not only reflect the true complexity of disease data but also represent a more realistic picture of a living system. Graphical models (inc. Graph Neural Networks) have shown promise in being able to ingest multidomain data and provide high precision predictions, more specifically using patient Electronic Health Records.[96, 97] National Health Service in the United Kingdom is trialing a patient symptom triaging system that sifts through billions of relationships between various biological, chemical and phenotypic entities to provide a rapid recommendation to patients regarding their symptoms.[98] While this platform requires more evidence to be published with concrete case studies, early results do seem promising.

As we generate more comprehensive datasets across biomedical domains, better the probability that predictive models will deliver clinically/biologically meaningful results for patient benefit. Transfer learning—where models trained on data from one domain are used to predict outcome on another—would become the norm. There has been a continuous debate over the validity of current preclinical models in providing an accurate enough snapshot of patient biology, and while this topic is beyond the remit of this chapter, AI could provide one of the solutions. Rather than a general definition of validity or preclinical models, AI could help to identify the relevance of specific models in certain contexts and not in other. Transfer learning signals from clinical datasets to identify similar signals in preclinical models has been attempted with some success.[99, 100] AI, and applications to healthcare, is evolving at such a rapid pace that algorithms, we describe here as "next wave of innovation" could become commonplace by the time of publication of this chapter. For example, dynamic monitoring of health data using fitness trackers has been a very popular field for the past 4–5 years. But using these data for decision-making by your doctor is something new and has revolutionized the entire field of patient monitoring. Algorithms are now able to predict strokes, onset of Alzheimer's, glucose metabolism, cancer progression, and others using continuous data such as blood readouts and metabolites from sweat. Also, the advent of new-age "home gyms" as we become more health aware has opened up a whole new industry bringing AI and healthy living together. These provide a second-by-second updates of workouts reading from fitness trackers, cameras monitoring how well and accurate the workout/posture is, and recommending future exercise routines and dietary advice on-the-fly.

Key challenges for AI in precision medicine

While these examples do suggest an exciting future for healthcare AI, there are quite a few challenges and roadblocks that will need addressing to realize the true potential of these methodologies. First, and probably the most significant of all in the context of data science, is the current state of biological/clinical data. FAIR data principles (Findability, Accessibility, Interoperability, and Reusability) are fundamental toward making our data "AI-ready," and we as a community are far from it.[101] This will need to be achieved in smaller steps, as forcing the entire scientific workforce to abandon current practices is unrealistic. This task is as much an in silico responsibility as it is for the bench-scientist. Clear definitions need to be provided for bench-scientists to annotate, store and share experimental data. The informatics teams need to build the right infrastructure for such a system to thrive.

This leads to the next major challenge in this domain, biological complexity. This is not something to solve given that the more we understand of biology the more we realize there is to learn. To deal with this, multimodal experimental readouts should be generated to give a more comprehensive picture of this variability. IBM Watson for Health's failure to deliver on the promises[102, 103] can be attributed, among other roadblocks, to the system's inability to decipher precise clinical signals from the complex landscape of medical literature.[104] The reality of clinical decision-making is not just driven by statistical significance, or frequency of a particular observation. It is driven by extensive "real world" evidence and experience, quantifying which is probably an impossible task. It is this scenario that interpretability/explainability of a predicted feature becomes critical. While Watson for Health did an impressive job of capturing medical terminology, capturing context resulting in novel predictions was a major limitation of the system.[105] Context-specificity defining the experimental setup is a key part of this discussion, as well as modeling uncertainty/noise within any given dataset should be the norm. Translatability of preclinical signals to patients is still a major stumbling block. This in turn results in any models trained on preclinical datasets not translating to patients as well. Another major issue we have today is lack of ground truth/gold standard datasets. For the model to pick out "confident" patterns, there is a severe lack in understanding of what is context-specific "confidence." Despite recent technological advancements and our understanding of ML models, quite a lot still remain the stereotypical "blackboxes." Using such systems for clinical recommendations would be highly contentious, and unethical. The FDA recently published a white paper on establishing a framework for an AI/ML tool to be approved as a medical device-driving decisions.[106] This will form the foundation of the standards we as a Healthcare AI community should adhere to, and feedback on. Finally, but surely not the least, is the questions of ethics and policies around data sharing and (re)use. While almost everyone involved in this discussion agrees that a patient owns their data, how this propagates to the rights of siblings/parents/progeny remain unclear.

The discussions put forward in this book highlight that the future of healthcare AI undoubtedly remains exciting and groundbreaking in delivering patient benefit, it is hence critical that the limitations put forward above need to be at the forefront of any implications of such technologies on humankind.

References

1. Yan S, Kwan YH, Tan CS, Thumboo J, Low LL. A systematic review of the clinical application of data-driven population segmentation analysis. *BMC Med Res Methodol* 2018;**18**:121.
2. Abrahams E. Right drug-right patient-right time: personalized medicine coalition. *Clin Transl Sci* 2008;**1**:11–2.

3. Morgan P, et al. Impact of a five-dimensional framework on R&D productivity at AstraZeneca. *Nat Rev Drug Discov* 2018;**17**:167–81.

4. Topalian SL, Taube JM, Anders RA, Pardoll DM. Mechanism-driven biomarkers to guide immune checkpoint blockade in cancer therapy. *Nat Rev Cancer* 2016;**16**:275–87.

5. Moscow JA, Fojo T, Schilsky RL. The evidence framework for precision cancer medicine. *Nat Rev Clin Oncol* 2018;**15**:183–92.

6. Drebin JA, Link VC, Stern DF, Weinberg RA, Greene MI. Down-modulation of an oncogene protein product and reversion of the transformed phenotype by monoclonal antibodies. *Cell* 1985;**41**:697–706.

7. Chakravarty D, et al. OncoKB: a precision oncology knowledge base. *JCO Precis Oncol* 2017. PO.17.0001.

8. Behan FM, et al. Prioritization of cancer therapeutic targets using CRISPR–Cas9 screens. *Nature* 2019;**568**:511–6.

9. Mullard A. Synthetic lethality screens point the way to new cancer drug targets. *Nat Rev Drug Discov* 2017;**16**:589–91.

10. Blume-Jensen P, et al. Biology of human tumors development and clinical validation of an in situ biopsy-based multimarker assay for risk stratification in prostate cancer. *Clin Cancer Res* 2015;**21**:2591–600.

11. Cullen J, et al. A biopsy-based 17-gene genomic prostate score predicts recurrence after radical prostatectomy and adverse surgical pathology in a racially diverse population of men with clinically low- and intermediate-risk prostate cancer. *Eur Urol* 2015;**68**:123–31.

12. Paik S, et al. A multigene assay to predict recurrence of tamoxifen-treated, node-negative breast cancer. *N Engl J Med* 2004;**351**:2817–26.

13. Yothers G, et al. Validation of the 12-gene colon cancer recurrence score in NSABP C-07 as a predictor of recurrence in patients with stage II and III colon cancer treated with fluorouracil and leucovorin (FU/LV) and FU/LV plus oxaliplatin. *J Clin Oncol* 2013;**31**:4512–9.

14. Khan FM, Bayer-Zubek V. Support vector regression for censored data (SVRc): a novel tool for survival analysis. In: *Proceedings—IEEE international conference on data mining, ICDM*; 2008. https://doi.org/10.1109/ICDM.2008.50.

15. Shiao HT, Cherkassky V. Learning using privileged information (LUPI) for modeling survival data. In: *Proceedings of the international joint conference on neural networks*; 2014. https://doi.org/10.1109/IJCNN.2014.6889517.

16. Harrell FE. *Regression modeling strategies with applications to linear models, logistic regression, and survival analysis.* New York: Springer; 2001.

17. Gordon L, Olshen RA. Tree-structured survival analysis. *Cancer Treat Rep* 1985;**69**:1065–9.

18. Segal MR. Regression trees for censored data. *Biometrics* 1988;**44**:35–47.

19. Mangasarian OL, Street WN, Wolberg WH. Breast cancer diagnosis and prognosis via linear programming. *Oper Res* 1995;**43**:548–725.

20. Brown SF, Branford AJ, Moran W. On the use of artificial neural networks for the analysis of survival data. *IEEE Trans Neural Netw* 1997;**8**:1071–7.

21. Burke HB, et al. Artificial neural networks improve the accuracy of cancer survival prediction. *Cancer* 1997;**79**:857–62.

22. Zupan B, Demšar J, Kattan MW, Beck JR, Bratko I. Machine learning for survival analysis: a case study on recurrence of prostate cancer. In: *Joint European conference on artificial intelligence in medicine and medical decision making*; 1999. https://doi.org/10.1007/3-540-48720-4_37.

23. Evers L, Messow CM. Sparse kernel methods for high-dimensional survival data. *Bioinformatics* 2008;**24**:1632–8.

24. Van Belle V, Pelckmans K, Van Huffel S, Suykens JAK. Support vector methods for survival analysis: a comparison between ranking and regression approaches. *Artif Intell Med* 2011;**53**:107–18.

25. Balogh EP, Miller BT, Ball JR. Improving diagnosis in health care. In: *Improving diagnosis in health care.* US: National Academies Press; 2016. https://doi.org/10.17226/21794.

26. Mitchell KJ. What is complex about complex disorders? *Genome Biol* 2012;**13**:237.

27. Moore WC, Bleecker ER. Asthma heterogeneity and severity-why is comprehensive phenotyping important? *Lancet Respir Med* 2014;**2**:10–1.

28. Schennach R, Riedel M, Musil R, Möller HJ. Treatment response in first-episode schizophrenia. *Clin Psychopharmacol Neurosci* 2012;**10**:78–87.

29. Cui JJ, et al. Gene-gene and gene-environment interactions influence platinum-based chemotherapy response and toxicity in non-small cell lung cancer patients. *Sci Rep* 2017;**7**:5082.

30. Hartford CM, Dolan ME. Identifying genetic variants that contribute to chemotherapy-induced cytotoxicity. *Pharmacogenomics* 2007;**8**:1159–68.

31. Erikainen S, Chan S. Contested futures: envisioning "personalized," "stratified," and "precision" medicine. *New Genet Soc* 2019;**38**:308–30.

32. Day S, Coombes RC, McGrath-Lone L, Schoenborn C, Ward H. Stratified, precision or personalised medicine? Cancer services in the 'real world' of a London hospital. *Sociol Health Illn* 2017;**39**:143–58.

33. Fröhlich H, et al. From hype to reality: data science enabling personalized medicine. *BMC Med* 2018;**16**:150.

34. Ahlqvist E, et al. Novel subgroups of adult-onset diabetes and their association with outcomes: a data-driven cluster analysis of six variables. *Lancet Diabetes Endocrinol* 2018;**6**:361–9.

35. van Smeden M, Harrell FE, Dahly DL. Novel diabetes subgroups. *Lancet Diabetes Endocrinol* 2018;**6**:439–40.

36. Harrell F. Statistical errors in the medical literature. In: *Statistical thinking*; 2017. Available at: http://www.fharrell.com/post/errmed/.

37. Dahly D. *Magical clusters*; 2018. Available at: https://darrendahly.github.io/post/cluster/. [Accessed 16 November 2020].

38. Spielmann L, et al. Anti-Ku syndrome with elevated CK and anti-Ku syndrome with anti-dsDNA are two distinct entities with different outcomes. *Ann Rheum Dis* 2019;**78**:1101–6.

39. Pinal-Fernandez I, Mammen AL. On using machine learning algorithms to define clinically meaningful patient subgroups. *Ann Rheum Dis* 2019;**79**:e128.

40. von Luxburg U, Williamson RC, Guyon I. Clustering: science or art? *Proceedings of ICML Workshop on Unsupervised and Transfer Learning.* vol. 27. Cambridge: MLR Workshop and Conference Proceedings; 2012. p. 65–79.

41. Pembrey L, et al. Understanding asthma phenotypes: the World Asthma Phenotypes (WASP) international collaboration. *ERJ Open Res* 2018;**4**:00013–2018.

42. Belgrave D, et al. Disaggregating asthma: big investigation versus big data. *J Allergy Clin Immunol* 2017;**139**:400–7.

43. Hornberg JJ, Bruggeman FJ, Westerhoff HV, Lankelma J. Cancer: a systems biology disease. *Biosystems* 2006;**83**:81–90.

44. Somvanshi PR, Venkatesh KV. A conceptual review on systems biology in health and diseases: from biological networks to modern therapeutics. *Syst Synth Biol* 2014;**8**:99–116.

45. Lopez C, Tucker S, Salameh T, Tucker C. An unsupervised machine learning method for discovering patient clusters based on genetic signatures. *J Biomed Inform* 2018;**85**:30–9.

46. Hughes GF. On the mean accuracy of statistical pattern recognizers. *IEEE Trans Inf Theory* 1968;**14**:55–63.

47. Röttger R. Clustering of biological datasets in the era of big data. *J Integr Bioinform* 2016;**13**:300.

48. Libbrecht MW, Noble WS. Machine learning applications in genetics and genomics. *Nat Rev Genet* 2015;**16**:321–32.

49. Colaco S, Kumar S, Tamang A, Biju VG. A review on feature selection algorithms. In: *Emerging research in computing, information, communication and applications.* Singapore: Springer; 2019. p. 133–53.

50. Hira ZM, Gillies DF. A review of feature selection and feature extraction methods applied on microarray data. *Adv Bioinforma* 2015;**2015**:1–13.

51. Budach S, Marsico A. Pysster: classification of biological sequences by learning sequence and structure motifs with convolutional neural networks. *Bioinformatics* 2018;**34**:3035–7.

52. Jurtz VI, et al. An introduction to deep learning on biological sequence data: examples and solutions. *Bioinformatics* 2017;**33**:3685–90.

53. Ronan T, Qi Z, Naegle KM. Avoiding common pitfalls when clustering biological data. *Sci Signal* 2016;**9**:re6.

54. Nugent R, Meila M. An overview of clustering applied to molecular biology. In: *Methods in molecular biology*. vol. 620. Clifton, NJ: Springer; 2010. p. 369–404.

55. Guzzi PH, Masciari E, Mazzeo GM, Zaniolo C. A discussion on the biological relevance of clustering results. In: *Lecture notes in computer science (including subseries lecture notes in artificial intelligence and lecture notes in bioinformatics). 8649 LNCS*, Berlin: Springer Verlag; 2014. p. 30–44.

56. Andreopoulos B, An A, Wang X, Schroeder M. A roadmap of clustering algorithms: finding a match for a biomedical application. *Brief Bioinform* 2008;**10**:297–314.

57. Wiwie C, Baumbach J, Röttger R. Comparing the performance of biomedical clustering methods. *Nat Methods* 2015;**12**:1033–8.

58. Aure MR, et al. Integrative clustering reveals a novel split in the luminal A subtype of breast cancer with impact on outcome. *Breast Cancer Res* 2017;**19**:44.

59. Mathews JC, et al. Robust and interpretable PAM50 reclassification exhibits survival advantage for myoepithelial and immune phenotypes. *npj Breast Cancer* 2019;**5**:30.

60. Alderdice M, et al. Prospective patient stratification into robust cancer-cell intrinsic subtypes from colorectal cancer biopsies. *J Pathol* 2018;**245**:19–28.

61. Yang H, Pizzi NJ. Biomedical data classification using hierarchical clustering. In: *Canadian conference on electrical and computer engineering*; 2004. p. 1861–4. https://doi.org/10.1109/CCECE.2004.1347570.

62. Pontes B, Giráldez R, Aguilar-Ruiz JS. Biclustering on expression data: a review. *J Biomed Inform* 2015;**57**:163–80.

63. Padilha VA, Campello RJGB. A systematic comparative evaluation of biclustering techniques. *BMC Bioinf* 2017;**18**:55.

64. Xie J, Ma A, Fennell A, Ma Q, Zhao J. It is time to apply biclustering: a comprehensive review of biclustering applications in biological and biomedical data. *Brief Bioinform* 2018;**20**:1449–64.

65. Jensen AB, et al. Temporal disease trajectories condensed from population-wide registry data covering 6.2 million patients. *Nat Commun* 2014;**5**:4022.

66. Beck MK, et al. Diagnosis trajectories of prior multi-morbidity predict sepsis mortality. *Sci Rep* 2016;**6**:36624.

67. Yang H, et al. Disease trajectories and mortality among women diagnosed with breast cancer. *Breast Cancer Res* 2019;**21**:95.

68. Giannoula A, Gutierrez-Sacristán A, Bravo Á, Sanz F, Furlong LI. Identifying temporal patterns in patient disease trajectories using dynamic time warping: a population-based study. *Sci Rep* 2018;**8**:4216.

69. Zhang Y, Horvath S, Ophoff R, Telesca D. Comparison of clustering methods for time course genomic data: applications to aging effects. *ArXiv* 2014;**1404**:7534.

70. de Jong J, et al. Deep learning for clustering of multivariate clinical patient trajectories with missing values. *Gigascience* 2019;**8**:giz134.

71. Chalise P, Koestler DC, Bimali M, Yu Q, Fridley BL. Integrative clustering methods for high-dimensional molecular data. *Transl Cancer Res* 2014;**3**:202–16.

72. Chauvel C, Novoloaca A, Veyre P, Reynier F, Becker J. Evaluation of integrative clustering methods for the analysis of multi-omics data. *Brief Bioinform* 2020;**21**:541–52.

73. Zeng ISL, Lumley T. Review of statistical learning methods in integrated omics studies (an integrated information science). *Bioinf Biol Insights* 2018;**12**. 1177932218759292.

74. Beaulieu-Jones BK, Greene CS. Semi-supervised learning of the electronic health record for phenotype stratification. *J Biomed Inform* 2016;**64**:168–78.

75. Charrad M, Ghazzali N, Boiteau V, Niknafs A. Nbclust: an R package for determining the relevant number of clusters in a data set. *J Stat Softw* 2014;**61**:1–36.

76. Brock G, Pihur V, Datta S, Datta S. ClValid: an R package for cluster validation. *J Stat Softw* 2008;**25**:1–22.
77. Arbelaitz O, Gurrutxaga I, Muguerza J, Pérez JM, Perona I. An extensive comparative study of cluster validity indices. *Pattern Recogn* 2013;**46**:243–56.
78. Ruiz Marin M, et al. An entropy test for single-locus genetic association analysis. *BMC Genet* 2010;**11**:19.
79. Von Luxburg U. Clustering stability: an overview. *Found Trends Mach Learn* 2009;**2**:235–74.
80. Yu H, et al. Bootstrapping estimates of stability for clusters, observations and model selection. *Comput Stat* 2019;**34**:349–72.
81. Van Der Maaten L, Hinton G. Visualizing data using t-SNE. *J Mach Learn Res* 2008;**9**:2579–605.
82. Wagner S, Wagner D. Comparing clusterings—an overview. *Universität Karlsruhe Technical Report 2006-04*; 2007.
83. García-Escudero LA, Gordaliza A, Matrán C, Mayo-Iscar A. A general trimming approach to robust cluster analysis. *Ann Stat* 2008;**36**:1324–45.
84. Liu Y, Wu S, Liu Z, Chao H. A fuzzy co-clustering algorithm for biomedical data. *PLoS One* 2017;**12**:e0176536.
85. Dennis JM, Shields BM, Henley WE, Jones AG, Hattersley AT. Disease progression and treatment response in data-driven subgroups of type 2 diabetes compared with models based on simple clinical features: an analysis using clinical trial data. *Lancet Diabetes Endocrinol* 2019;**7**:442–51.
86. NHGRI. *The human genome project*. Available at: https://www.genome.gov/human-genome-project. [Accessed 18 March 2020].
87. Riddick G, et al. Predicting in vitro drug sensitivity using random forests. *Bioinformatics* 2011;**27**:220–4.
88. López-Reig R, et al. Prognostic classification of endometrial cancer using a molecular approach based on a twelve-gene NGS panel. *Sci Rep* 2019;**9**:18093.
89. Toth R, et al. Random forest-based modelling to detect biomarkers for prostate cancer progression. *Clin Epigenetics* 2019;**11**:148.
90. Johann PD, Jäger N, Pfister SM, Sill M. RF_purify: a novel tool for comprehensive analysis of tumor-purity in methylation array data based on random forest regression. *BMC Bioinf* 2019;**20**:428.
91. Djuric U, Zadeh G, Aldape K, Diamandis P. Precision histology: how deep learning is poised to revitalize histomorphology for personalized cancer care. *npj Precis Oncol* 2017;**1**:22.
92. Campanella G, et al. Clinical-grade computational pathology using weakly supervised deep learning on whole slide images. *Nat Med* 2019;**25**:1301–9.
93. Coudray N, et al. Classification and mutation prediction from non-small cell lung cancer histopathology images using deep learning. *Nat Med* 2018;**24**:1559–67.
94. Sun R, et al. A radiomics approach to assess tumour-infiltrating CD8 cells and response to anti-PD-1 or anti-PD-L1 immunotherapy: an imaging biomarker, retrospective multicohort study. *Lancet Oncol* 2018;**19**:1180–91.
95. Kim D, et al. Knowledge boosting: a graph-based integration approach with multi-omics data and genomic knowledge for cancer clinical outcome prediction. *J Am Med Inform Assoc* 2015;**22**:109–20.
96. Rotmensch M, Halpern Y, Tlimat A, Horng S, Sontag D. Learning a health knowledge graph from electronic medical records. *Sci Rep* 2017;**7**:5994.
97. Jiang P, et al. Deep graph embedding for prioritizing synergistic anticancer drug combinations. *Comput Struct Biotechnol J* 2020;**18**:427–38.
98. Zhang W, Chien J, Yong J, Kuang R. Network-based machine learning and graph theory algorithms for precision oncology. *npj Precis Oncol* 2017;**1**:25.
99. Caravagna G, et al. Detecting repeated cancer evolution from multi-region tumor sequencing data. *Nat Methods* 2018;**15**:707–14.
100. Taroni JN, et al. MultiPLIER: a transfer learning framework for transcriptomics reveals systemic features of rare disease. *Cell Syst* 2019;**8**:380–394.e4.
101. Vesteghem C, et al. Implementing the FAIR data principles in precision oncology: review of supporting initiatives. *Brief Bioinform* 2020;**21**:936–45.

102. Miller A. The future of health care could be elementary with Watson. *CMAJ* 2013;**185**:E367–8.
103. Hatz S, et al. Identification of pharmacodynamic biomarker hypotheses through literature analysis with IBM Watson. *PLoS One* 2019;**14**:e0214619.
104. Strickland E. How IBM Watson overpromised and underdelivered on AI health care. *IEEE Spectr* 2019. Available at: https://spectrum.ieee.org/biomedical/diagnostics/how-ibm-watson-overpromised-and-underdelivered-on-ai-health-care. [Accessed 28 October 2020].
105. Zou FW, Tang YF, Liu CY, Ma JA, Hu CH. Concordance study between IBM Watson for oncology and real clinical practice for cervical cancer patients in China: a retrospective analysis. *Front Genet* 2020;**11**:200.
106. FDA. *Artificial intelligence and machine learning in software as a medical device*; 2021. Available at: https://www.fda.gov/medical-devices/software-medical-device-samd/artificial-intelligence-and-machine-learning-software-medical-device. [Accessed 25 January 2021].

Image analysis in drug discovery

9

Adam M. Corrigan[a]**, Daniel Sutton**[b]**, Johannes Zimmermann**[c,*]**, Laura A.L. Dillon**[d,*]**, Kaustav Bera**[e,f]**, Armin Meier**[c]**, Fabiola Cecchi**[d]**, Anant Madabhushi**[e,g]**, Günter Schmidt**[c,*]**, and Jason Hipp**[d,*]

[a]*Data Sciences and Quantitative Biology, Discovery Sciences, R&D, AstraZeneca, Cambridge, United Kingdom,* [b]*Imaging and Data Analytics, Clinical Pharmacology & Safety Sciences, R&D, AstraZeneca, Cambridge, United Kingdom,* [c]*Translational Medicine, Research and Early Development, Oncology R&D, AstraZeneca, Munich, Germany,* [d]*Translational Medicine, Research and Early Development, Oncology R&D, AstraZeneca, Gaithersburg, MD, United States,* [e]*Center for Computational Imaging and Personalized Diagnostics, Department of Biomedical Engineering, Case Western Reserve University, Cleveland, OH, United States,* [f]*Maimonides Medical Center, Brooklyn, NY, United States,* [g]*Louis Stokes Cleveland Veterans Administration Medical Center, Cleveland, OH, United States*

Across all stages of drug discovery and development, experimental assays are performed to understand the effect of a drug or drug candidate—at the molecular, cellular, organ, or organism level. Imaging is a key technology in this process, and an imaging assay consisting of sample preparation, image acquisition, and image analysis provides a quantitative readout of a system. Historically, chemical assays have been the workhorse of early discovery, screening millions of compounds for a simple endpoint, whereas imaging was primarily used in lower throughput mechanistic studies. However, with the development of high-throughput high-content microscopy platforms, the throughput of imaging assays now rivals chemical screens. Similarly, innovations in image analysis mean that robust quantitative conclusions can be derived from complex and multimodal image data, driving informed decision making later in the drug development process. For these reasons, imaging is widely used throughout the pharmaceutical industry and throughout multiple stages of the drug discovery process.

Systems where imaging is used range from cellular models typically used in early discovery, through organoids and preclinical animals, to clinical tissue and in vivo patient imaging. The choice of whether to design an imaging screen is often determined by the complexity of the endpoint. To understand the effect of a perturbation in situ, imaging is a good choice, as it is able to capture orthogonal information in addition to the primary endpoint, building a more detailed understanding. The modality and instrument used for the acquisition is always tailored to the system. Microscopy is widely used for cellular imaging; cell lines are typically optimized to have good optical properties, allowing biomarkers and compartments of interest to be fluorescently labeled and visualized at high magnification. In addition, cells can be directly visualized without fluorescent labels using bright-field or phase contrast imaging. Traditionally, because of the difficulty in interpreting the resulting images, bright-field microscopy has been used for simple endpoints such as cell confluency, however, as described later in this chapter, developments in machine learning and deep neural networks (DNNs) are allowing more quantitative insight to be extracted from these images.

---

*These authors contributed equally.

The Era of Artificial Intelligence, Machine Learning, and Data Science in the Pharmaceutical Industry. https://doi.org/10.1016/B978-0-12-820045-2.00010-6

Table 1 Description of different systems.

System	Description
Cells	Single cell types cultured in specific media. Well-characterized cell lines and genetically modified cells
Spheroids	Small bundles of cells (50–200 cells) of single or mix cell lineages cultured
Microphysiological systems	Cells organized into artificial organs in an enclosed physiological system
Ex vivo tissue culture	Viable tissue is removed from animals and cultured in vivo preserving as much of the natural tissue architecture as possible
Animal models	Mainly rodent but includes larger species like dog or pig. These are complex biological systems that are translated to human effects
Primate	Animals with the closest available system to humans are at times used to investigate safety before first in man dosing

In cellular systems, imaging is often used to perform phenotypic screening for target identification, where rather than screening compounds against a specific target, we are interested in identifying perturbations that elicit a specific cellular phenotype, which includes but is not limited to biomarker expression, differentiation status, morphology, cell cycle, or response to DNA damage. Here, imaging captures additional relevant information such as toxicity concerns or effects on proliferation, which we can use to group or prioritize the results of the screen.

More complex systems, such as organoids and organ-on-chip systems, are thought to be more physiologically relevant, with the trade-off of increased cost and reduced throughput. Imaging can capture this increased complexity, quantifying how the different microenvironments within the sample drive spatial heterogeneity in the response. Microscopy also provides the potential for noninvasive measurement, thereby enabling time lapse measurements. Taking advantage of the increased translatability into the clinic, these multicellular systems are often used to identify toxicity and safety concerns at the lead identification and optimization stages. Organoid systems are still amenable to fluorescence microscopy techniques, though the thickness of the sample becomes a limiting factor, restricting how well fluorophores can be excited and visualized. Techniques such as multiphoton and light-sheet microscopy are able to penetrate deeper into 3D samples, but the throughput of these machines has not yet reached the level needed for screening at scale.

There are a number of biological systems tools available to researchers before human dosing to investigate drug effects, interaction, and toxicology. These start simply at the single cell level up to primate testing with multiple options in complexity and sustainability in-between as described in Table 1.

Cells

Cell lines grown in culture allow the basic units of cellular biology to be studied. These are cells removed from an animal or human source and their subsequent growth in an artificial environment or culture.[1] The cells may be extracted directly from the tissue and disaggregated by enzymatic or removed by mechanical means before cultivation. They also may be derived from a cell line or cell strain that has already been established. The advantage of cell culture is that high-throughput techniques can be used because of their size and volume.

Spheroids

The next logical step up in complexity from cell culture is to arrange the cells into a simple system that begins to resemble organs or components of organs. Spheroid are so called because they tend to grow in a spherical fashion in culture; they are also referred to as organoids. The advantages of these spheroid are they allow a high-throughput approach and they also provide an insight on the interaction between cells.[2] The limitations of spheroids are they often grow in a disorganized manner and without vascularization resulting in necrosis. In practical terms, spheroids are challenging to work within a pathology-and-imaging context because of their size. Getting accurate focus can be challenging.

Microphysiological systems

These are enclosed biological systems often incorporating multiple organs included to model interactions[3]; they are often referred to as organ on a chip for this reason. The cells in the systems are seeded on to artificial constructs to stimulate organ structures; physical forces can also be stimulated to try to mimic their natural environment (e.g., fluid flow). Microphysiological systems are useful to model interactions in a more organ-specific setting without using animal models but are limited in their artificial nature of their composition.

Ex vivo tissue culture

Precision cut sliced tissue are taken from animals shortly after they are culled and cultured in vivo.[4] These have the advantage of maintaining the natural cellular composition and structural arrangement of the tissue and that a large number can be produced from a single animal. It is difficult to keep the tissue viable as the tissue can degenerate quickly after removal from the host animal due to the lack of biological infrastructure to support it. This has meant that it is of limited use in research.

Animal models

Animals make up the bulk of the tissue-based research.[5] They offer a complete biological system with all the associated interactions that can be experimented on in a controlled and consistent manner. The ability to stimulate diseases in an animal allows the study of treatments and also the disease mechanism. The 3R framework (reduction, refinement, and replacement of animals in research) dictates that were possible the previous methods are used where possible with the long-term goal to bring down the numbers of animals used in research without compromising on data quality.

- Disease models
  - **Induced disease models**: These will typically be models that involve the animal being dosed with a compound to elicit a disease-like response. For example, in the house dust mite model the animals inhale house dust mite partially to produce a model of asthma.
  - **Strain-specific spontaneous models**: Strains of animals which when crossed with each other produce animals that are genetically susceptible to developing diseases. For example, the NZB/NZWF1 mouse cross produces a spontaneous lupus model at 16 weeks.

- **Surgical and cell transplantation model**: Animals are transplanted with ex vivo tumor cells. These cells are grown in animals to form tumors and are used to model growth, development, and treatment. These syngeneic and xenograft models allow the study of human tumors and tumors with specific engineered mutations.
- **Genetically modified models**: Knock-out and knock-in models are used to target a specific mutation to a gene locus. These methods are useful if a single gene is shown to be the primary cause of the disease that is being treated or to investigate individual tumor mutations.

- In toxicology safety studies, animals are used to asses a wide range of compound effects and mechanisms of action. Specific studies examine absorption, distribution, metabolism, and excretion as well as tolerability and dosing. Toxicological studies form part of the regulatory package to get a drug approved.
- Primate studies are used because of their biological system resemblance of human physiology, reproduction, development, cognition, and social complexity.[6]

The use of animal models is lower throughput compared with the other systems and requires a higher level of infrastructure and expertise, but there is no substitute for the complexity that animal models provide in the alternative systems.

Moving further through drug development, imaging is used to study tissue sections of both preclinical and clinical samples. Here a key driver toward using imaging is the ability to directly visualize and quantify spatial relationships in the tissue, for instance whether a drug is enriched at the intended site, and how different cell types are interacting and infiltrating into a tumor.

Tissue pathology

The purpose of pathology is to understand the nature of changes occurring in tissue whether they are caused by disease, toxicity, or natural processes. In practice, this involves generating high-quality slides to view down a microscope or scanned. The majority of tissue pathology involves formalin-fixed paraffin-embedded (FFPE) tissues that are cut into 3–5 µm sections and mounted on a slide before being stained with hematoxylin and eosin (H&E) stain, producing a standardized substrate for a pathologist to view and diagnose.

Alternative fixation and processing methods are used depending on the objective of the experiment. Frozen tissue processing is the most common technique after FFPE that preserves the tissue without the need for formalin fixation but increases the tissue section thickness and reduces the quality of the morphology when the slide is viewed. There are a number of specialized processing techniques for preserving RNA/DNA, but they are time consuming and do not have specialist automation to enable scaling to higher throughput.

The visualization of tissue components can broadly be split into two categories: dye-based and molecular-based. The H&E stain was first described in 1865 by Franz Böhmer and is the most common dye-based stain making up the majority of the pathology slides produced in laboratories today. The dye-based staining makes use of the chemical properties of the dye molecules to bind to specific tissue components like mucin, fats, and proteins. Molecular techniques are newer and constantly being improved with more research. The most commonly used is immunohistochemistry (IHC), which uses antibodies to bind to protein epitope targets with a high degree of specificity and is visualized using 3,3'-diaminobenzidine, which produces a brown stain.[7] There are other visualization colors and methods and these can be combined to look at multiple markers. Antibodies can also be used in immunofluorescence (IF), which uses fluorescent dyes, bringing the advantage that a much higher number of targets can be multiplexed. Probe-based techniques can be used to visualize mRNA, miRNA, and DNA in the tissue using brightfield or IF techniques. There is also a push toward hyperspectral imaging, distinguishing tens or hundreds of different biomarkers labeled with different heavy metals and visualized using imaging mass cytometry (IMC), or using mass spectrometry imaging[8] directly on unlabeled samples and capturing a full spectrum at each pixel. Here, the benefits of machine learning approaches become clear, as it is extremely challenging to manually extract the full information present in the images.

In the clinic and in research settings the trend continues for greater complexity and lower throughput. In vivo imaging, such as magnetic resonance imaging (MRI) scans, is used to measure volumes of compartments, for example, to quantify the shrinkage of a tumor or the change in size of an organ.

For a number of reasons, image analysis has been at the forefront of the machine learning and artificial intelligence (AI) revolution. First, the proliferation of the internet, social media, and smartphone cameras has resulted in an explosion in the amount of annotated and searchable image data. This has resulted in the ImageNet dataset and challenge, which since its creation in 2010 has improved methodologies to the point of outperforming human classification accuracy. Along with large repositories of digitized images comes questions around the authenticity and quality of the images, an issue that has gained much attention because of artificially generated fake images that are difficult to distinguish from their real equivalents. Second, the development of convolutional neural networks (CNNs) has allowed models to process images containing millions of pixels/features without requiring astronomically large numbers of parameters. Finally, unlike the vast majority of other data types, images have, regardless of how or why they were acquired, a common structure consisting of a grid (usually two-dimensional, but three or more dimensions are possible) of pixels each with a number of color channels. For photographs the color channels are usually standard as red, green, and blue, whereas for microscopy the number of channels can be greater depending on the emission wavelengths that are captured, to the extreme case of hyperspectral imaging where the number of channels can number in the tens, hundreds, or even thousands. Nevertheless, the common structure has meant that machine learning methodologies and even trained models have been transferred from one domain to a completely unrelated area, to an extent far greater than was anticipated. This has meant, for example, that models that have been trained to accurately classify ImageNet categories such as types of vehicle and animal can be used to reliably differentiate toxic and nontoxic phenotypes, for example. There are many discussions of transfer learning in the literature; however, the main reason for this cross-applicability is generally attributed to the manner by which DNNs encode image information, using early layers to extract general features and textures from images, before combining these features in a task-specific way in the final layers (Fig. 1).[9]

Aims and tasks in image analysis

Image analysis aims to capture and quantify the information present in an image in a robust and reproducible way. Although the specific goal and endpoints vary with modality and stage of drug development, the measurements extracted and the methodologies for doing so are surprisingly consistent. As described above the ultimate aim of imaging and image analysis is to convert raw image data into a quantitative understanding of the system being imaged; classically, this has been performed through a relatively small number of tasks, primarily image preprocessing and enhancement, image segmentation and classification, which are described in more detail in the following sections. With the advent of AI in

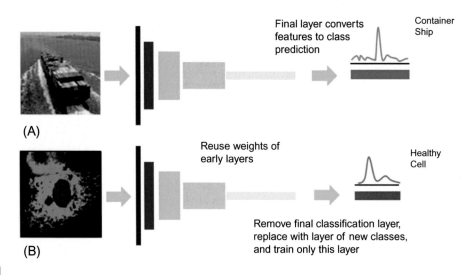

(A)

Final layer converts features to class prediction

Container Ship

(B)

Reuse weights of early layers

Healthy Cell

Remove final classification layer, replace with layer of new classes, and train only this layer

FIG. 1

Transfer learning. A model which has been trained on a task with a large amount of training data can be transferred to an application where less training data is available, reusing the knowledge gained in the first task. Here, a network trained to predict the classes from the ImageNET dataset (A), is repurposed to classify cellular phenotypes from fluorescence images (B). The early layers are reused with the same weights, and only the final classification layer is replaced and retrained with the cellular classes. Because only the weights from one layer are trained and the rest frozen, much less training data is required. The assumption made in transfer learning, which must be empirically tested, is that the features encoded in the early layers are general enough to be transferred to the new task.

image analysis, the simplest way in which machine learning methods are transforming image analytics is through the direct replacement or augmentation of these individual tasks. However, as methods continue to evolve, alongside commensurate increases in computational power and size and availability of datasets, machine learning is promising to fundamentally change the image analysis workflow, rather than replace individual parts. These directions are discussed at the end of this chapter.

Image enhancement

It is not always possible to acquire images of the optimal quality, free from artifacts and of the resolution required to easily resolve all the structures of interest. As such, image enhancement or preprocessing is often the first step in an analysis pipeline. In photography the types of correction include color-caste correction, where the apparent hue of the image is influenced by the ambient light at acquisition and correcting for spatial distortions introduced by lens aberrations. In microscopy applications, these steps also include correcting for uneven illumination across the image, attempting to remove noise introduced by the camera digitization, and correcting for the fundamental spread of the light as it passes through the microscope optics in a process known as deconvolution.

Traditionally, correction of these artifacts has relied on careful quantification of the source of the noise. For example, measuring the variation of the intensity across a field of view, either by averaging over a large number of acquired images or by capturing a blank or uniform reference image, allows the spatial illumination to be corrected. To deconvolve an image, the so-called point spread function (PSF) is calculated by capturing point-like sources of light such as fluorescent beads and measuring the spread in intensity due to the optics. Deconvolution is then an iterative process, which aims to calculate an image that once blurred by the PSF reproduces the acquired image.

As is the case in many disciplines, machine learning and AI tools are beginning to outperform and replace traditional methods. For example, neural networks are now used to remove camera speckle and background noise from fluorescence images. Some of these methods can train such a model without having access to the "ground truth" noiseless images (which may be difficult to acquire) by instead making more general assumptions about the distribution of the noise. A further improvement promised by deep learning (DL) is the ability to reconstruct a higher resolution image from a low-resolution input, in an approach referred to as superresolution imaging (contrast with superresolution microscopy, which is a solution, aimed at increasing the resolution at the point of acquisition, beyond the diffraction limit of microscopy). The applications made possible by these methods are many and varied, from the ability to visualize structures beyond the resolution of the microscope objective without expensive hardware and time-consuming localization methods to the possibility of capturing noisy short exposure images, which cause much less sample damage, and being able to extract the same information as longer exposure or higher magnification images.

Most straightforwardly the training data required for image enhancement tasks consist of a set of raw images and a corresponding set of "enhanced" images, capturing the same region with a longer exposure of a higher magnification, with the aim of predicting the high-quality image from the raw image. For denoising tasks, it has been proposed that a model can be trained using only raw, noisy images, whereby the model infers the common structure of the noise and how to remove it.[10,11] This approach has clear applications in live imaging, where it is not always possible to generate the high-quality target images.

Image segmentation

Classically, perhaps the most intuitive type of image analysis is image segmentation, the process of grouping pixels in the image into regions of known label. This is a universal imaging task, used for example in autonomous vehicles, to identify the location and extent of other vehicles, pedestrians, and road signage within the field of view. Within the pharmaceutical space the applications of image segmentation are many and varied. In early discovery, segmenting cells is almost always the first step in the image analysis workflow of a phenotypic screen. Identifying which pixels belong to which cells allows basic measurements of cell morphology such as cell area and cell shape. When measuring the intensity of a biomarker as an endpoint, it is important to distinguish whether the average channel intensity has reduced because the intensity of the biomarker has reduced, indicating potential activity, or because there are fewer cells in the image, indicating potential toxicity. Segmentation is an important step at all stages of drug development, including subclassifying regions in a clinical tissue slice and measuring the volume of a tumor from a patient scan.

Classically, segmentation is performed by designing a series of filtering and processing steps that separate foreground from background. The approaches vary widely with the image appearance and include thresholding of bright objects from darker background, watershedding to find boundaries and separate regions of the image, and active contours to find the optimal region boundary according to some imposed conditions on intensity and curvature. In all of these cases, construction and tuning of the algorithms requires significant experience in image processing. For tasks where the appearance of the image is sufficiently standardized, software packages and analysis platforms have been developed that implement commonly used algorithms and allow basic tuning of the main parameters.[12–17] In cellular imaging, where cells are commonly stained with a marker of the nucleus, open source software such as CellProfiler[14] allows analysis pipelines to be built to segment nuclei and cells and make measurements of intensity and shape, without requiring expertise in image analysis. These analysis platforms typically offer a small number of segmentation algorithms, based on image smoothing, thresholding, and separation of touching objects by a watershed method.[16] Each algorithm will often have tuneable parameters to improve the generality: for instance, by setting the typical object size to control the amount of smoothing or adjusting the threshold of detection to identify bright or faint objects. A key limitation of these platforms arrives when the appearance of the images is novel in some way, such that the prebuilt algorithms do not perform to the required standard. In these situations, it is possible to augment the built-in algorithms by bolting-on a preprocessing step such as image normalization or background subtraction; however, as the number of tuneable parameters increases, it becomes more challenging for the nonexpert to optimize the algorithm, and even then a bespoke algorithm designed for the task will likely outperform a prebuilt method. In the cellular domain, this can include cases where the shape, intensity, or spatial distribution of the cells have changed, and also cases where one does not want to include the standard markers that are used by the preset algorithms. For these tasks, machine learning approaches are beginning to replace the classical methods and achieve results that surpass what has previously been possible.

Considering image segmentation as a machine learning problem, we take the raw (or preprocessed) pixel data as the input and want to classify each pixel as belonging to background or foreground. The labeling scheme can be extended to include multiple foreground labels: for instance, different cell types, types of tissue, or tumor grades. Early machine learning methods for segmenting images grouped neighboring pixels into superpixels and then classified the superpixels based on their properties. Ongoing success in this area has been achieved by applying filters to the raw image to calculate a set of textural features for each pixel. The size of the neighborhood explored to make the classification is determined by the size of the filtering kernel used, and it is possible to combine multiple size scales in the feature vector. Training data are created by manually "painting" regions of images as classes of interest, and then a model is trained to predict the class of the pixels from the textural features. A clear benefit of this approach is that, although the textural features such as Gabor and Haralick features are based on decades of image analysis research, generation of a new segmentation model is as straightforward as painting regions of a new image and does not require any expertise. Therefore this approach is able to be deployed to end users, and examples of software implementations include Columbus from Perkin Elmer, which uses a linear classifier as the underlying machine learning model, Halo (Indica Labs), which is designed primarily for tissue segmentation, and the open-source framework *ilastik*,[17] which has been applied to many imaging domains and uses random forest (RF) as the underlying classifier. The application of region segmentation to tissue quantification is considered in detail in the next section.

DL methods generally work directly with the raw image data, without making assumptions about the type of textural features that capture differences in the image relevant for the specific task. A number of approaches are the subject of intensive research. In the autonomous vehicle field, driven by a requirement for real-time processing, models tend to focus on predicting the bounding boxes of objects within the image.[18] A DNN architecture, which has had success for pixel level classification, is the U-Net[19] that makes pixel-level predictions of the region classes. For application to cellular imaging, it is important not only to predict cell versus noncell, but also to distinguish densely packed and touching cells. One way to achieve this is to modify the loss function to weight background pixels that lie on the boundary between two or more cells more highly than other pixels, to ensure that individual cells are separated by a thin line of background pixels.

A valid criticism often leveled at DL models is the vast amount of training data required to create useable predictions. For the case of image segmentation, this problem is alleviated to some extent because predictions are made at the pixel level rather than the whole image. Therefore the annotation of 50 or even 20 training images can provide millions of training instances and allow reasonable model performance.

There is potential for using machine learning in segmentation tasks to perform tasks that would be difficult to achieve otherwise. An example of this is the segmentation of cells from phase contrast images (see Fig. 2). Phase contrast images do not require any staining or fluorescent imaging; however, because of the complexity of the images, it is highly challenging to segment areas of interest using a classical algorithm or even through manual annotation. As described above, training data for machine learning segmentation can be created through manual annotation. An alternative to this approach is to use an automated method, for instance a classical algorithm, to segment the image using the standard markers, for instance a nuclear marker and a cell marker. This allows far more training data to be created, with the caveat that errors will be present depending on the automated method used. A model is then trained to perform the segmentation, but without using the nuclear or cell markers in the input image, instead using the phase contrast image. Using one image modality to create training data for prediction using a

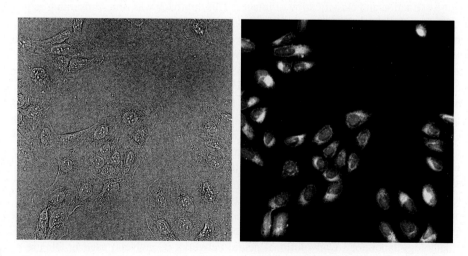

FIG. 2

Prediction of fluorescent labels from unlabeled bright-field images. A promising area is the use of deep learning to predict the appearance of fluorescent stains labeling cellular organelles, such as mitochondria, endoplasmic reticulum, and Golgi apparatus, and structures such as actin filaments *(right panel)*, directly from unlabeled bright-field images *(left panel)*. This approach exploits information that is present in the bright-field image but that is difficult for humans to access.

different modality is a promising area of image analysis, which has the potential to change the types of experiments that are performed within drug discovery and development[20] (see Fig. 2).

Another area where machine learning potentially outperforms classical segmentation is in terms of generality. As an example, images of spheroids in routine safety assays can be highly variable, because of both the natural heterogeneity of these complex samples, and the range of treatments and perturbations applied. As such the identification of a classical algorithm that is able to correctly process the full range of appearances is a challenge. A machine learning model can solve this problem, provided that the full range of behaviors is captured within the training data, and the model has sufficient learning capacity.

Region segmentation in digital pathology

Initially, digital pathology consisted of the ability to make simple measurements and basic manual annotations on to the images and return intensity and shape values. The breakthrough came from using the pixel values as 2D array data to allow computational analysis of the images. Often it is not required to identify, segment, and classify individual cells in a whole slide image to make a measurement and diagnosis. Instead, by identifying regions of the tissue that appear normal, diseased, or any other pathological phenotype, and quantifying the area and shape of these regions, computational methods can begin to replicate and automate the work of the pathologist.

Software was brought to market that used these features to allow tissue- and cell-based analysis of the images producing quantitative data from slides for the first time. There were software like the Definiens Developer software that enabled these features to be built up using a form of programming to allow feature extraction and classification of images. This approach was time consuming as it required the user to define all the features for classification and suffered due to the inflexibility of a rule-based approach. As the programming complexity increased, automated identification of some standard features like nuclei, vacuoles, blood vessels, and cells was added to improve workflow.

There has been the development of RF machine learning techniques[21] for classifying tissue. The difference from rule-based approaches is that machine learning can automatically extract features from the training regions quickly. The advantage of RF is that it isn't computationally demanding and can be done in real time on most computers even laptops. The main issues are the small sampling areas and the strong reliance on color to distinguish between the different tissue components. This means that it is of limited utility when the tissue is chromogenically homogeneous and the accuracy of the classification suffers as a result. It is not possible to build up features from simple shapes into complex shapes and structures.

In the last 3 years, there has been a migration from RF machine learning techniques to DL, driven by two main factors: the increase in graphics processing unit (GPU) computing power combined with the decrease in associated cost. Thus it has emerged out of specialist computer laboratories into the standard pathology laboratory. Convolutional neural networks (ConvNets) used in DL use convolutional layers to build up tissue features from simple edges and shapes into complex features. The fully connected neural network layers at the end of the ConvNet learn to combine and weigh these features to give the most accurate classification. The main advantage of DL is the ability to train the model to improve the accuracy; each iteration of the network is improved by backpropagation that updates and tries to improve the weight associated with the features. To create high-performing DL models, an accurate set of annotations are needed for training; here, the role of the pathologist and scientist is critical. The model is only as accurate as the training annotations, and in contrast, poorly trained models will result in inaccurate classification and errors. The most common models used for pathology DL models are VGG16[22] and UNET,[19] but they are constantly being developed and updated to improve performance.

Validation of the DL models is important to ensure that the data generated is accurate and consistent; the model, if possible, should be tested against an equivalent ground truth (Fig. 3). The validation study images set should be split into training and validation sets; the size of the training is roughly 10% of the overall set but this is at the discretion of the user. If you have heterogeneous data, then a larger training set maybe needed to ensure the model has enough examples to be accurate, the converse is true if the images are homogeneous. The validation set uses existing experts and processes (most likely pathologists) to produce a measure where possible of the same feature as the DL model. Comparison of the ground truth with the DL model data can be done using Pearson correlation coefficient,[23] with the R^2 giving an indication of the degree of correlation.

There are a large number of use cases for DL in digital pathology as the discipline is based on review and interpretation of images. Table 2 details some use cases.

For example, the genetically engineered mouse model[24] is used to test new drugs to assess their effectiveness against specific mutations in lung adenocarcinomas. One of the primary pathology readouts is tumor burden; this was calculated in the past by manually annotating the tumor regions on images. This was a labor-intensive process and would a take a whole week (37 h) of full time employee time per study. By training an AI model to recognize tumor regions and separate them from background normal tissue with almost equivalent accuracy, the same task was completed in 1 h, allowing results to be reported much more quickly (Fig. 4).

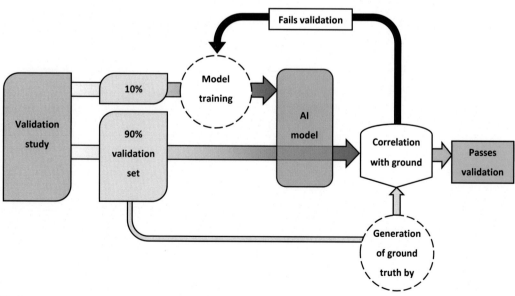

FIG. 3

Visualization of model validation steps.

Table 2 Description and examples of use cases.

Use case	Description	Example
Screening	Identifying changes in large pathology datasets	Large studies with high numbers of individual tumors to be segmented. Tumor burden change is the primary pathology data for the oncology studies
Diagnostic	Characterizing pathological changes in tissue	Automated Ashcroft scoring system for lung fibrosis reduces the burden on pathologists reading the slides
Modeling	Accurate disease/systems modeling	Accurate classification of tissue compartments and their cellular make up allows modelers to improve the predictive power of models
Toxicological pathology	Identifying compound-induced pathological changes	Identification of compound-related changes from normal pathology. These are regulatory studies that are required to bring a drug to market

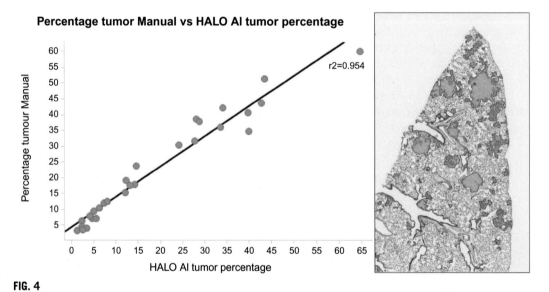

FIG. 4

Prediction of tumor regions and example tumor region image.

Why is it used?

The role of digital pathology has been used to provide quantitative data from digitized images. DL adds a powerful tool to analyze images by replicating some of the manual annotation and improving on previous ML models for classification and adding additional capabilities (Table 3).

The reduction in time to build acute models compared with rule-based solutions is significant

Before the introduction of DeepLearning networks into pathology image analysis software, there were limited options for tissue classification and analysis. RF techniques and rule-based transformations

Table 3 Example of uses.

Use	Example
Quantitative data for tissue sections	Quantitative rather than qualitative data from the images allows experimental hypothesis to be tested more rigorously. Categorical data from pathologist scoring system lack the granularity of continuous computer-derived data
Reduction in workload	Images that would take hours to annotate manually or be read by a pathologist can be processed in minutes. Work can be run in downtime overnight and over weekends to allow continuous data generation
Consistency in decision making	The software makes consistent decisions on repeated measurements with no biases. There are none of the human factors that affect consistency in decision making over time
Increase in the data extracted from image	Additional end points can be gathered with a greater degree of accuracy and efficiency. Measurements that would be discounted because the human interaction time cost was too great can now be gathered using automation
Increase in capabilities	AI is changing the way we work, from the use of annotation-free unsupervised learning to detect tissue changes or segmentation of multiomic data sets

were used for tissue classification. These could take large amounts of time to program and be inflexible when dealing with variations in tissue. The new generation of DL software can build rule sets with a high degree of accuracy in a relatively short period of time. The training regions can be added using simple annotation tools to feed training examples into the model; the model can be trained over the course of a day. There are a number of different DL offered in software allowing for the most appropriate model to be used; shallow DL models offer fast training but with a lower degree of accuracy for complex classifications; conversely, Dense or deep DL models are accurate for classification but take longer to train. There is potential to combine the strengths of both models by using shallow DL networks to identify ROIs (region of interest) for denser DL analysis.

Reduction in pathologist and scientist time doing manual aspects of annotation and analysis

By using DL methods to replicate the manual aspects of tissue analysis, there is a reduction in time spent with each image and there is a transition away from pathologist workload to scientists. Image analysis has long been a tool to reduce the burden on scientists and pathologists, but they are not an appropriate solution in every case, the time saved by creating a model must be weighed against the time to create and validate it. High-volume tasks benefit the most from a DL automation approach as they have the largest time saved to set up ratio.

Using DL methods allows analysis that would not have been attempted previously because of the high burden on the user. Accurate and comprehensive annotation of complex tissue ROI on large-scale experiments are now possible with correctly trained models. After a model has been established and validated, the time to process each image is a matter of minutes compared with the hours it might take a human. The software does not tire, get bored, or develop repetitive strain injury like humans and this has positive effect on wellbeing and health and safety implications that should not be undervalued.

The use of AI as a screening tool to identify pathological changes and focus pathologist time will allow further optimization of workflows. This could be using DL to identify rare events or changes that require additional urgent investigation or prioritizing treatment groups that see the largest drug effects. An interesting use of AI is to use an unsupervised approach to screening toxicology studies; the untreated group is used as a baseline and all the other groups are classified in their deviation from the normal tissue. This removes the need for annotation in the model building process, allowing further automation of the process.

Consistency of decision making (inter and intrauser error)

The ability to extract data from slides has previously relied on human interpretation from pathologists or trained scientists. These data were semiquantitative or qualitative data based on human interpretation and estimations. Humans are also poor at gauging percentages and large number of features in an image; computers can do this in a precise manner that gives absolute numbers in a robust way. A good example of this is H-scorning tissue; this is a score of percentage positivity of stained IHC cells.[25] The user estimates the percentage of weak, medium, and strong staining to generate the score. Humans are surprisingly good at this; however, there is a large amount of variation in scores between users. An image analysis software will produce highly accurate data with consistent scoring of the weak, medium, and strong staining.

Feature extraction

As described above the purpose of image segmentation is to identify objects of interest within the image. Sometimes, this step will be sufficient to provide the information required from the assay, for example, in cases where we want to measure cell confluency (the fraction of the field of view covered by cells) or cell count for use cases such as quantifying the rate of cell death or of cell proliferation, or to provide a binary readout of whether an object of interest is present in the image. More often, after segmenting the image, we want to make further measurements on the identified objects.

Depending on the type of imaging, these measurements can include morphological properties such as object area and elongation, quantification of the average image intensity within each object, or where regions of brightness or darkness are located within an object. Classically, the endpoints calculated have been designed to have biological relevance; for example, if we are looking for an inhibitor of a particular cellular pathway, then a common strategy is to fluorescently label a biomarker in the pathway and then quantify the fluorescence intensity within each cell. An advantage of imaging in this context is the ability to determine where in the image the signal is coming from—we can quantify the extent to which a protein is localized to the cell nucleus, the cell cytoplasm, or within other cellular compartments by using appropriate labeling. It is also possible to quantify whether a signal is localized to the edge of the cell, to the center, or the whole distribution in between, for applications where we are interested in how a cell internalizes a fluorescent cargo. Clearly, the choice of how to quantify the effect is key in determining the quality of the resulting readout; if we do not engineer the correct feature, we may not see a difference between images reflected as a change in the readout. Furthermore, there may be other measurements that are relevant to the biological interpretation, either as confounding factors or as additional orthogonal information, which are ignored unless a measurement has been designed to quantify it. Therefore an alternative strategy is to calculate a large number of general features that quantify as much as possible about the image. These include features, such as average intensity or cell area, that are directly interpretable, but also textural, morphological, and correlative features that are not straightforward to give a biological meaning. Texture features generally consist of small-scale filters that are designed to pick out particular textures, such as ridges or spots. Morphological features capture shape information by breaking a region into subregions between the boundary and the center and measuring the variation in intensity across these regions. The morphological regions can be used in combination with texture features to create further features. Finally, correlative features quantify the degree to which one color channel is correlated with another, to the extent that a high value in one channel is correlated with either a high or low value in the other. This approach of extracting a large number of features from images is known as image profiling.

A number of software platforms, including CellProfiler and Columbus, allow the calculation of textural and morphological features. Taken together, these features provide a fingerprint for a given cell, capturing as much information as possible. The features can be retained at single cell resolution or aggregated to the well level by median averaging to give a morphological fingerprint for a given compound or treatment.

As an alternative to textural and morphological features, DNNs can also be used as feature extractors. Typically, this will use a transfer learning approach, taking a network pretrained for a different task, inputting either a whole image or a single cell tile and taking the output of an intermediate layer (often the penultimate layer) as the features. The reasoning behind doing this is that the intermediate layers of a neural network have learned general features, and networks trained on the 1000 classes of ImageNet have often been shown to have generally useful features in their intermediate layers.

In tandem with the calculation of relatively unbiased features from cellular images, there is increasing interest in using unbiased assays to capture as much phenotypic diversity as possible. The most well-known example is the cell painting assay,[26] which images markers for many organelles and cellular subcompartments, including stains for actin filaments, nucleoli, endoplasmic reticulum, and Golgi apparatus in addition to the usual nucleus stain. Through this approach, rich phenotypic information is captured.

The advantage of this method is that the assay encodes more information than a standard assay that has been tailored specifically toward a clean readout of a single endpoint. This information includes mechanistic insight and potential off-target effects and toxicity that would otherwise have to be picked up when following up with secondary assays. Because of this potential to extract more than specific endpoints, efforts are now ongoing to standardize the cell painting approach, including accounting for batch and cell line effects, and to reinterrogate existing datasets with new questions.[27]

Once features have been chosen and calculated, they can then be used for the downstream application. This approach is often used for unsupervised clustering of cells or treatments based on their features, where annotations or labeling of the data is scarce or absent, because the feature generation step does not require any labeled data to perform. A key task to find relevant clusters is feature selection, to ensure that the groupings are based on biologically relevant features, rather than potential artifacts or confounding information.

Image classification

Often the purpose of image segmentation is to make measurements on the regions of interest to classify them into specific phenotypes. Therefore a natural application of machine learning approaches is to skip the segmentation step and make the classification directly from the raw image. This type of task is directly in line with the ImageNet challenge, where whole images are labeled with a single ground truth label.

A challenge of implementing this approach within the context of imaging in the pharmaceutical industry relates to a point noted in the introduction—often imaging is chosen as an approach because of the desire to extract more complex information than a simple bioassay endpoint. Nevertheless, it is an attractive proposition for a machine learning approach to directly learn how to extract all the relevant information from an image. Because of the requirement for large amounts of training data, a natural area of application for this approach is high-throughput screening, where compounds are screened at a scale up to millions, and on-board controls provide reasonable amounts of training data. The size of typical microscopy images is much larger than the images in the ImageNet dataset, presenting a further challenge to image classification approaches. Compressing millions of pixels down to a single number or vector per image requires deep networks with many parameters, and so here, the choice is often made to work with smaller regions of images, either tiled as a grid over a full image or centered around objects (typically cells) within the image. The classification is then made on a per-cell or per-tile basis, before averaging as necessary. A key challenge of whole image classification in a biological context is *explainability*; with end-to-end models, it is necessary to understand how and why a result has been obtained. As well as reverse engineering approaches to deconstruct how individual pixels contribute to the decision,[28] increasingly model architectures are being designed with interpretability built in from the outset.[29] It is held that there is a trade-off between model performance and interpretability, but this is a necessary compromise for drug discovery applications where there are human consequences to the decisions that are made.

Limitations and barriers to using DL in image analysis

The initial barrier to using DL in pathology is the setup cost, which is not insignificant. For an end-to-end solution, there is a need for hardware for digitalizing slides and the relevant IT infrastructure to store them. There are freeware DL software available for image analysis,[30] but most of the providers are commercial and charge for the software and the DL module. The DL software will need access to a GPU to run the iterations that build the model: this can be local or cloud based with the associated setup costs. The skills and expertise to build models will also need to be invested in, whether training existing employees or hiring new ones.

Not all pathology analysis will be appropriate for DL applications because of the weaknesses of the patch-based approach to analysis. The patch-based analysis is powerful for analyzing cells in a localized zoomed in way but struggles to link these together to allow large tissue features to be classified in relation to each other. For example, it could classify brain tissue as white or dark matter but would be unable to tell you if it came from the frontal lobe, primary motor cortex, or the central sulcus. It is important not to underestimate the amount of contextualization that occurs when humans look at images.[31] This limitation will hopefully be overcome in the future with newer models and increased computing power.

The status of imaging and artificial intelligence in human clinical trials for oncology drug development

A comprehensive understanding of tumor heterogeneity and immune contexture is a key determinant for developing and optimizing diagnostic and therapeutic approaches for oncology drug development (Fig. 5). The availability of imaging modalities at multiple spatial scales and functional depths provides

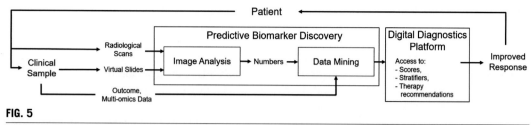

FIG. 5

From imaging to diagnostics: radiomics and computational pathology workflow.

real-world "big data" of exceptional value that can only be leveraged by AI, bypassing the limitations of human perception. Latest developments in AI for computational pathology and radiology are prominent examples of how general-purpose advances of computer hardware and computer sciences drive clinical applications, ranging from the understanding of mechanism of action of drugs in humans to digital companion diagnostics for precision medicine.

Computational pathology image analysis

Computational pathology establishes a relationship between the content of tissue slides acquired from formalin-fixed histochemically stained clinical tissues and the patient's disease state and progression. Although in routine workflows a limited set of options exists, such as functional staining with H&E, or IHC using chromogen-labeled, protein-specific antibodies, the discovery process for predictive biomarkers involves methods of higher conceptual sophistication highlighting the functional state of the cells in the tissue context, such as multiplexed IF and IMC. In both application scenarios, image analysis is used to identify regions of interest in the tissue, to detect and classify cells using their functional and morphological appearance, and to subsegment individual cell compartments within the cells. Clinically relevant examples for regions of interest are normal and neoplastic epithelium, tumor invasive margin, stroma, inflammation, and necrosis. Examples for cell populations that are central in targeted and oncology-related therapies are HER2-positive epithelial cells, PD-L1-positive epithelial cells, macrophages and immune cells, and various populations of immune cells such as T-cells, cytotoxic T-cells, B-cells, natural killer cells, and regulatory cells, to name a few. For targeted therapies, such as ENHERTU (trastuzumab deruxtecan), a precise quantification of target protein expression on the membrane of neoplastic epithelial cells is required. It is hypothesized that therapies based on checkpoint inhibitors benefit from a patient selection algorithm, which involves the spatial organization of the various cell populations, such as the average density of CD3- and CD8-positive cells in the tumor center and the invasive margin in ipilimumab-treated melanoma patients,[32] or the co-occurrence of PD-L1 and CD8-positive cells in durvalumab-treated lung cancer patients.[33]

Computational pathology starts with image acquisition, which had been historically one of the main bottlenecks in the pathology laboratory digitization. Although used for telepathology as the major use cases, early whole slide scanning systems had been characterized by prohibitive scanning times and storage costs.[34] Over the last 20 years that situation has improved markedly, with Philips UFS leading the industry, which provides a Food and Drug Administration (FDA)-approved primary read, 40× high-resolution scan in less than 3 min. Although automated image analysis has been used in research pathology since the 1980s,[35] the size of virtual slides, the heterogeneity of tissue, and the variability in staining and sample processing prevented its ubiquitous application in drug and diagnostic development. However, 2012 represented a turning point: with the revolutionary success of the AlexNet DL architecture for ImageNet image classification,[36] the computer vision community has seen an exponential growth in convolutional network methodologies, architectures, and applications. This growth is fueled by ever-increasing computational power (e.g., Nvidia GPUs), the availability of freely available open-source packages (e.g., TensorFlow, PyTorch), the availability of big data, and huge financial investments by institutional and commercial entities.

Digital pathology has been recognized as one of the most impactful application areas, and headlines such as "Artificial intelligence could yield more accurate breast cancer diagnoses" on the work of Mercan et al.[37] are being followed by medical products implemented in daily clinical routine. The first systems have been already approved by the FDA, showing performance comparable to well-trained clinicians. As an example, scientists at AstraZeneca developed a research prototype for an automated PD-L1 scoring solution for lung cancer immune therapy prediction using an Auxiliary Classifier Generative Adversarial Network (AC-GAN).[38] The AC-GAN enriches real datasets with artificially generated, but realistically looking images for which, by design, the classification results are known. In 2012, AlexNet relied on millions of human annotations to learn to classify natural images by optimizing the weights of its convolutional network layers so that the predicted classes match best the classes provided by humans. Because multiple human observers may classify an image in different ways, the AlexNet and its successors learned to generalize from those multiple sources of annotations to become superior to humans.[39] The same principle is observed in pathology—annotations from multiple pathologists enable a DL system to become "better," in the sense of being more consistent, than an individual pathologist.

In addition to the replication of pathologist scores for key predictive biomarkers such as PD-L1 and HER2, which are challenging to be consistently assessed by eyeballing, the improvement of classical pathologist grading schemes based on composite morphological criteria in H&E-stained tissue samples has moved into the focus of AI-based approaches in digital pathology. For prostate cancer detection and grading, more than 100 million tissue sections are evaluated each year globally, which highlights the need for automation in clinical routine use.[40–43]

DL models for assessing genomic status based on H&E-stained images are using deep residual learning to predict microsatellite instability,[44] and to classify molecular cancer subtypes.[45] The results indicate that the prediction of mutations of canonical cancer genes is possible, but the biological interpretation remains hidden within the complexity of the layer weights of the CNN. Similarly, a breast and colorectal cancer outcome prediction used a long short-term memory CNN, which again makes the interpretation difficult.[46]

Still, most digital pathology systems are based on thousands of cell and region annotations in histopathology images by human experts. The "annotation gold" is the bottleneck, and the currency we have to pay, to get to models that are as good as "consensus" human experts. To overcome the bottleneck, little biological knowledge can be used to predict patient survival[47] directly from the image content, which shifts the role of the pathologist from a provider of annotations to an interpreter of the diagnostic procedures invented by the machine. Deep survival learning enables the machine to associate image pixels in an end-to-end approach with the survival risk of the patient.[48] The survival risk may be modeled using a Cox proportional hazard model, so that the optimizer of the convolutional network layers minimizes the difference of the predicted risk and the observed risk (Fig. 6). As shown by Meier et al.,[49] such a system significantly identifies a subgroup of gastric cancer patients with high risk of disease-related death using CD20 (B-cells) and KI67 (proliferating cells) stained tissue microarrays.

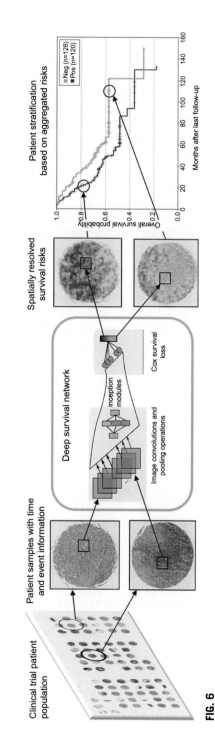

FIG. 6

In a first step, for each patient in the clinical trial cohort, individual cores are detected in a tissue microarray (TMA, H&E-stained). These core images are tiled and patients' survival times and events are assigned to their corresponding patches. Subsequently, a convolutional neural network is trained to minimize a Cox-model based loss. Applying the network results in spatially resolved survival risk maps, which allow to identify regions of low and high risk (*green and red* in the figure). All individual risk values for a specific patient are aggregated by taking their median. In the end the cohort is stratified into a low- and a high-risk group according to these patient risk scores.

Generative adversarial networks (GANs)[50] have been successfully used in multiple domains, such as image style transfer, resolution improvement, and generation of fake images.[51] In the field of computational pathology, generative methods are used to generate artificial but realistically looking "fake" images to augment limited training datasets or to transfer the staining information from one set of images to another set of images, stained with a different stain. In one remarkable application, a Joint Domain Adaptation and Segmentation GAN (DASGAN) was used to train an automated PD-L1 scoring system for nonsmall cell lung cancer (NSCLC).[52] With this approach, epithelial and nonepithelial PD-L1-stained image regions were discriminated using knowledge encoded in a convolutional network by adversarial training from cytokeratin (CK)-stained images. Because both image sets were acquired from distinct patient populations, the system may have learned autonomously the common characteristics of the morphology of hematoxylin-stained epithelial and nonepithelial nuclei. The DASGAN opens the door to "chemically annotate" image regions (e.g., CK-positive regions) with little pathologist monitoring, without having the need for carefully aligned serial sections[53] or complex "stain » scan » re-stain » scan" protocols and workflows.[54] Other weakly supervised approaches use clinical diagnoses as training labels[55] on training and validation sets of unprecedented scale or use attention models.[56] Despite these advances of DL-based approaches, significant open questions remain, namely the mentioned missing interpretation and opacity of DNNs,[57–59] which are beginning to be addressed by initiatives like the generation of explainable knowledge from unannotated images.[60] Another challenge is frequently the low volume of clinical datasets, where even with data augmentation, no training of a robust model can be achieved, and the transferability of models is frequently hard to predict. Furthermore, for the full phenomic exploitation of large histological sections on a cellular or subcellular level, the outcome of predictions in the form of probability maps must be transferred into tangible image objects. Here, heuristic approaches like Cognition Network Technology[61, 62] can be used to provide fractal-hierarchical solutions to challenging image-based biological problems. Following the principle of an eidetic reduction, the formulation of the essence, for example, the common denominator of specific cellular structures, helps to reliably identify the objects of interest, isolated from the noise of morphological variation. Through an iterative-evolutionary process, structures of interest are dynamically refined and semantically enriched, until an abstracted final state is reached.[63] The clear linguistic structure of such sets of rules provides a high explicability. Furthermore, heuristic approaches do not necessarily need large training sets if they are developed keeping the potential biological and preparational variability in mind, which can become essential once the size of a clinical cohort is limited, as in Phase I studies. For instance, Corredor et al.[64] showed that computer-extracted features of the spatial interplay of cancer and immune cells on H&E images of early stage NSCLC patients were strongly associated with the likelihood of recurrence. Barrera et al.[65] subsequently showed that this pattern is also associated with likelihood of response to check point inhibitors in the context of NSCLC. With the expected fractality of upcoming basket trials, including multiple cancer indications, for combinatorial therapies, the number of study subjects, and therefore samples on which solutions are trained, can become low. It is evident that combining the strengths of data- and knowledge-driven methods can significantly enhance the performance of digital pathology workflows.

Radiology image analysis

Radiomics is the high-throughput extraction and assessment of quantitative features from radiographic medical images, including computer tomography (CT), MRI, or positron emission tomography (PET).

The field acknowledges that medical images contain pathophysiological information beyond what is available for visual interpretation. Image features, which can include size, shape, intensity, and texture, provide information on the tumor itself as well as the tumor microenvironment (TME). By extracting high-dimensional data from digitized medical images, a range of bioinformatics methods can be applied to extract maximal amounts of information that, when combined with additional data types including clinical and genomic data, can be used to inform decision making. Data extracted from image features can be mined to create models that can be used to inform disease diagnosis and predictions of prognosis or response to treatment.[66] An example radiomics workflow, from Vaidya et al.,[67] is shown below (Fig. 7).

Radiomics has several advantages over other technologies that can be used to model tumor biology. Medical images contain extensive phenotypic information about a patient's disease, including its underlying molecular, histological, immunological, or genetic characteristics. Radiographic features can be used to characterize heterogeneity within and between tumors and with the surrounding TME,[68] and radiomics features can be compared over time for the same lesion to assess disease progression or measure response to treatment. Because the acquisition of images used for radiomics analysis is noninvasive, it has the potential to be integrated into clinical practice, and indeed, images used for radiomics analysis are often generated as part of standard-of-care practice for cancer patients. This makes radiomics a candidate for companion diagnostic development with the potential for broad applicability across the field of oncology.

Advances in AI, and in particular in machine learning and DL, have brought exciting new opportunities to the emerging field of radiomics.[69] Although radiomics has had a broad impact in oncology ranging from improving our understanding of tumor biology to providing diagnostic information (stage, prognosis) and disease monitoring and surveillance,[57, 70] a more burgeoning area involves the application of radiomics for the prediction of treatment response.[71, 72] Although there are no AI radiomics algorithms being used as a CDx to select for patients, retrospective clinical trial studies have been conducted to further support this exciting new area of oncology drug discovery.

In the following, we will provide an overview of the ways AI and radiomics has been applied to predict response to chemo/chemoradiotherapy, targeted therapy, or immunotherapy, followed by a detailed discussion of the many challenges associated with applying AI and radiomics to drug discovery.

AI-based radiomics to predict response to therapy

Oncologists continue to struggle with the identification of predictive biomarkers to differentiate between patients who will respond to a given therapy and those who will not. Companion diagnostics have been developed for targeted therapies, but even patients with a given gene mutation or those that express a given protein marker will not always respond to treatment. In addition, although immunotherapy has shown great potential to treat and even cure patients across a range of cancer indications, not all patients respond to therapy and predicting which patients will respond to immunotherapy remains a challenge.

Multiple biomarkers have been evaluated for their ability to differentiate between patients who will respond to both traditional therapies and immunotherapy and those who will not, including PD-L1 expression status[73]; mutational burden and characteristics,[74–77] transcriptomic signatures,[78] the presence of tumor-infiltrating lymphocytes,[79] and cytokine expression.[80] Although some of these biomarkers have been able to separate patients by response, none have yet been able to reliably predict response.

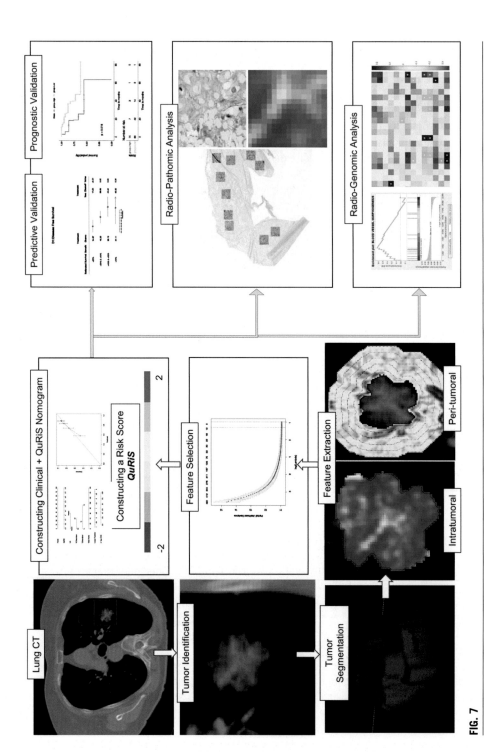

FIG. 7

The first step in radiomics analysis involves identifying and segmenting the region of interest on the CT scan. A standard radiomics pipeline would involve extracting texture, shape from within and outside the nodule and following feature selection to select the top features, building a risk score (e.g., QuRiS). The radiomic features could be combined with relevant clinical features to build a risk prediction nomogram (e.g., QuRNom) to individualize treatment decisions for each patient. Following that survival predictions are represented by Kaplan-Meier curves for prognostic validation, as well as comparing the survival between the control and intervention groups for predictive validation. Further association of the radiomics score could be done with computerized histopathologic analysis (Rad-Path association) as well as with Gene Ontology analysis for radiogenomic association (QuRiS, quantitative radiomic risk score; QuRNom, quantitative radiomic nomogram).

From Vaidya P, et al. CT derived radiomic score for predicting the added benefit of adjuvant chemotherapy following surgery in stage I, II resectable non-small cell lung cancer: a retrospective multicohort study for outcome prediction. Lancet Digit Heal 2020;**2**:e11-e128; used under CC BY 4.0.

Because medical images from oncology patients contain phenotypic information about tumors and the surrounding microenvironment, including information on gene signatures, protein expression, and intratumor heterogeneity, radiomics features mined from medical images using AI are increasingly being used to predict patient response to treatment.

Protein kinase inhibitors

Imaging-defined response assessment (RECIST) is a cornerstone of modern oncologic practice; however, it is limited in several targeted therapies, including treatment with multitargeted tyrosine kinase inhibitors (TKIs) such as sunitinib, sorafenib, and pazopanib. It is well established that conventional response evaluation criteria (RECIST)-defined response rates are lower than the proportion of patients who achieved disease control. Substantial tumor necrosis may occur because of the antivascular effect, resulting in a change in tumor attenuation but little overall size change. This heterogeneity in tumor morphology is not considered by RECIST criteria; however, radiomics features extracted from medical images could be used to predict response to TKIs regimens across a range of tumor types. Efforts include the application of MRI radiomics to predict risk stratification in glioblastoma-treated with antiangiogenic therapy[81] and PET/MRI radiomics to characterize early response to sunitinib therapy in metastatic renal cell carcinoma.[82] Upcoming radiomic feature-based imaging signatures allow prediction of survival and stratification of patients with newly diagnosed glioblastoma and advance the knowledge in the noninvasive characterization of brain tumors at low additional cost, as imaging is routinely and repeatedly used in clinical practice.[83]

Moreover, the use of radiomics-based response assessment tools could improve the stratification between TKIs-sensitive and TKIs-resistant patient populations. Misregulation of Epidermal Growth Factor Receptor (EGFR) expression is a hallmark of many cancers. Radiomics features can identify an EGFR-TKI response phenotype and distinguish between tumors with and without EGFR-sensitizing mutations at baseline and a quantitative change in these radiomic features at follow up.[84]

Chemotherapy/chemoradiotherapy

Similarly, despite the broad desire in the oncology field to identify which patients will benefit from chemotherapy/chemoradiotherapy, molecular biomarkers that can predict patient response remain elusive. To help fill this gap, radiomics features extracted from medical images are being identified to predict response to chemotherapy/chemoradiation therapy regimens across a range of tumor types. Efforts include the application of MRI radiomics to predict response to induction chemotherapy in nasopharyngeal cancer,[85, 86] neoadjuvant chemotherapy in breast cancer,[87, 88] and neoadjuvant chemoradiotherapy in rectal cancer.[89, 90] In addition, radiomics features extracted from CT scans from NSCLC patients have been used to predict response to chemotherapy,[91] chemoradiation,[92, 93] and neoadjuvant chemoradiation.[94, 95]

Although the specific approaches used in studies to date may differ, the utilization of radiomics in this space has generally involved the extraction of radiomics features followed by the application of regression analysis and/or machine learning to develop a radiomics signature; this signature is then applied to retrospective datasets to measure the association between the identified signature and patient response to therapy. Although these methods have had reasonable success in predicting patient response to chemotherapy/chemoradiotherapy in retrospective cohorts, it remains to be seen whether they will have similar success in independent cohorts or when applied prospectively.

Immunotherapy

CT-based radiomics biomarkers are also being developed to predict response to immunotherapy. Radiomics features have been analyzed to predict response to metastatic malignant melanoma in patients treated with anti-PD1 therapy,[96] to predict response of NSCLC patients previously treated with platinum therapy to anti-PDL1 therapy,[97] and to develop a predictor of rapid progression for NSCLC patients treated with anti-PD1 and/or anti-PD-L1 therapy.[98] A machine learning approach was used to develop a radiomics-based predictor of CD8 infiltration from CT scans and RNA-seq data, which was applied to predict response to anti-PD1 or anti-PD-L1 therapy in patients with solid tumors.[99] Machine learning was also applied to pretreatment CT images to predict response to anti-PD1 therapy at the individual lesion level in patients with advanced melanoma and NSCLC.[100] Khorrami et al.[101] showed that differences in texture features both inside and outside the nodule on CT scans were strongly associated with response to three different checkpoint inhibitors and that the signature was also strongly associated with overall survival; and Vaidya et al.[102] were able to identify radiomics features associated with hyperprogression in NSCLC patients receiving checkpoint inhibitor monotherapy. The early success of these efforts leads us to believe this will be an increasingly active area of research in the years to come.

Challenges in applying radiomics to drug discovery

Although radiomics biomarkers[71, 103, 104] have shown increasing promise in drug discovery, there remain challenges to be to overcome before they can be deployed in clinical practice.

Clinical trial validation

The gold standard for validating radiomic methodologies is in a clinical trial, but access to clinical trial datasets continues to be a major limitation for the development, testing, and validation of radiomics models for predicting drug response. Many of the imaging datasets are not easily accessible and to get access is a Catch-22 (before one approaches a cooperative group/pharmaceutical company to get access to the key imaging data, one needs promising results, which can only be obtained in a company that provides you with such data). In addition, clinical trial cohorts are also perfect to develop or construct a radiomic model because they consist of relatively clean patient cohorts with randomized arms, which limit variability compared with radiomics models developed traditionally on retrospective data that often are noisy. For generating true Level I evidence, radiomics needs to be embedded into a clinical trial with patients randomized according to radiomic assessment. However, such trials are usually expensive and time consuming.

Because radiomic methods are susceptible to preanalytic sources of variation including but not limited to scanner vendors, acquisition parameters, reconstruction kernels as well as existential pathologies in the background, there needs to be a way to minimize or eliminate these variations before radiomics can be truly reproducible. AI tools can be deployed for this purpose as well as to quality check the images before analysis to maximize performance. AI tools can also be used to increase the quality of low-quality or noisy images to maximize the signal to noise ratio as well.

Regulatory approval

Before clinical deployment, radiomics-based methods for patient selection or other applications would need to be approved. This could be through the model adopted by medical devices, by being approved by the FDA as a medical device. The FDA has categories of medical devices with Class I being the lowest risk that can be approved by the 510(k) pathway, which is relatively straight forward. The 510(k) pathway also allows a device to have fast-track approval if it is similar to a predicate device. Although 510(k) approvals are possible for AI applications in radiology, which help with segmentations or workflow, in case of AI methods that seek to risk stratify patients or predict response to treatment, it needs to be classified as a Class III or high-risk device and needs to submit a premarket approval, which is significantly more exhaustive and time consuming, needing clinical trial data to be approved. Instead, like in the case of several prognostic genomic assays[57] that have skipped FDA validation and are marked as CLIA (Clinical Laboratory Improvement Amendments)-based laboratory assays, AI tools can be marketed simply as CLIA tests without going through the hurdles of the FDA regulatory approval process. In this case the scans to evaluate/predict response would need to be sent to CLIA-approved central facility.

Distribution and reimbursement

Before radiomics can be used clinically, there needs to be a way to find the efficient way of deploying the tool that can maximize efficiency without disrupting the clinical workflow, as that would enable quicker adoption. Several prospective ways of deployment include (a) making it cloud based where the oncologist/radiologist uploads the scan to a cloud-based interface and a prediction is available to the provider immediately; (b) the AI tool could be part of a complete oncology or radiology workflow suite such as Picture Archiving and Communication System (PACS) and would be one of the features that could be invoked when required; and (c) the AI tool could be integrated with a radiology scanner. Although integrating an AI tool with a radiology scanner would minimize workflow disruption, it would not be vendor neutral, being confined to being used only with that particular scanner and patients scanned on that device. When it comes to reimbursement, there are no Current Procedural Terminology (CPT) codes for AI or radiomics assays. If radiomics tools would need to be deployed, there needs to be new CPT codes created for the assays, which would enable insurance/CMS covering the costs for the radiomics test.

Conclusion

In summary, although radiomics and computational pathology is an exciting emerging field, there is a relatively limited body of work on the application of AI to identify radiomics-based and computational pathology-based image biomarkers from oncology clinical trials. Although there are several challenges that have so far limited the broader implementation of radiomics and computational pathology in oncology drug development, we expect a growing number of AI studies to be conducted because of the exciting potential of this technology to transform the field.

Future directions
Imaging for drug screening

When analyzing screening data using the profiling approach described above, it becomes clear that not all the calculated features are directly relevant to the specific task, for example, hit identification. Indeed, there is a large amount of information present in the feature data, which can be reinterrogated for a different task. A well-known example of this approach is the work of Simm et al.,[105] where features extracted from a high-throughput compound screen were used to predict activity in a number of orthogonal assays. A small amount of data for each assay, using identified control compounds, was used to validate which assays could be accurately predicted from the high-throughput profiling feature data. For those selected assays, it is then possible to predict the activity across all the compounds in the high-throughput screen, without having to run them all in the assay. This approach, treating data as a resource that can be repurposed to answer a range of questions, is clearly of great promise to the pharmaceutical industry, potentially reducing the number of large-scale screening campaigns that need to be run, saving money and importantly reducing the time required to screen a large number of compounds for putative hits. Taking the idea of generally useful datasets further, Cell Painting assays (described above) are beginning to be used for morphological profiling,[106] with the hope that this unbiased approach will provide greater diversity in hit compounds compared with a traditional endpoint-based screening approach.

Computational pathology and radiomics

GANs[50] have been credited with the ability to generate realistic images and data, to the extent that the generated images are indistinguishable from the real target images. At the heart of a GAN is an adversarial training approach, which uses two networks training on competing tasks to replace a traditional human-designed loss function. One application of GANs in imaging is to create one imaging modality from another, for example, the prediction of fluorescent labels from bright-field imaging. The ability to generate realistic images is key to ensuring that relevant features are being learned within the model.

Another area where generative models are of interest is the augmentation of training data. The ability to increase the amount of data with which to train a model is always of interest in DL; the possibility of using a GAN to resample training space to produce realistic images that are different from the original training data is attractive. A key challenge of this approach is how to capture the complexity of an image in the sampling distribution, with the underlying assumption that a point sampled from outside of the existing training space will still be relevant. Another question around GAN augmentation concerns whether synthetic samples provide additional predictive value compared with the original data from which they are generated.

An important example of GAN augmentation is that of clinical data. Not only is clinical data typically of much smaller scale than in other areas of drug development, but consent is required for any application. In the era of machine learning and AI, it is difficult to predict how data will be found to be useful. Therefore one application of GAN augmentation in the clinical space is to anonymize the data so that it can be used in secondary tasks. Synthetically generated images, containing relevant features and information, but different to the original data and anonymized, could then be used to derive additional insights, in ways that could not have been foreseen at the time of original data generation. This approach is not limited to image data and perhaps would be improved by integration of multiple data types. In a similar manner to the challenge described above, a key question here is the data can be perturbed sufficiently to fulfill ethical criteria, while still retaining information useful for further analysis. Nevertheless, the adversarial approach, allowing the machine to learn how best to go about the task is of great interest in the pharmaceutical industry and, in future years, will likely be a part of more end-to-end machine learning workflows for analysis of image-based data.

References

1. Mather J, Barnes D. Animal cell culture methods. In: *Animal cell culture.* San Diego: Academic Press; 1998.
2. Cui X, Hartanto Y, Zhang H. Advances in multicellular spheroids formation. *J R Soc Interface* 2017;**14**. https://doi.org/10.1098/rsif.2016.0877.
3. Cirit M, Stokes CL. Maximizing the impact of microphysiological systems with: in vitro–in vivo translation. *Lab Chip* 2018;**18**:1831–7.
4. Van De Merbel AF, et al. An ex vivo tissue culture model for the assessment of individualized drug responses in prostate and bladder cancer. *Front Oncol* 2018;**8**:400.
5. Knoblaugh SE, Hohl TM, La Perle KMD. Pathology principles and practices for analysis of animal models. *ILAR J* 2018;**59**:40–50.
6. Phillips KA, et al. Why primate models matter. *Am J Primatol* 2014;**76**:801–27.
7. Dey P. *Basic and advanced laboratory techniques in histopathology and cytology.* Singapore: Springer; 2018.
8. Giesen C, et al. Highly multiplexed imaging of tumor tissues with subcellular resolution by mass cytometry. *Nat Methods* 2014;**11**:417–22.
9. Pan SJ, Yang Q. A survey on transfer learning. *IEEE Trans Knowl Data Eng* 2010;**22**:1345–59.
10. Krull A, Buchholz T-O, Jug F. Noise2Void – learning denoising from single noisy images. In: *Proceedings of the IEEE/CVF conference on computer vision and pattern recognition (CVPR)*; 2019. p. 2129–37.
11. Lehtinen J, et al. Noise2Noise: learning image restoration without clean data. In: *35th international conference on machine learning, ICML 2018*; 2018.
12. De Chaumont F, et al. Icy: an open bioimage informatics platform for extended reproducible research. *Nat Methods* 2012;**9**:690–6.
13. Jones TR, Kang IH, Wheeler DB, et al. CellProfiler Analyst: data exploration and analysis software for complex image-based screens. *BMC Bioinform* 2008;**9**:482. https://doi.org/10.1186/1471-2105-9-482.
14. McQuin C, et al. CellProfiler 3.0: next-generation image processing for biology. *PLoS Biol* 2018;**16**:e2005970.
15. Schindelin J, et al. Fiji: an open-source platform for biological-image analysis. *Nat Methods* 2012;**9**:676–82.
16. Cousty J, Bertrand G, Najman L, Couprie M. Watershed cuts: minimum spanning forests and the drop of water principle. *IEEE Trans Pattern Anal Mach Intell* 2009;**31**:1362–74.
17. Berg S, Kutra D, Kroeger T, et al. Ilastik: interactive machine learning for (bio)image analysis. *Nat Methods* 2019;**16**:1226–32. https://doi.org/10.1038/s41592-019-0582-9.
18. Redmon J, Divvala S, Girshick R, Farhadi A. You only look once: unified, real-time object detection. In: *Proceedings of the IEEE computer society conference on computer vision and pattern recognition*; 2016. https://doi.org/10.1109/CVPR.2016.91.
19. Ronneberger O, Fischer P, Brox T. U-net: convolutional networks for biomedical image segmentation. In: *Lecture notes in computer science. Lecture notes in artificial intelligence and lecture notes in bioinformatics*; 2015. https://doi.org/10.1007/978-3-319-24574-4_28.
20. Christiansen EM, et al. In silico labeling: predicting fluorescent labels in unlabeled images. *Cell* 2018;**173**:792–803.e19.
21. Bosch A, Zisserman A, Munoz X. Image classification using random forests and ferns. In: *Proceedings of the IEEE international conference on computer vision*; 2007. p. 1–8. https://doi.org/10.1109/ICCV.2007.4409066.
22. Simonyan K, Zisserman A. Very deep convolutional networks for large-scale image recognition. In: *3rd International conference on learning representations, ICLR 2015—conference track proceedings*; 2015.
23. Bravais A. Analyse mathematique sur les probabilités des erreurs de situation d'un point. *Mem Acad Roy Sci Inst France Sci Math Phys* 1844;**9**:255–332.
24. Chung WJ, et al. Kras mutant genetically engineered mouse models of human cancers are genomically heterogeneous. *Proc Natl Acad Sci USA* 2017;**114**:E10947–55.

25. Walker RA. Quantification of immunohistochemistry—issues concerning methods, utility and semiquantitative assessment I. *Histopathology* 2006;**49**:406–10.
26. Bray MA, et al. Cell painting, a high-content image-based assay for morphological profiling using multiplexed fluorescent dyes. *Nat Protoc* 2016;**11**:1757–74.
27. Way G, et al. *Predicting cell health phenotypes using image-based morphology profiling*. bioRxiv; 2020. https://doi.org/10.1101/2020.07.08.193938.
28. Zeiler MD, Fergus R. Visualizing and understanding convolutional networks. In: *Lecture notes in computer science. Lecture notes in artificial intelligence and lecture notes in bioinformatics*; 2014. https://doi.org/10.1007/978-3-319-10590-1_53.
29. Kraus OZ, Ba JL, Frey BJ. Classifying and segmenting microscopy images with deep multiple instance learning. *Bioinformatics* 2016;**32**:i52–9.
30. Bankhead P, Loughrey MB, Fernández JA, et al. QuPath: open source software for digital pathology image analysis. *Sci Rep* 2017;**7**:16878. https://doi.org/10.1038/s41598-017-17204-5.
31. Kaiser D, Inciuraite G, Cichy RM. Rapid contextualization of fragmented scene information in the human visual system. *NeuroImage* 2020;**219**:117045.
32. Galon J, et al. Immunoscore and immunoprofiling in cancer: an update from the melanoma and immunotherapy bridge 2015. *J Transl Med* 2016;**14**:273. https://doi.org/10.1186/s12967-016-1029-z.
33. Althammer S, et al. Automated image analysis of NSCLC biopsies to predict response to anti-PD-L1 therapy. *J Immunother Cancer* 2019;**7**:121.
34. Pantanowitz L, et al. Twenty years of digital pathology: an overview of the road travelled, what is on the horizon, and the emergence of vendor-neutral archives. *J Pathol Inform* 2018;**9**:40. https://doi.org/10.4103/jpi.jpi_69_18.
35. Wied GL, Bartels PH, Bibbo M, Dytch HE. Image analysis in quantitative cytopathology and histopathology. *Hum Pathol* 1989;**20**:549–71.
36. Krizhevsky A, Sutskever I, Hinton GE. ImageNet classification with deep convolutional neural networks. *Commun ACM* 2017;**60**:84–90.
37. Mercan E, et al. Assessment of machine learning of breast pathology structures for automated differentiation of breast cancer and high-risk proliferative lesions. *JAMA Netw Open* 2019;**2**:e198777.
38. Kapil A, et al. Deep semi supervised generative learning for automated tumor proportion scoring on NSCLC tissue needle biopsies. *Sci Rep* 2018;**8**:17343.
39. He K, Zhang X, Ren S, Sun J. *Deep residual learning for image recognition*. arXiv 1512.03385; 2015.
40. Bulten W, et al. Automated deep-learning system for Gleason grading of prostate cancer using biopsies: a diagnostic study. *Lancet Oncol* 2020;**21**:233–41.
41. Litjens G, et al. Deep learning as a tool for increased accuracy and efficiency of histopathological diagnosis. *Sci Rep* 2016;**6**:26286.
42. Nagpal K, et al. Development and validation of a deep learning algorithm for improving Gleason scoring of prostate cancer. *npj Digit Med* 2019;**2**:48.
43. Ström P, et al. Artificial intelligence for diagnosis and grading of prostate cancer in biopsies: a population-based, diagnostic study. *Lancet Oncol* 2020;**21**:222–32.
44. Kather JN, et al. Deep learning can predict microsatellite instability directly from histology in gastrointestinal cancer. *Nat Med* 2019;**25**:1054–6.
45. Sirinukunwattana K, et al. *Image-based consensus molecular subtype classification (imCMS) of colorectal cancer using deep learning*. bioRxiv; 2019. https://doi.org/10.1101/645143.
46. Bychkov D, et al. Deep learning based tissue analysis predicts outcome in colorectal cancer. *Sci Rep* 2018;**8**:3395.
47. Mobadersany P, et al. Predicting cancer outcomes from histology and genomics using convolutional networks. *Proc Natl Acad Sci* 2018;**115**:E2970–9.

48. Courtiol P, et al. Deep learning-based classification of mesothelioma improves prediction of patient outcome. *Nat Med* 2019;**25**:1519–25.

49. Meier A, et al. Hypothesis-free deep survival learning applied to the tumour microenvironment in gastric cancer. *J Pathol Clin Res* 2020;**6**:273–82.

50. Goodfellow I, et al. Generative adversarial nets. In: *NIPS'14: proceedings of the 27th international conference on neural information processing systems*; 2014. p. 2672–80.

51. Gui J, Sun Z, Wen Y, Tao D, Ye J. *A review on generative adversarial networks: algorithms, theory, and applications*. arXiv2001.06937 [cs, stat]; 2020.

52. Kapil A, et al. *DASGAN—joint domain adaptation and segmentation for the analysis of epithelial regions in histopathology PD-L1 images*. arXiv 1906.11118; 2019.

53. Harder N, et al. Segmentation of prostate glands based on H&E or IHC counterstain with minimal manual annotation in prostate cancer. In: *IEEE 16th international symposium on biomedical imaging*; 2019.

54. Nadarajan G, et al. Automated multi-class ground-truth labeling of H&E images for deep learning using multiplexed fluorescence microscopy. In: *Medical imaging 2019: digital pathology*. International Society for Optics and Photonics; 2019. https://doi.org/10.1117/12.2512991.

55. Campanella G, et al. Clinical-grade computational pathology using weakly supervised deep learning on whole slide images. *Nat Med* 2019;**25**:1301–9.

56. Tomita N, et al. Attention-based deep neural networks for detection of cancerous and precancerous esophagus tissue on histopathological slides. *JAMA Netw Open* 2019;**2**:e1914645.

57. Bera K, Schalper KA, Rimm DL, Velcheti V, Madabhushi A. Artificial intelligence in digital pathology—new tools for diagnosis and precision oncology. *Nat Rev Clin Oncol* 2019;**16**:703–15.

58. Campolo A, Crawford K. Enchanted determinism: power without responsibility in artificial intelligence. *Engag Sci Technol Soc* 2020;**6**:1–19.

59. Tizhoosh HR, Pantanowitz L. Artificial intelligence and digital pathology: challenges and opportunities. *J Pathol Inform* 2018;**9**:38.

60. Yamamoto Y, et al. Automated acquisition of explainable knowledge from unannotated histopathology images. *Nat Commun* 2019;**10**:5642.

61. Baatz M, Schäpe A, Schmidt G, Athelogou M, Binnig G. Cognition network technology: object orientation and fractal topology in biomedical image analysis. Method and applications. In: Losa GA, Merlini D, Nonnenmacher TF, Weibel ER, editors. *Fractals in biology and medicine*. Basel: Birkhäuser; 2005. p. 67–73.

62. Zimmermann J, et al. Image analysis for tissue phenomics. In: Binnig G, Huss R, Schmidt G, editors. *Tissue phenomics: profiling cancer patients for treatment decisions*. vol. 1. Singapore: Pan Stanford; 2018. p. 9–34.

63. Baatz M, Zimmermann J, Blackmore CG. Automated analysis and detailed quantification of biomedical images using definiens cognition network technology. *Comb Chem High Throughput Screen* 2009;**12**:908–16.

64. Corredor G, et al. Spatial architecture and arrangement of tumor-infiltrating lymphocytes for predicting likelihood of recurrence in early-stage non-small cell lung cancer. *Clin Cancer Res* 2019;**25**:1526–34.

65. Barrera C, et al. Computer-extracted features relating to spatial arrangement of tumor infiltrating lymphocytes to predict response to nivolumab in non-small cell lung cancer (NSCLC). *J Clin Oncol* 2018;**36**:12115.

66. Gillies RJ, Kinahan PE, Hricak H. Radiomics: images are more than pictures, they are data. *Radiology* 2016;**278**:563–77.

67. Vaidya P, et al. CT derived radiomic score for predicting the added benefit of adjuvant chemotherapy following surgery in stage I, II resectable non-small cell lung cancer: a retrospective multicohort study for outcome prediction. *Lancet Digit Heal* 2020;**2**:e116–28.

68. Lambin P, et al. Radiomics: extracting more information from medical images using advanced feature analysis. *Eur J Cancer* 2012;**48**:441–6.

69. Koçak B, Durmaz EŞ, Ateş E, Kılıçkesmez Ö. Radiomics with artificial intelligence: a practical guide for beginners. *Diagn Interv Radiol* 2019;**25**:485–95.

70. Limkin EJ, et al. Promises and challenges for the implementation of computational medical imaging (radiomics) in oncology. *Ann Oncol* 2017;**28**:1191–206.

71. Bera K, Velcheti V, Madabhushi A. Novel quantitative imaging for predicting response to therapy: techniques and clinical applications. *Am Soc Clin Oncol Educ Book* 2018;**38**:1008–18.

72. Thawani R, et al. Radiomics and radiogenomics in lung cancer: a review for the clinician. *Lung Cancer* 2018;**115**:34–41.

73. Antonia SJ, et al. Durvalumab after chemoradiotherapy in stage III non-small-cell lung cancer. *N Engl J Med* 2017;**377**:1919–29.

74. Samstein RM, et al. Tumor mutational load predicts survival after immunotherapy across multiple cancer types. *Nat Genet* 2019;**51**:202–6.

75. Keenan TE, Burke KP, Van Allen EM. Genomic correlates of response to immune checkpoint blockade. *Nat Med* 2019;**25**:389–402.

76. Hellmann MD, et al. Tumor mutational burden and efficacy of nivolumab monotherapy and in combination with ipilimumab in small-cell lung cancer. *Cancer Cell* 2018;**33**:853–861.e4.

77. Wolchok JD, et al. Overall survival with combined Nivolumab and Ipilimumab in advanced melanoma. *N Engl J Med* 2017;**377**:1345–56.

78. Prat A, et al. Immune-related gene expression profiling after PD-1 blockade in non-small cell lung carcinoma, head and neck squamous cell carcinoma, and melanoma. *Cancer Res* 2017;**77**:3540–50.

79. Tumeh PC, et al. Liver metastasis and treatment outcome with anti-PD-1 monoclonal antibody in patients with melanoma and NSCLC. *Cancer Immunol Res* 2017;**5**:417–24.

80. Bridge JA, Lee JC, Daud A, Wells JW, Bluestone JA. Cytokines, chemokines, and other biomarkers of response for checkpoint inhibitor therapy in skin cancer. *Front Med* 2018;**5**:351. https://doi.org/10.3389/fmed.2018.00351.

81. Kickingereder P, et al. Large-scale radiomic profiling of recurrent glioblastoma identifies an imaging predictor for stratifying anti-angiogenic treatment response. *Clin Cancer Res* 2016;**22**:5765–71.

82. Antunes J, et al. Radiomics analysis on FLT-PET/MRI for characterization of early treatment response in renal cell carcinoma: a proof-of-concept study. *Transl Oncol* 2016;**9**:155–62.

83. Grossmann P, et al. Quantitative imaging biomarkers for risk stratification of patients with recurrent glioblastoma treated with bevacizumab. *Neuro-Oncology* 2017;**19**:1688–97.

84. Song J, et al. A new approach to predict progression-free survival in stage IV EGFR-mutant NSCLC patients with EGFR-TKI therapy. *Clin Cancer Res* 2018;**24**:3583–92.

85. Wang G, et al. Pretreatment MR imaging radiomics signatures for response prediction to induction chemotherapy in patients with nasopharyngeal carcinoma. *Eur J Radiol* 2018;**98**:100–6.

86. Dong D, Zhang F, Zhong LZ, et al. Development and validation of a novel MR imaging predictor of response to induction chemotherapy in locoregionally advanced nasopharyngeal cancer: a randomized controlled trial substudy (NCT01245959). *BMC Med* 2019;**17**:190. https://doi.org/10.1186/s12916-019-1422-6.

87. Braman NM, et al. Intratumoral and peritumoral radiomics for the pretreatment prediction of pathological complete response to neoadjuvant chemotherapy based on breast DCE-MRI. *Breast Cancer Res* 2017;**19**:57.

88. Partridge SC, et al. Diffusion-weighted MRI findings predict pathologic response in neoadjuvant treatment of breast cancer: the ACRIN 6698 multicenter trial. *Radiology* 2018;**289**:618–27.

89. Liu Y, Wu S, Liu Z, Chao H. A fuzzy co-clustering algorithm for biomedical data. *PLoS One* 2017;**12**:e0176536.

90. Cui Y, et al. Radiomics analysis of multiparametric MRI for prediction of pathological complete response to neoadjuvant chemoradiotherapy in locally advanced rectal cancer. *Eur Radiol* 2019;**29**:1211–20.

91. Khorrami M, et al. Predicting pathologic response to neoadjuvant chemoradiation in resectable stage III non-small cell lung cancer patients using computed tomography radiomic features. *Lung Cancer* 2019;**135**:1–9.

92. Xu Y, et al. Deep learning predicts lung cancer treatment response from serial medical imaging. *Clin Cancer Res* 2019;**25**:3266–75.

93. Fave X, et al. Delta-radiomics features for the prediction of patient outcomes in non-small cell lung cancer. *Sci Rep* 2017;**7**:588.

94. Coroller TP, et al. Radiomic phenotype features predict pathological response in non-small cell radiomic predicts pathological response lung cancer. *Radiother Oncol* 2016;**119**:480–3.

95. Khorrami M, et al. Combination of peri- and intratumoral radiomic features on baseline CT scans predicts response to chemotherapy in lung adenocarcinoma. *Radiol Artif Intell* 2019;**1**:e180012.

96. Durot C, et al. Metastatic melanoma: pretreatment contrast-enhanced CT texture parameters as predictive biomarkers of survival in patients treated with pembrolizumab. *Eur Radiol* 2019;**29**:3183–91.

97. Nardone V, et al. Radiomics predicts survival of patients with advanced non-small cell lung cancer undergoing PD-1 blockade using Nivolumab. *Oncol Lett* 2020;**19**:1559–66.

98. Tunali I, et al. Novel clinical and radiomic predictors of rapid disease progression phenotypes among lung cancer patients treated with immunotherapy: an early report. *Lung Cancer* 2019;**129**:75–9.

99. Sun R, et al. A radiomics approach to assess tumour-infiltrating CD8 cells and response to anti-PD-1 or anti-PD-L1 immunotherapy: an imaging biomarker, retrospective multicohort study. *Lancet Oncol* 2018;**19**:1180–91.

100. Trebeschi S, et al. Predicting response to cancer immunotherapy using noninvasive radiomic biomarkers. *Ann Oncol* 2019;**30**:998–1004.

101. Khorrami M, et al. Changes in CT radiomic features associated with lymphocyte distribution predict overall survival and response to immunotherapy in non-small cell lung cancer. *Cancer Immunol Res* 2020;**8**:108–19.

102. Vaidya P, et al. Novel, non-invasive imaging approach to identify patients with advanced non-small cell lung cancer at risk of hyperprogressive disease with immune checkpoint blockade. *J Immunother Cancer* 2020;**8**:e001343.

103. Fleming N. How artificial intelligence is changing drug discovery. *Nature* 2018. https://doi.org/10.1038/d41586-018-05267-x.

104. Vamathevan J, et al. Applications of machine learning in drug discovery and development. *Nat Rev Drug Discov* 2019;**18**:463–77.

105. Simm J, et al. Repurposing high-throughput image assays enables biological activity prediction for drug discovery. *Cell Chem Biol* 2018;**25**:611–618.e3.

106. Mullard A. Machine learning brings cell imaging promises into focus. *Nat Rev Drug Discov* 2019;**18**:653–5.

Clinical trials, real-world evidence, and digital medicine

Jim Weatherall[a], Faisal M. Khan[b], Mishal Patel[c], Richard Dearden[d], Khader Shameer[b], Glynn Dennis[b], Gabriela Feldberg[e], Thomas White[f], and Sajan Khosla[g]

[a]Data Science and Artificial Intelligence, R&D, AstraZeneca UK Ltd, Macclesfield, United Kingdom, [b]AI and Analytics, Data Science and Artificial Intelligence, Biopharma R&D, AstraZeneca, Gaithersburg, MD, United States, [c]Imaging and Data Analytics, Clinical Pharmacology & Safety Sciences, R&D, AstraZeneca, Cambridge, United Kingdom, [d]Digital Health, Oncology R&D, AstraZeneca UK Ltd, Cambridge, United Kingdom, [e]Digital Health, Oncology R&D, AstraZeneca UK Ltd, Durham, NC, United States, [f]AI and Analytics, Data Science & Artificial Intelligence, Biopharma R&D, AstraZeneca, Cambridge, United Kingdom, [g]Oncology Data Science, AstraZeneca, Gaithersburg, MD, United States

Introduction

The development of new medicines through the clinical trials process offers countless opportunities for the application of approaches such as those offered by artificial intelligence (AI) and machine learning (ML). Broadly speaking, the application of these techniques is usually aimed at doing one of the following:

1. Increasing the speed and therefore reducing the time to develop a medicine and get it to patients who need it.
2. Becoming more accurate at picking the winners—those potential new medicines that will go on to succeed, rather than produce negative clinical trial results.
3. Generating productivity or efficiency gains to offset the large costs of running clinical trial programs.

These are all difficult problems to solve. However, the good news is that the clinical trials process tends to generate large volumes of data, some of which are of good quality—because it is required to be according to regulatory standards. These data are vital for using AI & ML techniques to achieve gains in one of the three opportunity areas outlined above. In addition, new data streams are maturing all the time—such as patient data from sensors and wearable devices, genomic testing, and real-world evidence (RWE).

In this chapter, we discuss applications of AI and ML in the operational and scientific aspects of running clinical trials, the use of RWE throughout that process and beyond, and the rapidly evolving area of digital medicine. Digital medicine is a broad term, encompassing a variety of areas such as tele-health and digital apps. For the purposes of this chapter, we will focus mostly on digital devices (sensors and wearables) as well as mentioning digital therapeutics.

The Era of Artificial Intelligence, Machine Learning, and Data Science in the Pharmaceutical Industry. https://doi.org/10.1016/B978-0-12-820045-2.00011-8

The importance of ethical AI

It is key to recognize that implementations of AI, ML, and data science in these settings, often involve dealing with personally identifiable information and/or algorithms or models that play a role in decision making related to health. For this reason, it is critically important to take ethical considerations into account. As an example, AstraZeneca has published its five principles for data & AI ethics, outlining that it is important to be

- explainable and transparent
- fair
- accountable
- human-centric and socially beneficial
- private and secure

Each area then expands into further details, including their impact in practice. For example, being aware of the limitations of AI systems, so they can be applied in the right context, or the need to respect the privacy rights of all relevant stakeholders. These ethical considerations are important to bear in mind as context, through the rest of this chapter.[1]

Clinical trials

Clinical trials are an essential part of the pharmaceutical R&D process. They ensure that a potential new medicine or other therapeutic intervention is tested thoroughly in humans before being considered for regulatory approval and its subsequent widespread availability for patients. Trials are required both for completely new medicines (i.e., they have never been licensed before) as well as for new indications (i.e., medicines that are currently licensed, but for a different disease or patient population). The phases of clinical trials that are followed vary somewhat based on the requirements of the development pathway in question—generally, they can be classified as:

- Phase 1: testing on a small group (few 10 s) of usually healthy people, to judge the safety of the medicine, and establish an idea of what the right dose might be.
- Phase 2: testing on a larger group (10 s to 100 s) of patients with the condition, to better understand safety, as well as establish efficacy and what the optimal dose might be.
- Phase 3: more information collected about both safety and effectiveness, in a large group of patients (100 s to 1000 s). Here, the medicine is compared against another treatment or placebo to see if it is better in practice. If a regulatory authority agrees that the results of phase 3 are positive, it will approve the medicine for use.
- Phase 4: postapproval trials, intended to test for safety and effectiveness in large, diverse populations, and over longer periods of time.

In addition, observational studies—which are not technically clinical trials per se—observe people in normal settings, grouping them and comparing changes over time. There are many different designs of clinical trial, based on the scientific question that is to be answered. For instance, a noninferiority trial would be designed to establish whether one therapeutic intervention is effectively equivalent to another, whereas a superiority trial would be designed to definitively establish whether one is clearly better than another. A bioequivalence trial will show—for two different routes of administration—that the same concentration of drug is present in the relevant tissue. A thorough-QT study will be structured to assess cardiovascular safety. These are just examples—modern clinical trials are becoming increasingly varied and innovative as drug developers respond to unmet medical needs and the expectation of personalized medicines.[2, 3] Some of the limitations of clinical trials data that can effect analysis and inference are issues such as missing data due to incomplete follow up of patients, patient nonadherence to the trial protocol, and the trial population representing only a subset of the potential ultimate target population.[4, 5]

Site selection

The increasing complexity of protocols coupled with the eroding recruitment performance of many investigative sites has pushed analytics-driven feasibility and site selection to the top of the list for areas that can be most highly impacted by leveraging ML and AI techniques. Although this area has seen a steady increase in the utilization of data and analytics to deliver better outcomes, there is still a long way to go to maximize the impacts of what can be achieved by a shift toward more predictive and prescriptive tactics. To improve site selection, one often needs to combine disparate data sources into a single, consistent, and comprehensive integrated dataset, improve predictions of key performance metrics through the application of ML, develop an optimization model to select the best sites from an integrated list according to complex study criteria, and forecast recruitment and site activation variables for the selected sites.

Traditionally, within the pharmaceutical industry, disparate internal and external data sources have been used to generate multiple candidate site lists for a study, often ranked according to different performance metrics. This has led to frustration for study teams, who have had to manually merge and interpret these lists to decide which sites to select.

One solution is to integrate these data sources to provide a single, consistent dataset from which to draw candidate site lists. Data sources range from internal and external clinical trial management system (CTMS) to external real-world databases, with each providing a piece of the overall site performance puzzle. It is crucial to select a broad range of data sources to ensure that one continues to strive toward a holistic view of site performance as well as gather an understanding of the environment in which they are operating. CTMS and other external clinical operational data sources provide detailed data points on site activation, patient recruitment, and retention and some quality measures around protocol deviations and data quality. Other data sources such as real-world data (RWD) provide views into the impact of the protocol design on the availability of targeted patient pools. Depending on the RWD source, one can use the provided information to identify hot spots of patients that match study design and then overlay site performance data to identify the highest performing sites near the most appropriate patients for the study. Other data sources included provide information around an investigator's scientific expertise and level of connectivity to peers in the industry by looking at their presentations, publications, and grants in combination with their clinical trial experience. In addition, by incorporating sources that provide long-term views into planned and ongoing clinical studies and new drug market launches that should be considered when selecting sites and countries in which to conduct clinical trials. To enable a comprehensive view of sites across these various sources, a large effort has gone underway to clean and merge data records as much as possible. Using various clustering techniques to identify and merge potentially duplicated site records with only a small portion needing manual curation results in a high-quality integrated database. This integration creates a common set of robust performance measures that allows a ranked site list to be generated and filtered according to common study parameters. A data extraction and integration pipeline is run regularly to incorporate updates to the source data.

To enable study teams to quickly achieve a level of comfort with the many performance measures in the integrated dataset, site ranking is created and based on a weighting across all key performance measures. Although this ranking is initially focused on core, site-level operational performance measures, it can be enhanced by pulling in data sources to enable the selection of key external experts (KEEs) who are best suited to bringing the highest quality science to the studies. The features considered for this unique set of individuals focus on measures of influence, clinical practice, and scientific expertise and include key metrics such as number of publications, impact factor of journals, grants, and scientific presentations. One can then develop a ranking system to identify and compare KEEs by scaling and weighting the features to create a comprehensive score similar to the operational performance ranking described earlier. In addition, models may be built to estimate the influence of these KEEs on the performance of other investigators selected for the study to better understand the network effect that highly respected investigators have on their peers. Another key consideration for this work is to identify new and up and coming KEEs using unsupervised ML techniques such as hierarchical cluster analysis, especially in more competitive disease areas where well-known KEEs may not be available.

At the site level, the use of natural language processing (NLP) techniques may be used to interrogate the large amount of quality data captured in reports and monitoring notes that are not readily available to use as performance measures but are highly valuable when assessing site quality. Understanding of the sentiment in these notes and adding them to the integrated data helps alert one to any sites that merit further scrutiny before selection. Other areas of focus regarding quality include the incorporation of an internally assigned site quality score. This quality score is a composite of multiple indicators, including protocol deviations, standing with various regulatory authorities, Good Clinical Practice flags as well as feedback from onsite monitoring staff. This information is then used to create the appropriate monitoring plan for any selected sites. Quality may include a view into past clinical results data to better understand if sites are recruiting the most appropriate patients to meet the inclusion and exclusion criteria of a protocol. This can be extremely helpful especially in disease areas where there are not discrete patient test results data for making the decision on patient eligibility and trial teams and are instead relying on more qualitative measures.

Generating a ranked site list is the right first step to aid site selection, but many studies can be large requiring hundreds of active sites to deliver the necessary number of patients. Candidate lists submitted to the study team for consideration will often consist of thousands of sites across the globe for a given study or program of studies. Expecting a study team to manually select an optimal list according to complex criteria and multiple performance measures is not realistic.

Often, a mathematical optimization model is developed to efficiently recommend the optimal group of sites from the input ranked site list according to critical study criteria using a site-first, country-second approach. Individual site performance drives selection, which guides teams toward an optimal set of countries, as opposed to selecting countries first and then identifying the best sites in those countries. This represents a fundamental change to the methodology the industry has used for many years. Success requires close partnership with study teams to ensure realistic output from the optimization model through incorporating critical business constraints, such as the minimum number of sites opened per country and must include or exclude countries. The model should apply these constraints while still following a site-first approach.

An initial optimization model may leverage linear integer programming to generate a list of sites that minimizes study duration while satisfying the provided constraints. Study duration is defined as the sum of two cycle times: site activation and recruitment duration. Constraints are combined with the objective function in linearized form, and the optimization model sets binary decision variables indicating whether a site is either selected or not to arrive at an optimal solution. In addition to the identification of the selected sites, the output also contains key derived variables such as the expected number of patients recruited for each selected site. All sites from the initial ranked site list are included in the output, allowing unselected sites to be used as a backup list. For example, if a selected site decides not to participate in the study, teams can review the list of unselected sites and find a candidate with a similar rank in the same country to use as a backup site, without potentially having to rerun the optimization model excluding the site that declined to participate.

Close collaboration with study teams to gather their feedback on the optimization model feeds their expert insight into the model as study constraints, inspiring confidence in this analytics-driven approach to site selection. As research expands in the development of optimization models to consider additional objective functions, such as minimizing cost, incorporating expert knowledge through intelligent implementation of further constraints will take on added importance due to multiplying interactions between cost, speed, and quality variables. As more objectives and constraints are considered, the decision of the optimal set of sites will become even harder to ascertain without the support of data science, and building confidence in the soundness of the optimization model through the inclusion of business best practices will be critical for study team buy-in.

Site ranking and the optimization model rely on performance measures developed over the last decade, which consist of summary statistics of historical performance or, at best, a normalized performance factor that enables a cleaner comparison of sites by removing the variability inherent in different protocol designs within the same disease area. These data-derived performance metrics represented a substantial improvement over selecting sites based on anecdotal experience alone, but do not extract maximum value from these datasets when it comes to assessing site performance.

ML may be used to better predict future performance at the site level and constrain uncertainty around these predictions. There are also efforts to build similar predictive models to accurately estimate key performance metrics where these are missing in the historical data, expanding the pool of candidate sites to potentially allow for better sites to be selected.

The optimal site list generated by the optimization model, along with improved predictions of performance for these sites using ML, is fed into statistical models developed to forecast site recruitment, with consideration to each country's recruiting limits. Such models are comprised of a thorough literature review of existing work, in addition to internal developments from a team of data scientists. These models are able to predict different scenarios of recruitment by (a) changing different constraints obtained from optimization and (b) fitting statistical distributions of key site metrics (randomization rate and site start-up time) to simulate patient recruitment at baseline and during the course of a trial. The models are then further adjusted to consider a trial's therapeutic area and each site's geographical region.

As the industry continues to leverage data science to change the way that sites for clinical trials are defined, identified, and ultimately selected, it will be imperative for data scientists to continue to partner with clinical study teams. This partnership enables the data scientist to understand the nuances of the underlying data and operational practices and gather expert insight into which features are considered most relevant to performance, which will improve the predictive models enabling them to provide the most meaningful impact and business value.

Recruitment modeling for clinical trials

Predicting the recruitment of participants onto clinical trials is critical for trial design and planning, for monitoring ongoing trials to ensure they are collecting the data they need, and for making changes to trials when they deviate from the original plan. Typically, a trial will need an initial prediction, often referred to as a *baseline*, of its expected recruitment rate before the trial starts, and then an updated prediction during the recruitment period of the trial to check if the original plan is being followed. The baseline prediction may be used to evaluate the feasibility of a trial, determine how many and which hospitals or clinics (referred to as sites or *centers*) to use in the trial, and to plan the drug supply for the trial. After the trial has begun, the updated prediction (we will refer to this here as the *ongoing prediction*) will be compared against the baseline to check whether the trial is going as originally planned. It may also be used to refine the drug supply plan and to decide what to do (e.g., to add new hospitals, pay for local recruiting efforts, or change the trial protocol to increase the number of potential recruits) if the trial deviates significantly from the plan.

Because large drug trials are typically performed in multiple countries, with multiple centers per country, prediction of recruitment at all these levels is often needed. Center-level predictions are used to select the hospitals to use in a trial and to plan shipments of drugs to each center. Country-level or region-level predictions are needed for international drug shipment planning (typically, there is a single depot per country and shipments to the depot and then from the depot to the centers in that country must all be planned) and to ensure requirements from country regulators are satisfied. Study-level predictions are needed to determine the amount of time required for recruitment, to monitor overall performance, and to plan the quantity of drug that must be manufactured for the trial. To meet all these requirements, it is sensible to model the recruitment rate at the level of the individual centers and then aggregate the data to generate predictions at the country, region, or study level. However, this brings a data scarcity challenge: in an analysis of 100 recent late stage trials, 85% of centers were used in only one trial and just over 5% of centers were used in five or more trials. In contrast, less than 15% of countries were used in five or fewer trials and 22% of countries were used in more than 30 of the 100 trials.

Among other factors, recruitment rates at centers depend on the size and catchment area of the hospital or clinic, so country-level data for recruitment rates cannot reliably be used to predict center-level recruitment, but in the absence of sufficient center-level data—particularly for centers that have not been used previously—they may be the best available estimator.

Recruitment rates on trials depend on many factors other than the centers that are used. Most obviously, they depend on the disease, or more accurately the patient segment, the trial is studying. However, the recruitment rate also depends on what other trials are going on in the same centers that may be recruiting from the same pool of participants, whether the drug being trialed has already been approved for other diseases, the drug's side effects and other aspects of the trial that make it hard for participants, etc. This again reduces the quantity of useful, relevant data to train recruitment models. The lack of data makes the use of relatively simple statistical models to predict recruitment common.

Recruitment start dates

Centers do not start recruiting at the same time on trials, so models of recruitment need to consider both the start-up time for each center and the rate of recruitment after centers have started. Center start-up depends on, among other things, the administrative and regulatory processes of the center and the country, the earliest time at which the required drugs can be delivered to the center, and the capacity of the organization performing the trial.

Center start-up dates, or *ready-to-enroll* dates, are often assumed to be known for trials that have already started, as they are often provided by the centers themselves, and we will make that assumption here. However, these are not available when studies are in early planning stages. In addition, analysis of actual first enrollment dates for centers shows they typically occur significantly later than ready-to-enroll dates. Predictions could be significantly improved if a reliable predictive model of center start-up was developed, particularly one that could be updated as actual start-up rates were observed.

The Poisson Gamma model of trial recruitment

Because of the sparsity of data for training models of recruitment rates, most approaches use relatively simple statistical models with few parameters. The most widely used and cited recruitment model for multicenter trials, first introduced by Anisimov and Fedorov,[6] is the *Poisson Gamma model*. In this, participants are recruited at centers according to independent Poisson processes with time-constant rates. Consider a trial recruiting n participants at C centers. Each center c is assumed to have a constant rate of recruitment λ_c and starts recruiting at time t_c. For any time t after t_c:

$$n_c(t) \sim Pois(\lambda_c, t - t_c)$$

where $n_c(t)$ is the total number of participants recruited by time t at center c. The overall recruitment rate of the trial is a nonhomogeneous Poisson process because of the different ready-to-enroll dates of the centers. In the Poisson Gamma model, we assume that the recruitment rates for each center are sampled from a Gamma distribution:

$$\lambda_c \sim Gamma(\alpha, \beta)$$

where $Gamma(\alpha, \beta)$ is Gamma-distributed with probability density function:

$$p(x; \alpha, \beta) = \frac{e^{-\beta x} \beta^\alpha x^{\alpha-1}}{\Gamma(\alpha)}$$

To predict recruitment rates for a trial given a set of historical data of recruitment rates at centers in multiple trials (and assuming that the center ready-to-enroll dates are given), fit a Gamma distribution to the historical data for each center c, giving prior parameters α_c and β_c, and then sample a recruitment rate λ_c from the Gamma distribution. Monte Carlo sampling from the Poisson process then generates a recruitment prediction for that center, and these can be summed over all centers in a country or the whole trial to generate a country-level or trial-level prediction. Repeating this process can be used to estimate the uncertainty in the prediction.

Given data from a trial in which center c has already been recruiting for time t_c and has recruited k_c participants, Anisimov and Fedorov show that these prior parameters can be updated in a Bayesian approach to give a posterior $Gamma(\alpha_c + k_c, \beta_c + t_c)$, thus allowing predictions of ongoing trials to learn from the data seen so far in the trial.

Nonhomogeneous recruitment rates

The assumption made in the model described above is that recruitment rates are constant throughout the recruitment period of each center (as the centers start at different times, the *overall* trial recruitment rate will not be constant). This assumption is often poor in trials due to several factors. First, there may be other trials that are competing for the same pool of participants, and recruitment rates can change significantly when these start or finish recruiting. Publicity about the drug can also change recruitment, as can the actions of the study team, for example, by advertizing for participants. The most significant effect is often that there is an initial high recruitment rate at a center as participants who have been waiting for a trial for their condition join the trial but then the rate slows as the only new recruits are people newly diagnosed.

A number of approaches have been proposed for modeling nonhomogeneous recruitment rates at individual sites. Tang et al.[7] have used an inhomogeneous Poisson model similar to that described above but where the intensity parameter λ_t is allowed to change piecewise linearly over time, with either fixed change points or by estimating them from the data. Zhang and Long[8] and Deng et al.[9] have used cubic B-splines as a nonparametric model of recruitment rate, in the first article with the assumption that overall rates are continuous and smooth, and later allowing the overall rate to be discontinuous, although still requiring individual centers to be smooth. Other approaches have been discussed, including Gaussian processes and ensembles of different curves. The challenge with all these approaches is the much greater data requirements to estimate models with larger numbers of parameters, a problem which is even worse for nonparametric approaches. As a result, they are often restricted in application to modeling at the level of countries or whole trials.

A typical recruitment curve for a whole trial is S-shaped, with a slow initial recruitment rate as the number of centers recruiting is relatively small, a faster rate in the middle of the recruiting period when most centers are active, and then a slowing of recruitment at the end. This slowing at the end is due in part to the reduction in recruitment as the initial pool of potential participants is exhausted, but the largest effect is due to centers stopping recruitment as they, their country or the trial reach their target number of participants. To accurately predict when recruitment will finish and produce a realistic overall curve, this *shutdown* process needs to be modeled, which requires that at least the study target is known at modeling time.

One approach to modeling center shutdown is to explicitly model the rules used to determine when a center stops recruiting. The difficulty with this approach is that the rules are often poorly defined and may even be different in different countries or regions. Studies may have recruitment targets for all of centers, countries, regions (e.g., Europe), and overall, and these targets may not all be followed. For example, some countries that have reached their target may keep all their centers recruiting to compensate for another country that is behind its target, whereas in other countries, the targets cannot be shared in this way. In theory, it may be possible to infer the rules that are actually used from data, but this is complicated by the fact that decisions about whether to continue recruiting or shut a site down may not be centrally recorded. The alternative is expert elicitation of the rules, but in our experience, this is rarely consistent from one expert to another.

A possible alternative approach to modeling center shutdown is to simply use a nonhomogeneous recruitment model and allow the model to fit all the data, including data from after recruitment ends at each center. The challenge with this approach is that recruitment targets at a center will be different from trial to trial, so historical trial data will require some kind of scaling based on the target to align the datasets. We are not aware of any work in the literature on this approach.

Applications of recruitment modeling in the clinical supply chain

An efficient clinical supply chain relies crucially on having good estimates of patient numbers at each center on a trial. Manufacturing, purchasing of competitor's comparator drugs where needed, packing and labeling, international shipping, and shipping from the country depot to individual centers all need to be planned to ensure there is always drug available for every participant in a trial, and many of these steps are time-consuming so often decisions need to be taken based on expectations of future recruitment rather than just on patient already enrolled in the trial. For this reason, accurate recruitment predictions, at least at the country level and ideally at center level, need to be made.

Without good predictions, there is a significant risk of delays to trials as recruitment at a center is frozen due to lack of drug, out-of-stock situations where no drug is available for a participant, or of significant wastage of drug if it exceeds its shelf-life. Clinical trial supply chains tend to err on the side of wastage rather than out-of-stock situations as the latter can invalidate the data from that participant. A recent survey of 200 clinical trials found that 62% of the material packaged was never used.[10]

Although recruitment prediction is a significant part of supply chain planning, it is by no means the only predictive model that is needed. Retention of participants on a trial also has a significant impact on supply costs, and models may also be needed of the likely dose participants will receive. If a trial includes *titrations*, where the dosage is varied from one visit to the next to discover what is effective or can be tolerated by a participant, this may be a complex model with priors based on distributions of body weights, for example, and stochastic transitions between doses.

The standard approach to supply chain planning is to treat it as an optimization problem and cast it as a linear program, where the objective is to choose shipment sizes and dates, plus resupply strategies for individual centers so as to minimize the cost of supply given that all the demand is met, and any other constraints are not exceeded. Although this is a sensible approach given the complexity of the optimization problem, it requires a good estimate of demand and cannot explicitly consider uncertainty, which is important in the clinical supply chain because of the high cost of out-of-stock situations. What is needed is an upper bound or high quantile (for the purposes of this description we will assume the 99th percentile) of the number of participants to enable a demand for each center to be calculated that minimizes out-of-stock situations. Monte Carlo simulation of the recruitment model as described in the previous section can be used to estimate demand for each center.

The last link in the supply chain—from the depot that supplies each country to the centers—is typically fast, so demand at individual centers can be aggregated to calculate demand at depots. However, because the probability that *all* the centers in a country will recruit at the 99th percentile of their expected recruitment is much lower than 1%, this will result in a significant overestimate of the depot demand. A better approach is to use the 99th percentile of the aggregated recruitment predictions at the country level to estimate demand at the depot, but to optimize the resupply strategy from the depot to the centers with the center-level predictions. The same approach can be used at the trial level to determine the volume of drug that must be manufactured or procured.

As we said above, after the demand has been calculated, we can cast the supply chain optimization problem as a linear program and solve to compute the shipment sizes and dates, the depot-to-center resupply strategy, and the manufacturing and procurement volumes—we refer to this as the *supply plan*. To evaluate the overall likelihood of an out-of-stock situation or other supply problem, the supply plan needs to be simulated again, starting with the individual center recruitment prediction model to compute the final likelihood of an out-of-stock or other undesirable situations.

Clinical event adjudication and classification

Clinical event classification (CEC), both from a proactive risk assessment as well as post hoc diagnostic perspective, is a critical function for all parties interested in the delivery of high-quality healthcare and improvement of the human condition. CEC is typically accomplished by skilled clinicians with

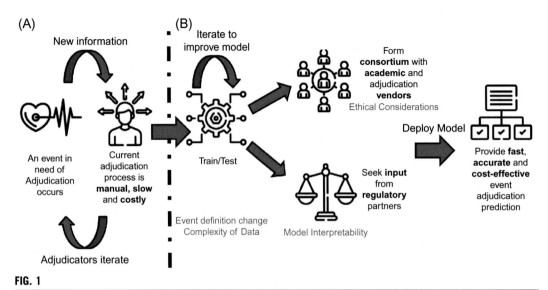

FIG. 1

Digital transformation of a clinical event classification process using AI-based approach: (A) current, manual approach; (B) digital transformation approach enabled using seamless data integration and automated algorithms.

years of training examining patient medical records to investigate whether established criteria for a particular adverse event or medical condition are met—for example, integrating multiple diverse pieces of information to adjudicate whether the patient has experienced a major adverse cardiovascular event. ML methods can be applied to CEC at a number of levels, including assessing data quality control (QC), using RWE data from patient registries and electronic medical record (EMR) data to establish and inform patient trajectories, extracting semantic information from clinical notes, and performing classification of an event or clinical review triage based on diverse data sources (see Fig. 1).

Application of ML to CEC as a field is still in its infancy. Some clinical diagnoses, such as image-based classification methods for melanoma diagnostics or diabetic retinopathy, are further along the path toward clinical care. Promising approaches to the automation of EMR interpretation and diagnosis based on longitudinal health data do exist but more development needs to occur before they reach maturity.

Digital transformation of a complex process like CEC has its own challenges. Here, we discuss two major technological issues (model transparency and bias) associated with design, development, and implementation of an AI-based approach for clinical decision making. From both a regulatory perspective and a provider, payer, or sponsor's risk tolerance, complex black-box models are not acceptable when it comes to CEC. Model transparency is typically addressed in the ML research community through assessment of model interpretability. A model is interpretable if the set of features that resulted in a particular classification can be identified, leading to a "transparent" model. "Opaque" models, such as neural networks, present difficulty when it comes to attributing the set of features that lead to a classification. One recent approach involved training one neural network to interrogate another, leading to an "attribution" approach that is able to identify specific features in sequence that result in a classification. Some models, such as rule-based Expert Systems, are completely transparent, and while lacking the capabilities of neural networks when it comes to complex data, these approaches do provide interpretability for regulatory and legal purposes. In summary, there is a critical requirement to be able to "step into" the model and identify what features result in classification of an event. ML algorithms that demonstrate bias are unable to perform accurately under certain conditions, leading to undesirable and sometimes dangerous outcomes. Bias in ML algorithms can result from imbalanced training data, changes in practice that result in training data that is no longer relevant, poisoning attacks where an adversary seeks to intentionally alter model performance, or systemic bias captured in training data that does not necessarily reflect the desired outcome of the model. Training data must be carefully selected and curated to prevent bias. In addition, model performance can periodically be interrogated for bias, although this will not necessarily catch underrepresented or edge cases.

Identifying predictors of treatment response using clinical trial data

In the course of analyzing the results of a randomized controlled trial (RCT), various questions are often posed. Which factors predicted efficacy? Are there subgroups that behave differently? Can responders vs nonresponders be identified? What would the impact of changes in trial design (inclusion/exclusion) criteria be? And many more.

The relationships between different predictors and efficacy can be complicated. An RCT can often have few data points which capture that relationship (most trials have fewer than 1000 enrolled subjects). Furthermore, there is no guarantee that the "true" predictors were measured. It may be difficult to distinguish between prognostic relationships (which predict outcome regardless of treatment) and predictive relationships (which predict outcome with significant interaction with treatment assignment). Prognostic relationships are a dominant signal and by definition exist in all arms of a trial.

Understanding and untangling these analytical factors is a nontrivial endeavor. There are many well-known issues when interpreting subgroups (even if they are prespecified). These include a high risk of false positives, low power of interaction tests, biased estimates in selected subgroups, and more. A nonideal approach would be an unorganized "data dredging" exercise, determined ad hoc and resulting in unknown performance, low power, and complicating the detection of multiple factors interacting together.[11,12]

There are developments to appropriately structure these analyses and consider them as a systemic and special case of model selection. Most of these approaches rely on tree-based algorithms such as random forests (RFs), as those are inherently designed to partition the population. One core approach relies on an analytical structure called "Virtual Twins." The key idea is to simulate how patients on one arm of a trial would perform on a different arm (as if a virtual twin had been enrolled). Comparing and evaluating the differences in outcomes for a patient and its "twin" across different arms can then lead to deeper understanding of predictive and prognostic factors, as well as relevant subgroups.[13–15]

The virtual twins approach can be described using a simple two-arm RCT example, a treatment arm A and a control arm B. Although a variety of ML algorithms may be leveraged, for the purposes of this illustrative example, let's assume a RF is being used. Two RF models are built separately, one per each arm. Model RFA is built using the cohort in treatment arm A and their outcomes, and model RFB is built using the cohort in treatment arm B and their outcomes. Both models are then used to predict outcomes for the full RCT population from both arms. Suppose Outcome RFA is the result of applying model RFA to both arms, and Outcome RFB is the result of applying model RFB to both arms. Then, Outcome Difference would be the difference of Outcome RFA—Outcome RFB. A third RF model, Model 3 can then be built using the full (A + B) trial population and Outcome Difference as the supervised target.

In essence, Model 3 is a multivariate model of the difference in outcomes for a patient depending on which treatment arm they were on. This can then lead to an understanding of prognostic and predictive factors, along with the definition of particular subgroups.

The virtual twins approach fits a separate conventional model per trial arm to predict outcome, and a final model to predict the difference in those estimates. An alternative approach is SIDES, which fits a single model where the splitting criterion for a decision tree is chosen based on the significance of the difference in outcomes by trial arms. There are other approaches to consider as well. An article compared 13 different approaches in their ability to distinguish prognostic and predictive relationships using both real RCT and synthetic trial data. Their evaluation identified no clearly superior method in all scenarios, echoing the conclusion of many previous comparisons of this type.[16] In general, it is important to understand the strengths and weaknesses of the various approaches and explore the usage of a few different ones given the particular context and situation of the RCT at hand.

These structured approaches to explore RCT subgroups and understand trial factors are invaluable for a variety of purposes. However, this is an emerging field of research, and practitioners are advised to tread carefully. Beyond the well-known risks of post hoc RCT analyses (multiple testing, data fishing), regulatory agencies and payers do not yet formally recognize many of these approaches. Some of these risks may be mitigated through proper planning and predeclaration of the intent to conduct structured exploration in advance.

Structured exploration represents both the best and risk of ML in analyzing RCTs. It can be leveraged to answer crucial questions from complicated data in a challenging setting. This area of research has emerged as a true collaboration between ML experts, data scientists, statisticians, and other analytical experts, with participation from academia, industry, and regulatory bodies. Although there is great promise, it is however a new area of research attempting to solve longstanding and complex problems and requires careful execution, extensive study, and further research.

Real-world data: Challenges and applications in drug development

The application of ML and AI in the space of RWD is growing in maturity and scale. To explain the approaches, this section provides an overview of how the pharmaceutical industry generates RWE, its purposes, and how ML is being used to advance the field and application.[17]

The healthcare environment is constantly evolving and the adoption of new medicines into the marketplace is increasingly reliant on ever-more sophisticated evidence-based criteria. For a long time, RCTs have been considered the gold standard for generating clinical data on efficacy and safety, to inform product registration, and subsequent prescribing. However, because of their inherent limitations and the characteristics of new and innovative medicines, it is not always possible to obtain all relevant data through an RCT methodology. Moreover, advances in technology, increased investment in digital solutions, and the creation of more powerful sophisticated analytics are all contributing to more pragmatic approaches to collect and use RWD in drug discovery and development. In the framework for its RWE program, the US Food and Drug Administration (FDA) draws a distinction between RWD and RWE. RWD is defined as data relating to patient health status or the delivery of healthcare routinely collected from a variety of sources. RWE is defined as clinical evidence about the usage and potential benefits or risks of a medical product derived from analyses of real-world data.[18] The framework makes this distinction because evaluation of RWE will need to take into consideration both the methodologies used to generate evidence and the reliability and relevance of the RWD used in analyses. One of the key objectives of RWE is to understand observations and events in patients in routine clinical practice. Moreover, RWE use spans clinical decision making, regulatory understanding and decision making, and payer decision making for reimbursement. Within the pharmaceutical industry, RWE is ubiquitously used to support a variety of internal decisions on how to progress pipelines ensuring drugs are developed for those with the most unmet clinical need, evaluating market potential for new products and much more.

Many collaborations in the space of innovative analytics include and not limited to the collaborations observed by the likes of Google DeepMind, UK Health Data Research—Digital Innovation Hubs, which are examples of private/public partnerships to evaluate the use of ML and AI in health.[19–21]

The RWD landscape

RWD is generated in the course of normal clinical practice or administrative claims processing or may be reported directly by patients. Examples include data from: patient charts, laboratory reports, prescription refills, patient registries, patients treated on- and off-label, patients treated through expanded access, pragmatic clinical trials, surveys, and mobile health devices, as well as other data from existing secondary sources used to support decisions concerning safety, quality, care coordination, coverage, and reimbursement.[22]

The current RWD landscape is characterized by enormous variety and complexity extending beyond traditional sources such as chart reviews, prescription, or claims data to include both structured and unstructured data from a variety of heterogeneous sources. This has been largely facilitated by growth in the adoption of electronic health records (EHRs) and the proliferation of consumer digital technologies, including mobile devices, wearables, sensors, adherence tools, online patient networks.[23]

EHRs have become a pervasive healthcare information technology for storing data associated with each individual's health journey (including demographic information, diagnoses, medications, laboratory tests and results, medical images, clinical notes, and more).[24,25] Although the primary use of EHR was to improve the efficiency and ease of access of health systems, it has found a lot of applications in clinical informatics and epidemiology.[25,26] In particular, EHRs have been used for medical concept extraction,[27,28] disease and patient clustering,[29,30] patient trajectory modeling,[31] disease prediction,[32,33] and data-driven clinical decision support[34,35] to name a few.

The generation of evidence from these RWD relies upon robust methods of: designing the study objectives, defining the patient cohort for establishing the evidence, choosing the correct methods for testing hypotheses, and negating any internal or external bias coming from the secondary use of RWD.[36] Many of these elements of good RWE generation practice are covered in the basics of epidemiology and some of these are emerging fields of innovation, especially as the sophistication of methods increases with more complex use cases of RWE becoming clear.

Barriers for adoption of RWD for clinical research

Although RWD, and specifically EHR, offer significant opportunities for improving healthcare research, innovation, and decision making, as with any rapidly evolving field there are challenges to leveraging its full potential. These challenges range from technical (e.g., claims data are collected for administrative billing purposes but are envisioned now to support drug development decisions), to ethical (a risk to privacy as health information and patient datasets are aggregated), and to analytical (selecting which RWD are appropriate to informing a particular decision and reducing the risk of bias). The growing emphasis on the use of RWD is expected to improve our understanding of these datasets as well as fill existing knowledge gaps.

Data quality

Because most RWD are not collected for research purposes, the data collection is episodic, reactive, and at best offers a partial picture. As a result, RWD are in general messy and sparse and require statistically rigorous and valid methods to clean the data and correct inconsistencies. Careful data curation, using both structured and unstructured data, is especially important for precision therapeutics, where often crucial information related to molecular biomarkers or endpoints data can be missing. Moreover, these data are also subject to selection bias, as cohort selection and treatment decisions in clinical practice are not random.[37]

Interoperability

Owing to the rapid evolution of RWD, the standards for the development and maintenance of data assets are lagging behind. A lack of interoperability between real-world databases creates difficulties for combinatorial analysis and collaboration between data holders. Even within individual organizations, there is often a lack of consolidated or centralized data storage, leading to difficulties in analyzing data across different datasets. In general, there is a need to implement standardization and maintain robust quality assurance/QC practices to support data robustness. Organizations and collaboratives such as OHDSI—Observational Health Data Sciences and Informatics (https://www.ohdsi.org/) and Health Level 7 (http://www.hl7.org) are creating standards for electronic health data and promoting interoperability among systems. In the future, advances in data standardization, interoperability, and linkage techniques are anticipated to further enable disparate data sources to converge into a single platform for more seamless and efficient analytics.

Even though the adoption and use of EHRs has grown significantly, extracting meaningful data from EHRs in an accurate and efficient manner remains challenging. This is because a significant portion of high-value clinical information in EHRs is often stored in unstructured, free-text clinical documents that are inaccessible to algorithms and requires layers of preprocessing. NLP methods provide one approach for extraction and conversion of unstructured information from clinical text data to structured observations; this can become powerful when used in conjunction with human expertise such as Flatiron's technology-enabled abstraction.[38]

Use of RWE/RWD in clinical drug development and research

Historically, there has been significant use of RWD and RWE by the pharmaceutical industry during the peri-launch period just before and just after the marketing approval of a drug.[36] It is used to describe patient populations as well as contribute to understanding and knowledge of patient safety as well as derive judgments on the comparative effectiveness between drugs.[36]

However, with the proliferation and availability of richer and deeper data, early stages of the clinical drug development pipeline are now starting to use RWD to support ever more critical pipeline decisions. RWD-based simulations are now accepted as a reasonable way to inform clinical study design, modeling the impact of different study eligibility criteria, the timing of endpoint assessments, and study timelines in the FDA.[18] RWD can also inform study site selection as well as identify patients potentially eligible for trials, making it possible to design clinical trials faster and in a more "realistic" (how patients present themselves in a true medical setting) manner that includes creating a feasible set of eligibility criteria.[39] Simulation of study control arms, potentially replacing the need to randomize a patient to a control arm in some scenarios,[40,41] has the potential to reduce the size, duration, and cost of clinical trials and could also make a particular difference in cases of rare disease where patient recruitment is especially challenging.

The use of health data for secondary purposes, that is, beyond the care of the individual, for the delivery of high quality of care delivery, is complex from legal and ethical perspectives.[42,43] One key element to this is the mechanism of consent, something that federated networks have the ability to support. By taking models that are run centrally and then running them in a federated approach on local branches allows for the models to push consent notifications. Should patients consent, their data becomes usable or they are contacted for clinical trial screening.

The advanced applications that can build complex models and then push them to federated data sources require more than links to the federated datasets, but also common data models to ensure data elements are ready for model application. Common data models, such as the Observational Medical Outcomes Partnership common data model, create a structure for a majority of routinely collected health data and have also driven a series of tools and data science capabilities that use the common data model.[44-46]

The early analyses of EHR relied on simpler and more traditional statistical techniques.[47] More recently, however, statistical ML techniques, such as logistic regression,[48] support vector machines,[49] Cox proportional hazard model,[50,51] and RF,[52] have been successfully used for mining reliable predictive patterns in EHR data. Although the simplicity and interpretability of such approaches are desirable for medical applications, they have a number of limitations, depending on the method: for instance, dealing with high-dimensional input, reliance on strong assumptions, both statistical and structural, and their need for handcrafted features/markers, guided by a domain expertise.[53-55] A health history is almost always complex, a rich patina of life experiences, symptoms, encounters, and treatments. The number of potential predictor variables in the EHR may easily number in the thousands, particularly if free-text notes from doctors, nurses, and other providers are included.

One needs to analyze each individual's entire medical history,[54] using modeling techniques that can discover and take into account complex nonlinear interactions among variables.[56-58]

Deep learning and artificial neural networks offer a way to cut this Gordian knot, allowing us to analyze irreducibly complex data where we have little idea of the important variables or the underlying model. Deep learning techniques have achieved great success in many domains through deep hierarchical feature construction and capturing long-range dependencies in data in an effective manner. These systems are known for their ability to handle large volumes of relatively messy data, including errors in labels and large numbers of input variables. A key advantage is that investigators do not generally need to specify which potential predictor variables to consider and in what combinations; instead, neural networks are able to learn representations of the key factors and interactions from the data itself.[54] Given the rise in popularity of deep learning approaches and the increasingly vast amount of patient data, there has also been an increase in the number of publications applying deep learning to EHR data for clinical informatics tasks,[57,59-65] which yield better performance than traditional methods and require less time-consuming preprocessing and feature engineering. A comparative review put together by Ayala Solares et al. details the main approaches for deep learning on EHRs,[66] which has been detailed in Table 1.

Table 1 Some of the main approaches for deep learning on EHRs.[66]

Model	Architecture	Learning process	# Patients considered	Predictors considered	Outcome	Performance metrics
eNRBM[67]	AE	Modular	7578	Diagnoses (ICD-10), procedures (ACHI), Elixhauser comorbidities, diagnosis related groups, emergency attendances and admissions, demographic variables (ages in 10-year intervals and gender)	Suicide risk prediction	F-score recall precision
Deep patient[59]	SDA	Modular	704,587	Demographic variables (i.e., age, gender, and race), diagnoses (ICD-9), medications, procedures, lab tests, free-text clinical notes	Future disease prediction	AUROC accuracy F-score precision
Deepr[55]	CNN	End-to-end	300,000	Diagnoses (ACS), procedures (ACHI)	Unplanned readmission prediction	AUROC
DeepCare[53,66]	RNN	End-to-end	7191	Diagnoses (ICD-10), procedures (ACHI), medications (ATC)	Disease progression, unplanned readmission prediction	Precision F-score recall
Med2Vec[64]	FFNN	End-to-end	ND	Diagnoses (ICD-9), procedures (CPT), medications (NDC)	Medical codes in previous/future visits	AUROC recall
Doctor AI[64]	RNN	End-to-end	263,706	Diagnoses (ICD-9), procedures (CPT), medications (GPI)	Medical codes in future visits, duration until next visit	Recall R2
RETAIN[68]	RNN	End-to-end	32,787	Diagnoses (ICD-9), procedures (CPT), medications (GPI)	Heart failure prediction	AUROC
RetainVis[69]	RNN	End-to-end	63,030 (heart failure) 117,612 (cataract)	Diagnoses (KCD-9), procedures, medications	Heart failure prediction, cataract prediction	AUROC AUPRC
Ensemble model[54]	RNN/FFNN	End-to-end	216,221	Demographics, provider orders, diagnoses, procedures, medications, laboratory values, vital signs, flowsheet data, free-text medical notes	Inpatient mortality, 30-day unplanned readmission, long length of stay, diagnoses	AUROC

Concluding thoughts on RWD

Advanced analytics using ML on longitudinal RWD has the potential to inform and reframe clinical drug development and clinical trial design strategy, through patient stratification into subgroups based on disease subtypes, drug treatment efficacy, progress, side effects, and toxicity profiles. Currently, the type of patient data collected in routine clinical practice is growing ever rapidly ranging from simple demographics to genomics and imaging. Moreover, improved means of capturing, storing, and analyzing longitudinal RWD on patients allows for greater insights to be extracted to aid drug development. As ML algorithms and frameworks continuously advance, there will be improvements in the ability of these models to learn continuously as new information emerges either in the form of additional data sources or updated treatment guidelines.

Sensors and wearable devices

Wearable technologies span a broad spectrum of sensor-containing devices that collect activity, biological, environmental, or behavioral information. One of the key objectives for wearable device continuous monitoring is to augment medical decision making by detecting symptoms and adverse events in near real time. Today, this objective is made possible by synergistic technologies, including the Internet of Things (IoT) framework, cellular and WiFi connectivity, cloud and machine, which collectively make it possible for devices to continuously stream data to a centralized location from nearly anywhere in the world.

Continuous monitoring of vital signs using wearable technology is not as new as recent hype around Digital Health might suggest. c.1968, as part of the NASA Apollo program, thousands of hours of biotelemetry data were continuously transmitted from outer space and monitored on Earth by NASA physicians.[70] Only when astronauts were in lunar orbit on the far side of the moon was there an interruption in the steady transmission of vital sign data. Medical monitoring during these operations permitted real-time adjustments in activity timelines formulated before flight as such alterations were needed. Fast-forward 50 years and this same type of medical monitoring and real-time adjustment embodies the foundation of today's digital health revolution. Much like the NASA bioinstrumentation system of the 1960s, modern remote health monitoring systems similarly consist of body-worn sensors that acquire data from a patient, transmit data to the cloud continuously or in micro-batches, and process it using data analysis techniques ranging from business rules to deep neural networks. In clinical practice, real-time monitoring of patients combined with automated recognition of a specific health condition could greatly improve patient survival by significantly reducing or altogether eliminating the need for human intervention. For example, the cardiac implantable pacemakers and defibrillators detect arrhythmias and respond with life-saving interventions without ever consulting human practitioners.[71]

The electrocardiograph (ECG) is the most commonly performed cardiovascular diagnostic procedure, with > 100 million ECGs obtained annually in the United States,[72] including use in 21% of annual health examinations and 17% of emergency department visits. Although computerized interpretation algorithms for ECGs have existed for decades,[73] they have been constrained in that they aim to replicate the rules-based approach to ECG analysis used by human readers. Although the 12-lead ECG trace contains a large amount of information produced from cardiac electrical and structural variations, the standard approach aims to detect the presence or absence of disease by evaluating fairly simple criteria on only a small subset of the total information contained in the

ECG.[74] The application of ML ECG analysis could potentially perform a wide range of novel ECG-based tasks, including improving accuracy, estimating quantitative cardiac traits, performing longitudinal tracking of serial ECGs, and monitoring disease progression and risk. Using a combination of ML methods, including convolutional neural networks and hidden Markov models, studies have shown an ability to perform detailed longitudinal tracking and personalized ECG vector profiles to estimate continuous measures of cardiac structure and function such as left ventricular mass and mitral annular e′ velocity.[75]

Several studies have reported the single-lead ECG data from wearable devices is comparable to 12-lead ECG data and appropriate for measuring pathological rhythms.[76–78] Modern ML applications to real-time streaming ECG data range from adaptive filtering and least mean squares for noise reduction.[79] Another interesting application uses single-lead ECG recordings to detect obstructive sleep apnea with 90% classification accuracy, thereby eliminating the need for a full sleep study.[80] The authors consider a set of features either extracted directly from heart rate variability RR-tachogram or from the surrogate ECG-derived respiration signal to combine for a total of 111 features.

The importance of remote cardiac function monitoring is reflected in the FDA's recent approval of Verily's Study Watch, classifying it as a Class II medical device for its on-demand ECG feature, which records, stores, transfers, and displays single-channel ECG rhythms.[81] The FDA approval of Verily's Study Watch adds to the growing market of cardiovascular devices, such as AliveCor's KardiaBand that was approved by the FDA in 2017. KardiaBand was the first medical-grade accessory for Apple Watch to quickly detect normal sinus heart rhythms and atrial fibrillation via ECG measurements. With the KardiaBand approval, AliveCor also introduced SmartRhythm, a new feature that combines Apple Watch's heart rate and physical activity sensors to continuously evaluate the correlation between heart activity and physical activity. When the feature detects an abnormal heart beat, the device notifies users to capture ECG data.

Similar to markers of cardiac health, the organization Healthy People 2020 has recognized physical activity as one of the leading health indicators for a nation's population.[82] Tri-axial accelerometers provide a low-power and high-fidelity measurement of force along the x, y, and z directions, and thus, provide a view into the physical movement of the person wearing the device. Although many commercial and medical grade activity trackers adopt wrist-mountable form because of the ease of accessibility, no consensus has been made regarding the positioning of sensors and their data acquisition details. Many open problems exist because of the sheer volume of reasonable configurations available. Although many device vendors provide only summarized activity data, the raw, high-frequency accelerometer data serve as a common denominator across all activity monitors and thereby provide an opportunity to develop standardized device configurations for transmitting data. The exploitation of such high-frequency, real-time telemetry data is an ideal task for ML. In particular, deep learning is renowned for its autonomous feature learning through neural networks, which accepts a large amount of raw data for training and subsequent classification of unseen data. As a derivative of recurrent neural networks (RNNs), long short term memory (LSTM) networks are known to perform well on data that span over long sequences, especially in time-series prediction domain. An LSTM architecture represents memory states that model local dependencies on features that last for a certain period of time when they occur. Human activity recognition is regarded as a time-series classification problem with temporal dependency, where input data that are temporally close have higher interdependencies than do those that are temporally distal. Thus, LSTM network models are renowned for state-of-the-art performance in human activity recognition domains.[83]

Both physical activity and cardiovascular health are essential for assessing overall health and wellness. And both of these physiological dimensions are often combined with monitoring of blood glucose (BG) for the management of diabetes, which is globally recognized as the seventh most common cause of death according to WHO.[84] BG dynamics are influenced by numerous factors, including food intake, insulin intake, previous BG level, pregnancy, drug and vitamin intake, smoking, and alcohol intake. ML application to continuous glucose monitoring has been a major research focus in recent years to help relieve patients of the self-management burden. In a recent literature review of ML applications in BG monitoring between 2000 and 2018, conventional feedforward artificial neural networks were the most frequently used technique in hypoglycemia (26%) and hyperglycemia (34%) classification from continuous glucose monitoring devices.[85] Generally, all of the studies have relied on either indirect indicator variables such as heart rate, QT interval, and others or a subset of input parameters that affect BG dynamics. The patient's contextual information, for example, meals, physical activity, insulin, and sleep, has a significant effect on BG dynamics, and a proper anomaly classification and detection algorithm should consider the effects of these parameters. In this regard, however, the individual patient is expected to record meal, insulin, and physical activity data. One of the main limitations is meal modeling, where most of the algorithm depends on the individual estimation of carbohydrate, which is prone to errors and further aggravates the degradation to detection performance. With regard to physical activity, there are various wearables and sensors that can record the individual's physical activity load and durations. However, there is the lack of a uniform approach among the studies with certain limitations on the way these signals are used in the classification and detection algorithms. For example, there are some studies that consider levels of activity as low, moderate, and high and others consider descriptive features by summarizing the number, intensity, steps, exercise durations, and others to better quantify the effect of physical activities. Moreover, recording insulin dosage has its inherent limitations, which might affect the detection performance.

Sample case study: Parkinson's disease

Parkinson's disease (PD) is a degenerative disorder of the central nervous system, which affects the motor control of the patients it inflicts. Three of the most common symptoms are hand and leg tremors, and dyskinesia, an uncontrollable spasming/movement of the patient's upper body.

Medication for PD attempts to control these symptoms, but the frequency and dosage as well as the appropriate type of medication is often difficult to determine. Patients frequently keep track of their symptoms in inaccurate self-maintained handwritten diaries, from which physicians attempt to learn about and manage their symptoms. Assessment of PD is difficult with paper diaries as they are labor-intensive, requiring patients to self-report every half-hour for several days in a row. For this reason, compliance tends to fall markedly overtime. In addition, the self-assessment is frequently imprecise.

More accurate digital records of the patient's symptoms, including time, duration, and intensity of onset, could facilitate better disease management, as well as permit potentially dynamic adjustment of the treatment regimen. The use of accelerometers is a potential solution; however, the detection of PD movement from normal signal is nontrivial. There are challenges of accurately detecting signal from background noise. Symptoms such as dyskinesia may be easily identifiable because of their drastic movements, but those such as hand tremors can be nearly indistinguishable. Finally, if digital sensors are obtrusive (in number and bodily location) and interfere with patients' daily routines, they will not be used and discarded.

There has been significant work accomplished in analyzing time series data from wearable accelerometers.[86–90] Preece et al.[91] provide a nice review of advances in the literature along with different classification methods that have been used. Other research provides detailed explanations on feature extraction methods. Notable works in assessing accelerometer signals for analysis of Parkinson's disease include[90,92,93]. Commercially available device with sensor configurations on the hand, write, finger, arms, trunk, back, and/or waist[94] can detect gain and postural impairments as well as tremor and dyskinesia severity. However, it has been observed that[95] such studies often are result of short monitoring periods as subjects are often required to wear cumbersome sensor configurations that are impractical in a daily life setting.

Khan[90] presents a system that is based on a single accelerometer worn on the waist. In addition, although different classification methods are assessed in the literature, most of the focus has been on developing robust and informative feature extraction methodologies. Accelerometer data are often noisy and difficult to work with. This reference provides a comparison of some of the core classification techniques that has not really been present in the literature. Some work[96] has mentioned the use of neural networks in a 10-fold cross-validation framework.

Standards and regulations and concluding thoughts

Wearable devices need to meet international quality standards to cross the boundary from consumer gadgets to medical devices. The regulations imposed by FDA are applied to mainstream medical devices and, therefore, many wearable sensors are classified as wellness/lifestyle tracking devices to circumvent FDA's rigid standards. FDA has recently issued guidelines for wireless medical devices to ensure they address major safety critical risks associated with radio-frequency wireless systems, short- and long-range communication, and secure data transmission. Over the next few years, it is anticipated that the FDA and other agencies dealing with medical device regulations will define wearable devices in terms of their regulatory compliance and their use for medical interventions.

As the internet evolves, it continues to pose issues of privacy and security. Not unlike internet-connected desktop PCs that are prone to cyber attacks, mobile phones and wearable devices are also now under constant threat of highly skilled, organized hackers. To mitigate the risk of cyber attacks on wearable devices, we need strong network security infrastructure for short- and long-range communication. In each technology layer, from the wearable sensors to the gateway devices to the cloud, careful precautions are desired to ensure the users' privacy and security.

To live up to the hype and hope, wearable technology needs not only to overcome the technical challenges of generating a flexible framework for networking, computation, storage, and visualization, but also to consolidate its position in designing solutions that are clinically acceptable and operational. Over the next few years, there is likely to be a trend toward siteless trials, where all the study materials and drugs are boxed up and shipped directly to the patient, and the patients access their study team directly via telecommunication apps in Smarthome and mobile devices. This model allows a trial to be centered around the patient in the most literal sense—the patient can participate from their own home and essential data are passively collected during normal daily living. Under patient-controlled privacy and sharing of data, medical decisions and alerts are augmented by patient-specific AI algorithms and indication-specific digital assistants that continuously monitor essential data stream and offer subject area expertise.

Conclusions

In summary, the outlook for the application of AI and ML to clinical trials, RWE, and digital medicine is positive. There are already multiple successful examples of where there has been benefit for the speed, efficiency, or accuracy of clinical development. At the same time, there are many areas being tested and that will come to fruition in the future, which promise to enhance the development of medicines still further. Some of the more mature applications include the use of ML for the identification of patient subgroups, using NLP to provide additional structure to textual information, and convolutional neural networks to perform automated image analysis. These use cases are valuable and important, although it should be recognized that in most cases they do not yet drive all the core critical pathways in drug development—but tend to be more peripheral and supportive to that main activity.

In the future, we will see more AI & ML applications begin to drive the core trial process, partly propelled by increasing regulatory acceptance. For instance, the use of ML to select patients who are likely to experience a differential medical outcome is rapidly gaining traction. Automation will be an important area, including for instance AI-driven adjudication of certain critical medical events, relating to both efficacy and safety. New data types, such as streaming data from sensors and wearable devices, as well as RWD, will gradually mature as AI/ML approaches enable them to be more effectively processed and interpreted in the clinical context. We will also see more joint development of ML algorithms with a new medicine, aiming to provide a more holistic package of "pill plus algorithm" to patients and healthcare providers seeking ever better outcomes through "digital therapeutics." In short, the future is bright for the use of AI and ML in this area.

References

1. *AstraZeneca data and AI ethics.* Available from: https://www.astrazeneca.com/sustainability/ethics-and-transparency/data-and-ai-ethics.html. [Accessed 12 November 2020].
2. *Clinical trials—NHS.* Available from: https://www.nhs.uk/conditions/Clinical-trials/. [Accessed 6 March 2020].
3. *What are clinical trials and studies?.* Available from: https://www.nia.nih.gov/health/what-are-clinical-trials-and-studies. [Accessed 6 March 2020].
4. Collet J. Limite des essais cliniques [Limitations of clinical trials]. *Rev Prat* 2000;**50**:833–7.
5. Weber D. *The limitations of randomized controlled trials.* Association of Integrative Oncology and Chinese Medicine (AIOCM); 2012. Available from: http://aiocm.org/uncategorized/the-limitations-of-randomized-controlled-trials/. [Accessed 12 November 2020].
6. Anisimov VV, Fedorov VV. Modelling, prediction and adaptive adjustment of recruitment in multicentre trials. *Stat Med* 2007;**31**(16):1655–74.
7. Tang G, Kong Y, Chang CCH, Kong L, Costantino JP. Prediction of accrual closure date in multi-center clinical trials with discrete-time Poisson process models. *Pharm Stat* 2012;**11**:351–6.
8. Zhang X, Long Q. Stochastic modeling and prediction for accrual in clinical trials. *Stat Med* 2010;**29**:649–58.
9. Deng Y, Zhang X, Long Q. Bayesian modeling and prediction of accrual in multi-regional clinical trials. *Stat Methods Med Res* 2017;**26**:752–65.
10. Dürr S. *Meeting the clinical supply challenge with IRT.* Available from: https://www.appliedclinicaltrialsonline.com/view/meeting-clinical-supply-challenge-irt. [Accessed 20 March 2020].
11. Bornkamp B, Ohlssen D, Magnusson BP, Schmidli H. Model averaging for treatment effect estimation in subgroups. *Pharm Stat* 2017;**16**.

12. Brookes ST, et al. Subgroup analyses in randomized trials: risks of subgroup-specific analyses; power and sample size for the interaction test. *J Clin Epidemiol* 2004;**57**:229–36.

13. Foster JC, Taylor JMG, Ruberg SJ. Subgroup identification from randomized clinical trial data. *Stat Med* 2011;**30**:2867–80.

14. Dmitrienko A, Millen B, Lipkovich I. Multiplicity considerations in subgroup analysis. *Stat Med* 2017;**36**:4446–54.

15. Lipkovich I, Dmitrienko A, D'Agostino RB. Tutorial in biostatistics: data-driven subgroup identification and analysis in clinical trials. *Stat Med* 2017;**36**:136–96.

16. Loh WY, Cao L, Zhou P. Subgroup identification for precision medicine: a comparative review of 13 methods. *Wiley Interdiscip Rev Data Min Knowl Disc* 2019. https://doi.org/10.1002/widm.1326.

17. Crown WH. Real-world evidence, causal inference, and machine learning. *Value Health* 2019;**22**:587–92.

18. US Food and Drug Administration. *Framework for FDA's real-world evidence program.* US Food and Drug Administration; 2018. Available from: www.fda.gov. [Accessed 19 March 2020].

19. Connell A, et al. Evaluation of a digitally-enabled care pathway for acute kidney injury management in hospital emergency admissions. *npj Digit Med* 2019;**2**.

20. Connell A, et al. Implementation of a digitally enabled care pathway (part 1): impact on clinical outcomes and associated health care costs. *J Med Internet Res* 2019;**21**:e13147.

21. Varadarajan AV, et al. Predicting optical coherence tomography-derived diabetic macular edema grades from fundus photographs using deep learning. *Nat Commun* 2020;**11**.

22. Sherman RE, et al. Real-world evidence—what is it and what can it tell us? *N Engl J Med* 2016;**375**:2293–7.

23. Swift B, et al. Innovation at the intersection of clinical trials and real-world data science to advance patient care. *Clin Transl Sci* 2018;**11**:450–60.

24. Birkhead GS, Klompas M, Shah NR. Uses of electronic health records for public health surveillance to advance public health. *Annu Rev Public Health* 2015;**36**:345–59.

25. Botsis T, Hartvigsen G, Chen F, Weng C. Secondary use of EHR: data quality issues and informatics opportunities. *AMIA Jt Summits Transl Sci Proc* 2010;**2010**:1–5.

26. Jensen PB, Jensen LJ, Brunak S. Mining electronic health records: towards better research applications and clinical care. *Nat Rev Genet* 2012;**13**:395–405.

27. Jiang M, et al. A study of machine-learning-based approaches to extract clinical entities and their assertions from discharge summaries. *J Am Med Inform Assoc* 2011;**18**:601–6.

28. Meystre SM, Savova GK, Kipper-Schuler KC, Hurdle JF. Extracting information from textual documents in the electronic health record: a review of recent research. *Yearb Med Inform* 2008;128–44. https://doi.org/10.1055/s-0038-1638592.

29. Doshi-Velez F, Ge Y, Kohane I. Comorbidity clusters in autism spectrum disorders: an electronic health record time-series analysis. *Pediatrics* 2014;**133**:e54–63.

30. Li L, et al. Identification of type 2 diabetes subgroups through topological analysis of patient similarity. *Sci Transl Med* 2015;**7**. 311ra174.

31. Ebadollahi S, et al. Predicting patient's trajectory of physiological data using temporal trends in similar patients: a system for near-term prognostics. *AMIA Ann Symp Proc* 2010;**2010**:192–6.

32. Zhao D, Weng C. Combining PubMed knowledge and EHR data to develop a weighted Bayesian network for pancreatic cancer prediction. *J Biomed Inform* 2011;**44**:859–68.

33. Austin PC, Tu JV, Ho JE, Levy D, Lee DS. Using methods from the data-mining and machine-learning literature for disease classification and prediction: a case study examining classification of heart failure subtypes. *J Clin Epidemiol* 2013;**66**:398–407.

34. Kuperman GJ, et al. Medication-related clinical decision support in computerized provider order entry systems: a review. *J Am Med Inform Assoc* 2007;**14**:29–40.

35. Miotto R, Weng C. Case-based reasoning using electronic health records efficiently identifies eligible patients for clinical trials. *J Am Med Inform Assoc* 2015;**22**:e141–50.

36. Khosla S, et al. Real world evidence (RWE)—a disruptive innovation or the quiet evolution of medical evidence generation? *F1000 Res* 2018;**7**.

37. White R. Building trust in real-world evidence and comparative effectiveness research: the need for transparency. *J Comp Eff Res* 2017;**6**:5–7.

38. Liede A, et al. Validation of international classification of diseases coding for bone metastases in electronic health records using technology-enabled abstraction. *Clin Epidemiol* 2015;**7**:441–8.

39. Claerhout B, et al. Federated electronic health records research technology to support clinical trial protocol optimization: evidence from EHR4CR and the InSite platform. *J Biomed Inform* 2019;**90**:441–8.

40. Carrigan G, et al. Using electronic health records to derive control arms for early phase single-arm lung cancer trials: proof-of-concept in randomized controlled trials. *Clin Pharmacol Ther* 2020;**107**:369–77.

41. Patel K, Ouwens M, Shire N, Khosla S. The application of electronic medical records (EMRs) as a virtual comparator arm in a lung cancer clinical trial: a case study. *J Clin Oncol* 2017;**35**:e18098.

42. Becker MY. Information governance in NHS's NPfIT: a case for policy specification. *Int J Med Inform* 2007;**76**:432–7.

43. Liaw ST, et al. Optimising the use of observational electronic health record data: current issues, evolving opportunities, strategies and scope for collaboration. *Aust Fam Physician* 2016;**45**:153–6.

44. Candore G, et al. Can we rely on results from IQVIA medical research data UK converted to the observational medical outcome partnership common data model? *Clin Pharmacol Ther* 2020;**107**:915–25.

45. Glicksberg BS, et al. ROMOP: a light-weight R package for interfacing with OMOP-formatted electronic health record data. *JAMIA Open* 2019;**2**:10–4.

46. Lamer A, et al. Transforming French electronic health records into the observational medical outcome partnership's common data model: a feasibility study. *Appl Clin Inform* 2020;**11**:13–22.

47. Harrell FE, Lee KL, Califf RM, Pryor DB, Rosati RA. Regression modelling strategies for improved prognostic prediction. *Stat Med* 1984;**3**:143–52.

48. Kurt I, Ture M, Kurum AT. Comparing performances of logistic regression, classification and regression tree, and neural networks for predicting coronary artery disease. *Expert Syst Appl* 2008;**34**:366–74.

49. Carroll RJ, Eyler AE, Denny JC. Naïve electronic health record phenotype identification for rheumatoid arthritis. *AMIA Ann Symp Proc* 2011;**2011**:189–96.

50. Hippisley-Cox J, Coupland C. Predicting risk of emergency admission to hospital using primary care data: derivation and validation of QAdmissions score. *BMJ Open* 2013;**3**.

51. Bottle A, Aylin P, Majeed A. Identifying patients at high risk of emergency hospital admissions: a logistic regression analysis. *J R Soc Med* 2006;**99**:406–14.

52. Rahimian F, et al. Predicting the risk of emergency admission with machine learning: development and validation using linked electronic health records. *PLoS Med* 2018;**15**:e1002695.

53. Pham T, Tran T, Phung D, Venkatesh S. Predicting healthcare trajectories from medical records: a deep learning approach. *J Biomed Inform* 2017;**69**:218–29.

54. Rajkomar A, et al. Scalable and accurate deep learning with electronic health records. *npj Digit Med* 2018;**1**.

55. Nguyen P, Tran T, Wickramasinghe N, Venkatesh S. Deepr: a convolutional net for medical records. *ArXiv* 2017;**21**:22–30.

56. Lecun Y, Bengio Y, Hinton G. Deep learning. *Nature* 2015;**521**:436–44.

57. Ravi D, et al. Deep learning for health informatics. *IEEE J Biomed Heal Informatics* 2017;**21**:4–21.

58. Goodfellow I, Bengio Y, Courville A. *Deep learning*. MIT Press; 2016.

59. Miotto R, Li L, Kidd BA, Dudley JT. Deep patient: an unsupervised representation to predict the future of patients from the electronic health records. *Nat Sci Rep* 2016;**6**.

60. Jagannatha AN, Yu H. Structured prediction models for RNN based sequence labeling in clinical text. In: *EMNLP 2016—conference on empirical methods in natural language processing, proceedings*; 2016. p. 856–65. https://doi.org/10.18653/v1/d16-1082.

61. Jagannatha AN, Yu H. Bidirectional RNN for medical event detection in electronic health records. In: *2016 conference of the North American chapter of the association for computational linguistics: human language technologies, NAACL HLT 2016—proceedings of the conference.* Association for Computational Linguistics (ACL); 2016. p. 473–82. https://doi.org/10.18653/v1/n16-1056.

62. Nie L, et al. Disease inference from health-related questions via sparse deep learning. *IEEE Trans Knowl Data Eng* 2015;**27**:2107–19.

63. Choi E, Schuetz A, Stewart WF, Sun J. *Medical concept representation learning from electronic health records and its application on heart failure prediction*; 2016.

64. Choi E, et al. Multi-layer representation learning for medical concepts. In: *Proceedings of the ACM SIGKDD international conference on knowledge discovery and data mining 13–17 August.* New York, NY, United States: Association for Computing Machinery; 2016. p. 1495–504.

65. Ma F, et al. Dipole: diagnosis prediction in healthcare via attention-based bidirectional recurrent neural networks. In: *Proceedings of the ACM SIGKDD international conference on knowledge discovery and data mining*; 2017. https://doi.org/10.1145/3097983.3098088.

66. Ayala Solares JR, et al. Deep learning for electronic health records: a comparative review of multiple deep neural architectures. *J Biomed Inform* 2020;**101**, 103337.

67. Tran T, Nguyen TD, Phung D, Venkatesh S. Learning vector representation of medical objects via EMR-driven nonnegative restricted Boltzmann machines (eNRBM). *J Biomed Inform* 2015;**54**:96–105.

68. Choi E, et al. RETAIN: an interpretable predictive model for healthcare using reverse time attention mechanism. In: *Advances in neural information processing systems*; 2016. Arxiv; https://arxiv.org/abs/1608.05745.

69. Kwon BC, et al. RetainVis: visual analytics with interpretable and interactive recurrent neural networks on electronic medical records. *IEEE Trans Vis Comput Graph* 2019;**25**:299–309.

70. Luczkowski SM. *s6ch3*. Available from: https://history.nasa.gov/SP-368/s6ch3.htm. (Accessed 19 March 2020).

71. Jeffrey K, Parsonnet V. Cardiac pacing, 1960-1985 a quarter century of medical and industrial innovation. *Circulation* 1998;**97**:1978–91.

72. Drazen E, Mann N, Borun R, Laks M, Bersen A. Survey of computer-assisted electrocardiography in the United States. *J Electrocardiol* 1988;**21**:S98–S104.

73. Pipberger H, Stallman F, Berson A. Automatic analysis of the P-QRS-T complex of the electrocardiogram by digital computer. *Ann Intern Med* 1962;**57**:776–87. https://doi.org/10.7326/0003-4819-57-5-776.

74. Casale PN, et al. Electrocardiographic detection of left ventricular hypertrophy: development and prospective validation of improved criteria. *J Am Coll Cardiol* 1985;**6**:572–80.

75. Tison GH, Zhang J, Delling FN, Deo RC. Automated and interpretable patient ECG profiles for disease detection, tracking, and discovery. *Circ Cardiovasc Qual Outcomes* 2019;**12**:e005289.

76. Bumgarner JM, et al. Smartwatch algorithm for automated detection of atrial fibrillation. *J Am Coll Cardiol* 2018;**71**:2381–8.

77. Haverkamp HT, Fosse SO, Schuster P. Accuracy and usability of single-lead ECG from smartphones—a clinical study. *Indian Pacing Electrophysiol J* 2019;**19**:145–9.

78. Haberman ZC, et al. Wireless smartphone ECG enables large-scale screening in diverse populations. *J Cardiovasc Electrophysiol* 2015;**26**:520–6.

79. Thakor NV, Zhu YS. Applications of adaptive filtering to ECG analysis: noise cancellation and arrhythmia detection. *IEEE Trans Biomed Eng* 1991;**38**:785–94.

80. Bsoul M, Minn H, Tamil L. Apnea MedAssist: real-time sleep apnea monitor using single-lead ECG. *IEEE Trans Inf Technol Biomed* 2011;**15**:416–27.

81. *FDA approves ECG device, verily study watch, HCPLive.* Available from: https://www.mdmag.com/medical-news/fda-approves-ecg-device-verily-study-watch. (Accessed 19 March 2020).

82. *Healthy people 2020*. Available from: https://www.healthypeople.gov/2020/Leading-Health-Indicators. (Accessed 19 March 2020).
83. Alsheikh MA, et al. Deep activity recognition models with triaxial accelerometers. In: *AAAI workshop—technical report WS-16-01*. AI Access Foundation; 2016. p. 8–13.
84. *The top 10 causes of death*. Available from: https://www.who.int/news-room/fact-sheets/detail/the-top-10-causes-of-death. [Accessed 19 March 2020].
85. Woldaregay AZ, et al. Data-driven blood glucose pattern classification and anomalies detection: machine-learning applications in type 1 diabetes. *J Med Internet Res* 2019;**21**:e11030.
86. Gafurov D, Helkala K, Søndrol T. Biometric gait authentication using accelerometer sensor. *J Comput* 2006;**1**:51–9.
87. Godfrey A, Conway R, Meagher D, ÓLaighin G. Direct measurement of human movement by accelerometry. *Med Eng Phys* 2008;**30**:1364–86.
88. He Z, Liu Z, Jin L, Zhen LX, Huang JC. Weightlessness feature—a novel feature for single tri-axial accelerometer based activity recognition. In: *Proceedings—19th international conference on pattern recognition*; 2008. p. 1–4. https://doi.org/10.1109/icpr.2008.4761688.
89. Kwapisz JR, Weiss GM, Moore SA. Activity recognition using cell phone accelerometers. *ACM SIGKDD Explor Newsl* 2011;**12**:74–82.
90. Khan FM, et al. A wearable accelerometer system for unobtrusive monitoring of Parkinson's diease motor symptoms. In: *Proceedings—IEEE 14th international conference on bioinformatics and bioengineering, BIBE 2014*. Florida, USA: Institute of Electrical and Electronics Engineers; 2014. p. 120–5. https://doi.org/10.1109/BIBE.2014.18.
91. Preece SJ, et al. Activity identification using body-mounted sensors—a review of classification techniques. *Physiol Meas* 2009;**30**:R1–R33.
92. Bonato P, Sherrill DM, Standaert DG, Salles SS, Akay M. Data mining techniques to detect motor fluctuations in Parkinson's disease. In: *Annual international conference of the IEEE engineering in medicine and biology—proceedings*; 2004. p. 4766–9. https://doi.org/10.1109/iembs.2004.1404319.
93. LeMoyne R, Coroian C, Mastroianni T. Quantification of Parkinson's disease characteristics using wireless accelerometers. In: *2009 ICME international conference on complex medical engineering, CME*; 2009. p. 1–5. https://doi.org/10.1109/ICCME.2009.4906657.
94. Maetzler W, Domingos J, Srulijes K, Ferreira JJ, Bloem BR. Quantitative wearable sensors for objective assessment of Parkinson's disease. *Mov Disord* 2013;**28**:1628–37.
95. Merello M, Antonini A. Evaluation of motor complications: motor fluctuations. In: Sampaio C, Goetz C, Schrag A, editors. *Rating scales in Parkinson's disease: clinical practice and research*. New York: Oxford University Press; 2012. https://doi.org/10.1093/med/9780199783106.003.0202.
96. Matzinger P. Tolerance, danger, and the extended family. *Annu Rev Immunol* 1994;**12**:991–1045.

Beyond the patient: Advanced techniques to help predict the fate and effects of pharmaceuticals in the environment

Stewart F. Owen and Jason R. Snape
Global Sustainability, AstraZeneca, Cambridge, United Kingdom

Overview

Pharmaceuticals enter the aquatic environment and can be found in low concentrations in almost every environmental compartment worldwide, but especially water. The hazard and risk these pharmaceuticals pose in the environment is a complex question. Although regulations in some regions are protective, not all parts of the world have legislation or infrastructure to manage micro-pollution from the patient use of pharmaceuticals. Innovative approaches to unbiased data analysis such as machine learning (ML) and other artificial neural networks (ANNs) offer new avenues to better understand the processes and key features that determine the impact of pharmaceuticals in the environment. We are at an early stage of using ML in environmental toxicology and exposure assessment, but early developments are offering insight and understanding that previously were not evident from traditional approaches.

As these innovative methods become general purpose technologies (GPTs), ANNs and ML will have significant benefits for environmental protection and conservation. We anticipate a shift from chemical by chemical in vivo exposures and assessment to in silico techniques predicting real-world effects from combined chemical exposure coupled with other environmental stressors and stimuli. This needs to be safe and trusted technology with a high level of transparency. We expect to be able to go from understanding binding affinities and molecular docking to an integrated systems approach through absorption, distribution, metabolism, and excretion to confidently predict and compare toxicology across species to understand sensitivity and vulnerability. Based on the science, we foresee the future use of task-driven artificial intelligence (AI) approaches to support an expert global registration and risk assessment of all chemicals with less inherent uncertainty than the current manual and limited processes. Together, these approaches are new to environmental protection and promise a rich and rapid growth of knowledge and understanding.

The Era of Artificial Intelligence, Machine Learning, and Data Science in the Pharmaceutical Industry. https://doi.org/10.1016/B978-0-12-820045-2.00012-X

Background

Before 2006, there was little legislation concerned with pharmaceuticals entering the environment; as a result, most active pharmaceutical ingredients (API) launched before this have little or no environmental information available.[1] Because of their constant use and pseudo-persistent nature, pharmaceuticals are found in the aquatic environment at low concentrations almost everywhere we look, typically low microgram (μg) or nanogram (ng) per liter concentrations.[2] They are also found inside fish and invertebrates in low nanogram per gram (ng/g) concentrations,[3] but the risk that their presence poses to wildlife is a complex question and a field of much scientific research.[4–6]

Pharmaceuticals reach the environment through a number of routes such as manufacturing waste[7] and incorrect disposal, but most (> 80%) of the pharmaceuticals that reach the aquatic environment is via patient use. APIs leave the patient through multiple routes: eliminated as unabsorbed and unchanged active ingredient in the feces; metabolites of the original medicine in urine and feces; or excreted unmetabolized parent compound in the urine. There are a few APIs that are completely metabolized in the patient. But given the advances in analytical chemistry and technologies to measure drug residues at low concentrations, we can expect most APIs will be found in the environment at some point. In more developed regions with sanitation infrastructure the urine and fecal wastes from patients will usually go to a wastewater treatment plant where the waste will be exposed to a variety of physical and biological procedures. These will determine how the API behaves and what it may or may not breakdown to. Once discharged, these breakdown products and API residues are usually diluted in surface waters where they can be further degraded or partition to suspended or benthic sediments. Consequently, aquatic and benthic taxa are exposed to these micro-pollutants in combination with many other compounds and a wide range of biological, chemical, and physical variation (Fig. 1).

There are cases of high concentrations or "hot spots" of some pharmaceutical discharges associated with manufacturing effluents that are unacceptable and controllable. Many of these monitoring studies are focused on antibiotic manufacturing in poorly managed or underregulated situations.[7] In some cases the concentrations in the receiving waters were measured in milligrams per liter: similar to the concentrations in human blood used to treat disease.[9] These pollutants add to the overall chemical burden on these environments and require effective management.[10] It is unlikely that these elevated concentrations are representative of all pharmaceutical manufacturing practice and typical concentrations that could be found downstream of drug production in most countries. However, more data are needed to confirm that this is the case. This is preventable pollution and approaches to defining safe limits for pharmaceutical production have been identified.[11–13] Our focus for environmental risk management is on the diffuse source of pharmaceuticals excreted by patients as part of their necessary treatment. This is the part of the micro-pollution that by its nature is difficult to eliminate, and so, we focus on understanding and better predicting these real-world risks to the global environment.

There are examples of pharmaceuticals causing significant environmental harm. For example, a nonsteroidal antiinflammatory drug called diclofenac was used as a veterinary medicine to treat cattle near their end of life within India and Pakistan. The practice of allowing the dead carcase of cattle (treated and untreated) to be cleared by vultures resulted in a secondary poisoning event where raptors received a toxic dose from treated cattle that led to renal damage. Gout-like symptoms resulted in the widespread deaths of these birds and almost total loss of three species of *Gyps* vultures.[14] Steps to reduce diclofenac exposure in this region seems to have reduced the unintended population-level impact on these species, but it remains a real risk that we could lose these fabulous bird species in the near

FIG. 1

Cities are reliant on high-quality water. The River Thames (London, United Kingdom) is typical of industrialized cities. It is the primary source of drinking water, with abstracted and extensively treated river water supplying about 80% of the city's need equating to approximately 13 million people across the whole catchment and about 7 million people in London itself. Groundwater aquifers support about 20% of the need. There are at least 32 wastewater treatment works (Source: www.thameswater.co.uk), and there are a wide range of pollutants and other factors that challenge the wildlife reliant on this catchment.[5] Pharmaceuticals have been measured along its length and concentrations vary day-to-day and are affected by flow/rainfall, season, and diurnal cycles.[8]

Courtesy: Jack Owen (2019).

future.[15] Predicting this hazard would be difficult even if there was prior knowledge of that unusual exposure route to these birds, and further, no current regulatory framework for the environmental risk assessment (ERA) of human or veterinary medicines globally would have predicted this. As a result of this impact, some approaches now consider these more vulnerable species through a comparative physiology approach[16] and ecopharmacovigilance.[17]

Ecopharmacovigilance is a process akin to how adverse effects of pharmaceuticals in patients are continuously monitored to ensure patient safety; the environmental scientific literature is continuously reviewed, and relevant findings are incorporated into ERAs of medicinal products.

Despite the example of impact on the vultures, a waterborne exposure route remains the most likely route for pharmaceutical risk. Another example of environmental toxicology from pharmaceuticals that is more directly related to aquatic exposure is the impact of the active ingredients within the birth control pill, ethinylestradiol (EE2), which can enter fish from the water. Although found in low nanogram per liter concentrations in rivers, the worldwide exposure is extensive and coexposure to other natural and synthetic estrogens and endocrine-disrupting chemicals compounds the issue. In male fish exposed to low nanogram per liter concentrations of EE2 in the water, uptake can feminize fish sufficiently to produce oocytes (first stage of the egg) in the testis. In the laboratory, changes in behavior, secondary sexual characteristics (such as male color), and potentially quality and quantity of sperm significantly impacts the breeding potential of these affected male fish.[18, 19] Effects and their recovery have been demonstrated in fish populations in experimental lakes at exposures above the typical environmental concentrations,[20, 21] and feminized males are commonly found in impacted waters around the world.[18] However, linking effects on populations directly to a single cause such as EE2 has proven difficult in the wild,[5, 22, 23] although likely when considered in mixtures.[24–26]

Given that pharmaceuticals are designed to interact with pharmacological (biological) targets, and that these targets can be widely conserved across a range of wildlife,[27] there is significant concern that some other classes of pharmaceuticals may also impact wildlife.[1, 18] It is therefore critically important that we understand the fate (where they go) and effects (what they do) of pharmaceuticals in the environment. We suggest that advanced/innovative techniques such as ML and similar statistical algorithms based on ANNs are potentially new and disruptive technologies that may help predict the hazard and risks of pharmaceuticals in the environment; an essential tool that is needed in this field.[28]

Current European and US legislation for environmental assessment of pharmaceuticals

An ERA is required as part of a marketing application for a new medicinal product within the European Union (EU) and the United States. Unlike preclinical toxicology studies, the ERA is conducted late in the development cycle of a new medicinal product. ERAs are typically conducted in parallel to Phase III clinical trials after patient safety and efficacy have been established. In the EU, long-term aquatic effects studies are triggered if the predicted surface water concentration is expected to exceed an action limit of 10ng/L.[29] In the United States, an ERA is required if the expected environmental concentration exceeds 100ng/L.[30] Within the EU, an environmental hazard assessment is required irrespective of the exposure-based action limit to determine whether the API is persistent, bioaccumulative, or toxic if the API is highly lipophilic ($logD \geq 4.5$). A tailored ERA is also required irrespective of environmental exposure, in both the EU and United States, for APIs that have an endocrine-mediated mode of action that impacts reproduction.[29–32] Aquatic toxicology studies focus on assessing impacts on primary producers (green and/or blue-green algae), an invertebrate primary consumer (typically the freshwater crustacean *Daphnia magna*), and a vertebrate secondary consumer (fish). For APIs that adsorb to sewage sludge solids (Koc > 10,000), a terrestrial risk assessment is required. APIs with a $logD > 3$ need to have their bioconcentration potential in fish determined and those that partition to sediment within the water-sediment transformation study need to have a sediment effects assessment and sediment ERA conducted.[31]

Ecological protection within any ERA is directed at ensuring environmental safety at the population level to maintain biodiversity and ecosystem integrity. The typical endpoints assessed within regulatory ERA frameworks are focused on growth, mortality, and reproduction. Effects at a molecular, biochemical, cell, or tissue level or impacts on behavior are not routinely considered within regulatory ERAs unless they have been linked to an adverse outcome pathway or the endpoint has been validated for regulatory use (see below).

Within the EU, maximum daily doses and default market penetration rates of 1% are assumed, unless reliable disease prevalence data exist, together with a per capita daily water use of 200L per day and an environmental dilution of 10-fold. These assumptions coupled with physicochemical, biodegradation, and partitioning data are used to understand the fate and behavior of an API in the environment to determine predicted environmental concentrations (PECs) in a range of environmental matrices (surface waters, sediments, ground waters, soil, etc.). Chronic environmental toxicity studies identify no observed effect concentrations (NOECs) for algae, daphnia, and fish. An assessment factor (typically a factor of 10) is applied to the lowest NOEC, derived from the most sensitive species, to determine a predicted no effect concentration (PNEC). The ratio of the PEC to PNEC is calculated for each environmental compartment of concern to determine whether a potential environmental risk exists; a PEC/PNEC > 1 is indicative of a potential risk. Given the conservative exposure assessment within an ERA, that is, the 1% market penetration overestimates consumption for 95% of medicinal products,[29] environmental risks can be overestimated. An article by Gunnarsson et al.[1] conducted consumption-based risk assessments for > 120 APIs. Gunnarsson et al.[1] showed that 95% of active ingredients posed low or insignificant risk even based on worst-case environmental exposure, that is, no removal or degradation within sewage treatment. Actives with the highest potential risk all had endocrine-mediated modes of action and high lipophilicity.

Currently, environmental hazard and risk is not a barrier to patient access to medicines. It is not part of the patient benefit-risk assessment within the marketing application. However, concerns about the environmental hazards and risks of pharmaceuticals are bringing this into question. Many publications have advocated the inclusion of environmental hazard and risk within the marketing authorization.[33] This poses a challenge to the pharmaceutical industry given its current model of testing the environmental hazard and risks late within drug development. Either regulatory ERA testing has to be brought forward or reliable, robust, and valid tools to predict environmental hazard and risk at a much earlier stage need to be developed to (i) screen candidate drugs within development and (ii) prioritize legacy APIs authorized for use before the introduction of the 2006 EMA ERA guidelines that lack full environmental datasets. Significant progress on all fronts would need to be made, not least in developing trusted predictions to anticipate issues, a key area for investment in ML.

One challenge for environmental toxicology is the relevance of the limited model species that are used within regulatory approaches for ERA. Within preclinical toxicology, animal models are used to inform clinical safety, that is, data from a few animal models such as mice, rats, dogs, and nonhuman primates are extrapolated to humans. Within environmental toxicology, we conduct a similar process; however, data from algae, daphnia, fish, and sediment invertebrates are extrapolated to protect all life in the biosphere.

Animal testing for protecting the environment

ERA is a tiered testing cascade that for many pharmaceuticals results in highest tier testing of freshwater invertebrates and fish to measure the effect on apical endpoints such as growth, survival, and reproduction. These are standardized tests used in assessing other chemical classes such as industrial chemicals and pesticides. Similarly, pharmaceutical assessment includes a battery of effect testing on algae and cyanobacteria as well as the fate testing on sorption and impact on microbial communities in terms of degradation and transformation to establish the degree of risk to the environment. Regulatory authorities ask for pharmaceuticals to be assessed using the guidance of practical ring-tested methods from the OECD and similar authorities. For pharmaceuticals, these typically might be OECD 202,

211, 218, 225, and 235 (all references to OECD tests throughout this chapter are available at: www. oecd.org/env/ehs/testing/). Invertebrates such as the crustacea *Daphnia* sp. (water flea), the larvae of a midge *Chironomus* sp., and sediment-dwelling organisms such as oligochaetes like *Lumbriculus* sp. are used to understand the lowest observed effect concentration (LOEC) and establish an NOEC that is used in the assessment against PECs based on patient use. Fish of various life stages and species are used to understand the risk to these vertebrates. Bioaccumulation (OECD 305) studies generally use several hundred fish and expose them to relatively high concentrations to facilitate analytical chemical quantification. Because of the high exposure concentration and long period needed to reach steady-state internal concentrations in some situations, these are generally of moderate severity, but can include range-finding studies of the highest severity[34] with a high risk of mortality or lasting harm. This provides an ethical challenge in terms of balancing the severity imposed on laboratory fish to protect those in the environment. Few pharmaceuticals have bioaccumulative properties in mammals as this is also a difficult property to manage in the patient, so are rarely selected. However, the regulatory environmental trigger to test this in pharmaceuticals is conservative, resulting in likely more tests than might otherwise be necessary. Better predictions based on scientific understanding are urgently needed in this area.[35–37]

Issues for database creation

The standard OECD 305 bioaccumulation test assumes a first-order (linear) uptake and depuration rate that calculates a ratio between the two.[38] This ratio is a number used by regulators with thresholds set as to whether the chemical is of concern. For example, a chemical with a ratio above 2000 is considered to be bioaccumulative. This is a kinetic measure and the assumptions of linear rates may not stand for all pharmaceuticals.[38] Further, a confusion exists within the scientific literature where academic researchers tend to measure a bioaccumulation factor as multiples of the water concentration, whereas regulatory studies are kinetic and use a rate of depuration to derive a ratio and call it a bioconcentration factor. These factors mean that just generating datasets to train new models is fraught with difficulty, and preprocessing is a critical step, especially when these datasets are likely to be relatively small. Further issues lie in finding robust molecular descriptors or physiochemical properties of pharmaceuticals. Often these are themselves modeled rather than measured and those that are measured rarely report the accuracy or variability of that measurement. This question of reliability of measurement also applies to the experimental data contained within regulatory or scientific reports. For example, a risk assessor may report a LOEC as a single concentration. However, there will be variation in the analytical chemistry and accuracy within that measure, and from the practicalities of maintaining a constant exposure concentration in an ecotoxicology laboratory where water is continuously flowing through a fish tank for weeks or months, makes these numbers more of an estimate of a geometric mean than an absolute value. Yet, we routinely see academic studies report effects at say 1 or 2 ng/L in the laboratory without any estimate of the variability. The accepted variability in this branch of environmental toxicology is compensated for by using large safety factors (typically factors of $10\times$, $100\times$, $1000\times$). Intuitively, this means that outputs from innovative methods such as ML or AI have the potential to be used with some confidence. Development of safe and trusted models and tools to predict these effects with similar ranges of accuracy seems achievable.

Opportunities to refine animal testing for protecting the environment

General toxicity to egg, larval, and juvenile fish is probably the most commonly requested vertebrate study for ERA; known as a fish Early Life Stage Test (OECD 210). This test aims to understand the risks for what is considered a vulnerable life stage of early development, and the endpoints are considered apical (relevant to the populations) such as growth and survival. The test uses newly fertilized eggs from species such as the zebrafish (*Danio rerio*), medaka (*Oryzias latipes*), or fathead minnow (*Pimephales promelas*) that are exposed to a range of concentrations while they develop and grow into small fish over several weeks; this typically can be many hundreds of fish per compound. Much larger and more complex chronic studies may also be conducted to understand potential impact on reproduction, life cycles (egg to adult to egg) and multigeneration (egg to adult to egg to adult), and abbreviated (adult to egg) studies may also use thousands of fish in regulatory guideline studies such as the OPPTS 850.1500: Fish Life Cycle Toxicity Test[39] and the OECD 240. Fish are the focus here for several reasons. Ecologists regard them as keystone species, representing higher trophic levels and essential positions in food webs. The presence and health of a fish population is seen as a good indicator as to the health of the ecosystem. Further, physiologists see fish as most closely related to mammals and therefore more likely to share pharmacological targets[27] and therefore likely of highest risk of impact from low concentrations of pharmaceuticals. We recently provided some evidence for this long-held hypothesis that if the pharmacological target is present in fish alone, then they are the most sensitive species, but if the target was in both fish and daphnia, then we need to test both because at present we do not know the driver of the toxicity is the intended pharmacology of the API or some other pathway. The internal concentration is likely to be involved, so the uptake, metabolism, and excretion are likely key.[1]

Fish studies exist for relatively few available pharmaceuticals, but from the data we do have, for most classes of pharmaceutical, we would expect the environmental risk to be low.[1] There are significant exceptions. Hormones and reproductive endocrine-modulating pharmaceuticals generally have a higher risk than other small molecules. We are currently unable to predict the impacts of antineoplastics and similar oncology therapies.[1] There are other classes for which we currently have little information to conclude, and this may be an opportunity for innovative analysis techniques to leverage across the chemical characteristics and allow us to predict which classes are priorities for further animal testing and which are not.[40]

Although these testing requirements have only been required in Europe since 2006, and through the tiered approach many new pharmaceuticals do not even trigger a full assessment based on their likely low volumes and small patient populations, there are some 229 plus pharmaceuticals with environmental toxicology data on at least one of the recommended test species.[1] There are about 975 pharmaceuticals of which 88% have little or no environmental data. To test all of these is practically difficult. It typically takes 2 years to complete a full environmental assessment with associated tests and reviews. There is insufficient capacity in certified contract laboratories to address these ~1000 legacy compounds in a reasonable timescale. But most importantly, conducting full testing would require likely hundreds of thousands of fish and other animals to be exposed and killed, an ethically complex situation when we currently see so little evidence of environmental hazard or risk.[1, 5] It seems likely that for most pharmaceuticals being used by patients in developed countries there is little environmental risk at the concentrations we measure now or are likely to measure in the future even with significant population growth or demographic change (i.e., aging populations are expected to use more pharmaceuticals).

However, in developing countries where the investments in healthcare and access to medicines are outpacing investments in basic sanitation and environmental infrastructure despite rapid urbanization, the potential risks associated with pharmaceuticals in the environment could be more significant.

Currently, the standard regulatory risk assessments in Europe are conservative, making assumptions of no metabolism and large patient populations that provide safety margins typically of many hundreds if not thousands of orders of magnitude.[1] However, as populations change and global access to medicines increases, we can expect more pharmaceuticals to reach the environment, and particularly, in regions that currently have lower access and to a restricted range beyond the World Health Organization essential medicines. To protect the environment and most importantly maintain patient access to essential medicines, we clearly need to direct our efforts toward a prioritized set of those pharmaceuticals that represent the highest risk. However, it is complex as to how that list should be set.[41] Should we focus on a particular class, physiochemical property, volume used, or potency? Or should we concentrate on the measured prevalence in the environment, or some other combination of factors? How might innovative methods like ML approaches help us here? Currently, few in the environmental field have tried to apply these more advanced computational methods to these issues,[37] but we are hopeful of significant developments in the near future.[28, 40]

Current approaches to predicting uptake of pharmaceuticals

The universe of chemicals that are created by humans is diverse, and we do not even have a single reliable list of what exists, but the latest estimate is upward of 350,000 unique chemicals.[42] Beyond simple linear relationships for closely related chemicals, it has been relatively difficult to accurately predict uptake of even general chemicals. Even more difficult has been to predict their likely impact on fish and other aquatic organisms using simple correlation, and regression to develop quantitative structure-activity relationships (QSARs) has proven a challenging field.

Predicting uptake of some classes of chemicals is relatively simple, and factors such as the attraction to lipids (lipophilicity) or concentration in water or the food can directly correlate in linear or log linear relationships.[43] But the range of molecular characteristics combined with the variety of environmental factors and differences in species makes for a complexity that does not lend well to derivation of linear correlations.[35] In fact, regulatory studies that determine bioaccumulation make a fundamental assumption of first-order (linear) uptake and depuration of chemicals in their calculations. This does not necessarily hold true for pharmaceuticals and could lead to false positives and negatives.[35] Indeed, many factors are interrelated, so correlations probably become more of a complex algorithm that has evaded much of the research field.

For issues such as acute toxicity from high concentrations, correlations have focused on the environmental concentration (in the water) rather than specifically looking at the internal concentrations within animals. We then go on to compare the water concentration with biological endpoints (such as death). For a good range of general chemicals (including a few that have a double role as pharmaceuticals, e.g., caffeine), we have reached a point where general models (e.g., Bayesian) are relatively good at predicting say the mortality of fish based on the concentration in the water. Importantly, this correlates well with media concentrations in cell culture that result in cell death; put simply, what kills a fish in vivo, kills a cell in vitro. The result is relatively accurate mortality predictions for live fish exposure.[44] Predicting chronic effects has been far less successful. Perhaps, this is because of thresholds of internal concentrations required to drive specific modes of action. Or perhaps baseline toxicity (narcosis) makes separating specific from nonspecific action difficult.[45] Indeed, which particular chronic endpoint we are most concerned with as the focus is critically important, we might expect the concentration that impacts growth to be different to the concentration that impacts reproduction. There is further complication added in that the experimental data often hold a significant degree of variation across the scientific literature. Replicability and reliability are significant issues in ecotoxicology,[46] and so the basic data available from which to derive models are subject to some degree of approximation. Making predictions based on these data is irrevocably tied to the biology, and although new statistical methods like ML are good at finding relationships, they are inevitably more complex than say chemical relationships. How we deal with this replicability issue will be key to the successful application of these technologies to make safe and trusted predictions. Indeed, to move from the hype of expectations for AI and ML to the delivery of GPT, the key is in the training datasets. Curated and relevant data are the most valuable resource, and in comparison, developing the AI/ML is simple and cheap.

What makes pharmaceuticals special?

Pharmaceuticals are a unique class of chemicals that are generally regarded as different from other chemicals (even pesticides, personal care, and plant protection products). The domain they occupy in the chemical universe is an almost unique and restrictive zone. Indeed, there is a relatively simple description that can be applied to most drugs, the Lipinski rule of five,[47] which is more an approximation concerning orally active pharmaceuticals that generally describe their properties.

Typically, Lipinski suggested that we see five or fewer hydrogen bond donors (oxygen-hydrogen and nitrogen-hydrogen bonds); fewer than 10 hydrogen bond acceptors (total number of nitrogen or oxygen atoms); a lipophilicity ($\log P$) less than five and a mass less than 500 a.u. Although not exhaustive, these general principles do place pharmaceuticals into a specific chemical space, and other properties tend to widen this domain only a little. The majority are ionizable, which makes their behavior in a complex aquatic environment even more difficult to predict.[48] They may or may not be susceptible to metabolic transformation. They cross cell membranes via physiochemical properties, but also via active transporter routes.[49] Efforts to predict uptake and bioaccumulation in fish from simple correlations have had some success,[50–52] but there are significant exceptions where this has not worked well,[53–55] and this is probably because the complexity of the studies (pH, concentrations, species, temperature, water chemistry) add a level of complexity not considered by simple models.

Only small datasets have been readily available to the academic community; the cost of material and difficulty of analysis have been largely restrictive until recently. Limited access to regulatory datasets has proven difficult to extract pertinent information.[1] Even within industry, data sharing across companies has been a complex issue with legal liabilities and intellectual property considerations confounding perceived commercial advantages, making transparency and availability of data almost nil. AstraZeneca was the first company to have shared high-level results and environmental assessments on their websites since 2012,[17] but even these data are insufficient for any real modeling approaches due to a low number of products and restrictive range of chemical diversity making correlation and generalized predictions difficult if not impossible. Other companies have now shared their data online, but it remains difficult to build a useable database. The relatively limited diversity in chemical properties of pharmaceuticals remains a barrier for traditional linear methods of developing rules and principles, especially when the empirical data do not distinguish the cause of toxicity, making traditional QSARs complex.[45] A recent collaboration across the European pharmaceutical companies in collaboration with academic and small enterprises has better addressed this situation of small and disjoint data under the Innovative Medicines Initiative (IMI). In that program, the fate and effects data from 13 companies were compiled into an initial database (http://i-pie.org/ipiesum/), and European waters modeled.[56] Approximately 200 APIs provide a resource with potential to develop and train algorithm approaches to find patterns and predict future compound liabilities.

Why do pharmaceuticals effect wildlife?

A central tenet for understanding the risk of any pharmaceutical in the environment is that to affect a response it must reach a target and in sufficient quantities to elicit the pharmacological response by driving enough receptors[1, 50, 52] (see references for explanation). Essentially, the concentrations in human plasma that are associated with pharmacological effect are likely to be those needed in other animals. We would intuitively expect toxicity to occur at higher concentrations than those for therapeutic effect, making the assumption inherently conservative. Given that other animals likely have similar numbers but less specific targets and receptors, then this intuitively feels like a reasonable conservative assumption. Verbruggen et al.[27] have combined multiple databases to connect APIs to their protein targets and their likely presence in animal species from their genome sequence; the application software enables a ranking of the species against the likely presence of the target (www.ecodrug.org). If we could follow the model of the medicinal chemists who design the API to fit the target, and reverse engineer this process to find the specificity of the targets in wildlife we will have a powerful tool to better determine which species are at risk, and which are not. Currently we are in a position to know we might have say 60% homology of target between say a daphnid and a human, or any other degree of similarity. Even if we restrict the homology consideration to the active features of the gene, we don't know if this degree of similarity means that the API will cause an effect or not, or indeed at what level of homology we should be looking for. Fundamentally, exactly what does 60% homology mean for functionality? Iterative ML approaches will be essential to derive this innovation, predicting binding of APIs will be the essential step between prediction of uptake and prediction of effect.

When we consider the risk of pharmaceuticals in the environment the focus in ecotoxicology has typically been on the concentrations in the water or soil; probably because it is difficult to measure low concentrations in small volumes such as inside small aquatic wildlife.[3, 35, 36] However, some laboratory studies on fish have tied cause and effect in fish with measured internal concentrations that are at or above those in humans.[54, 57–59]

Pharmaceuticals are often complex and designed to cross membranes and interact with the proteins. This is even more complicated at the cellular level by the potential for membrane-based transporter proteins that appear to have significant effect at low concentrations.[49] While we suspect this is the case for many pharmaceuticals and many species, most experimental (and regulatory) work continues to look to high (many-fold above environmental levels) concentrations in the laboratory to determine accumulation at more easily measured levels. These data are then used to project back in a linear fashion to low concentrations that were significantly lower than the tested range. This may not be a conservative approach since transporters and factors such as pH may drive higher than expected uptake at these critically important environmental concentrations (ng/L). In short, we just do not yet know if this is a significant risk; our experience from measured environmental concentrations, and those in aquatic wildlife would suggest that there is poor correlation.[60] Concentrations inside wildlife such as freshwater shrimps do not necessarily reflect concentrations in the surrounding water.[6] This is probably because the concentrations in the water are variable, and the time to reach equilibrium in an animal takes a little longer and or may be subject to metabolism and excretion. Few pharmaceuticals seem to have significantly accumulative properties, but those that do should be of higher priority for risk assessment,[1] and if we compare the internal concentration with that which elicits pharmacological effects in humans then we may be able to better prioritize these compounds[50, 52, 61] and this has recently been attempted for a range of pharmaceuticals found in river shrimps in the United Kingdom.[6]

What happens in the environment?

Pharmaceuticals can have a range of molecular properties, often designed to perform a function or interaction. For example, multiple and variable surface molecular charges drive interactions with proteins and while in some environments and specific pH conditions they may well follow specific relationships like other chemicals, the range where this is the case can be somewhat restricted. The pH in aquatic environments is rarely constant. For example, there are changes in lakes with algal and plant life driving lower pH during daylight due to net photosynthesis and high pH at night with net respiration. There are micro-climates within rivers and other water bodies that change temperature and redox on a scale commensurate with the animals and plants that is missed by average monitoring and typical reporting by scientists.

Typically, we might see a pH of environmental water reported as 6.5 ± 1.0; we need to remember that pH is a log scale; the variation is usually a standard deviation so there are likely more extremes than 5.5–7.5 in this case, and the conditions change significantly with these ranges. For example, the carbon dioxide concentrations and solubilities vary with pH, and membrane charges on respiratory surfaces such as gills will also likely be impacted, and hence, could affect uptake. Changes in carbon dioxide can change respiration rate, and it determines the volumes of water that pass over the gills and therefore the exposure to potential changes. Temperature has similarly complex effects and has been demonstrated in the laboratory to significantly impact the scale and direction of effect of pollution.[62] We need to understand this to help us incorporate the exacerbating effects of climate change on pollution, and we know little about this. For example, at the surface of a fish gill (at micrometer scale), the pH could be as low as pH 3 due to the secretion of protons to drive excretion and oxygen uptake, whereas millimeters away from that it could be neutral or even alkali in the main bulk water. The gill surface glycoprotein mucus adds to the complexity. Exactly what this means for the uptake of pharmaceuticals into fish (and invertebrates) we are not yet sure. What we do know is that ionized compounds such as pharmaceuticals are likely to be significantly affected by these conditions. To add to the complexity, some (not all) freshwater fish do not regulate the pH of their blood, seeing "alkali tides" after feeding, and acidosis on exercise, yet this phenomenon has not been extensively studied,[63] and we have no evidence as to whether this changes the uptake of ionized compounds. We suspect this is likely the case, and it may go some way to explaining the wide range of internal concentrations of pharmaceuticals such as ibuprofen measured in fish in the same tank[54] and is probably at least part of the reason uptake is difficult to predict. Although we are investigating these factors and trying to explain the impacts on physiology, we could apply basic ML to the problem.[37]

Predicting uptake using ML

ANNs likely represent the simplest route for the environmental toxicologists to move their field forward to a new paradigm. However, ecotoxicologists in general are unlikely to possess the skills to conduct this work de novo, collaboration and training is key, and our community needs to build those capabilities quickly. In simple terms the ANN is a model that maps observed data across networks with hidden layers (called neurons) to experimental outcomes. The path through these hidden layers is complex, but ML lets the weight of each pathway change to better describe the route to the observed data through each iteration. Essentially, we let the machine find the best explanation and correlation between the information we think might affect the outcome (such as mathematical descriptors that describe the molecule) and fit the outcome we have observed (such as the measured internal concentrations). The machine runs until it finds the best fit. But we must be mindful of the algorithms being overtrained. The datasets remain the key; we need sufficient data to derive a training set, a verification set, and test subsets. Overtrained or overfitted models describe the training data well but are unable to fit the test set. The best way we can manage this is to use the largest dataset we can and the fewest hidden layers (neurons) in the architecture of the ANN. We can also add to the regularization (a method to limit overfitting) by minimizing the weight we add to the neurons and their mathematical decay functions. This means extremes in the observed data should have less impact on the overall model; a regularized model is less likely to learn from noise in the data. This is because the model runs through iterations gradually reducing the error in the verification data prediction to a minimum, before increasing on overtraining. The machine stops learning at this minimum, and we then try the fit to the test data.

Using ANNs and mapping input to output data through hidden layers, we can approach the issue of asking which physiochemical factors of a pharmaceutical are most important in explaining the observed data in vivo. As we have highlighted extensively above, our field is data poor and we have serious concerns as to the uncurated quality and replicability. To provide confidence to a wider and nonspecialist audience, we would like to use Transfer Learning, where we would train models on thousands of data points to make predictions on smaller datasets and combine these as multimodal models. It is human nature to see these to be inherently convincing based on the huge volumes of data. But we do not have (and likely never will have) appropriate volumes of data in environmental toxicology. Datasets such as the TOXCAST program (https://www.epa.gov/chemical-research/exploring-toxcast-data-downloadable-data) offer probably the largest such databases on chemical toxicity, and ML has been applied to examine this across several hundred scientific manuscripts. But relatively few pharmaceuticals are within that dataset, raising concerns for the domain of applicability. Instead, our field is more likely to succeed with efforts based on relatively small but highly curated datasets. We then take an integrative approach and examine the data with multiple models to develop a multimodal approach to increase our confidence in the predictive power. Although we might not yet be ready to make wide reaching predictions to better protect the planet at this point in time, we can learn much from the approach.

Miller et al.[37] successfully predicted bioaccumulation of pharmaceuticals in fish and invertebrates from a relatively small dataset of 352 chemicals in fish and invertebrate tests. More importantly, they opened the door to an unbiased approach to transpose the key molecular descriptors associated with accumulation that had not previously been considered as important. A four-layer multiperceptron model used 14 input descriptors to map the bioaccumulation. In an ideal world, we would be able to make reasonable predictions based on one easily measurable input, transpose learning where we model on many input descriptors, and then identify the key one for prediction indication. Miller et al.[37] demonstrated that as might be expected the log$D$ was a key indicator. This is reassuring as it is the pH-dependent equivalent for ionizable compounds of the Log$P$ (lipophilicity/hydrophobicity) that is used in the regulatory triggers for testing. But by no means was this the only factor, and indeed, we were surprised to find that the number of nitrogen atoms and the topological polar surface area were also important among others. This insight is critically important, and we can now go back to the laboratory and look more closely at the mechanisms in vitro.[48] Probably more importantly the seminal work by Miller et al. demonstrated that for understanding uptake/bioaccumulation of pharmaceuticals into aquatic life, fish are not the same as invertebrates and different models are needed.[37, 38] Further, there seems to be species differences even within each of these groups; it seems although both are clearly fish, a carp might be different to a trout when it comes to taking up pharmaceuticals. If this is correct, then it follows that we may not be able to rely on a regulatory bioaccumulation on one species being protective of all wildlife as we currently do. Testing a wider range of species is impractical and unlikely to protect everything and so must be considered an unethical path. Modeling and multimodal (in silico/in vitro) innovative technologies may be the only appropriate route to help.

Regional issues and the focus of concern

Legislation and effort have generally focused ERAs on developed (western) markets. There is an inherent bias toward estimating the risk for a conservative and worst-case environmental exposure in these regions, with only some companies anticipating risk in other markets (developing and lower middle income). Europe and the United States have significant amounts of information concerning populations of patients, the environments and locations they live, the connectivity to infrastructure such as wastewater treatment, water flows, temperatures and rainfall; all factors that can impact the likely concentrations in rivers. For example, detailed models of European rivers down to 1-km resolution exist, and models have been developed to anticipate and predict specific risks based on use, flow, and population.[56] There are similar models for other regions like China among many others.[64] There is a fundamental right of access to medicines and increasingly programs to support access to medicines in more deprived regions are increasing the number of people who can benefit.[65] However, these communities are typically poor and have significant socioeconomic, political, and other complex issues in parallel. Often, there is little connectivity to infrastructure, low access to clean water and waste treatment. Indeed, of 7.8 billion people on the planet in 2020, 2 billion do not have a toilet,[66] and typically most of those who do, have a pit latrine rather than centralized water connectivity. This changes the risk situation from that estimated by regulations in Europe, the United States, Canada, and Japan. Established cities have evolved water supply and treatment systems over time (Fig. 1), but the rate of population growth and mega-city development in some low- and middle-income countries seems to be outstripping the rate of development of their infrastructure. Therefore groundwater becomes a potential sink for pharmaceuticals,[67] and little scientific evidence is available to help understand that risk. In this situation the standard tests conducted to protect surface waters using invertebrates, algae, and fish data are largely irrelevant as they do not occupy this groundwater environment. The potential risks in groundwater may be more significant than in surface waters. In regions where latrines are used, we are not yet clear on the extent of the risk. Combined with high populations and limited surface water dilutions, we know pharmaceuticals can travel significant distances, and latrine inputs will likely add to that picture.[68] We need the data to better determine these risks. Some of the earliest examples of using ANNs in this field were to model how pharmaceuticals partition to soils[69] and structural QSARS to predict sorption to the sludge in wastewater treatment.[70, 71] We clearly urgently need advanced techniques to help predict the fate and effects of pharmaceuticals in the environment.

Historical datasets are sparse and often poorly curated. Even compilations of scientific literature are prone to small errors. If a paper cites the occurrence of a pharmaceutical as 1,000 ng/L, misreading the comma as a full stop or the n as an m can have profound effects on a database, and significant time is needed for curation and checking anomalies against original reports and indeed the analytical credibility of those original reports.[3, 72] The key for developing innovative analysis, ML, or even AI is the quality of the data we use for training. Fewer high-quality, directly relevant data whereas not ideal, are better than more, lower quality uncurated data. This field is not a big data field with millions of data points. And this is the key challenge for our community.[28] However, historical data have already been used successfully in identifying the likely causes of drinking water contamination and discoloration using a ML approach,[73] and it seems likely that this field will rapidly expand. If we are to better protect the global environment, then this needs to happen, and soon. Indeed, if we are to make sustainability a key tenet of providing pharmaceuticals to patients, then we must address how innovative analytics such as ML and AI are integrated into the processes described throughout this book, and critically must include the principles of ethical AI approach. AI and ML are currently shaping how we manage manufacturing facilities, distribute supplies, deliver medicines to patients and work toward lower environmental impacts of our global environmental footprint. Adoption of this information stream has been rapid and disruptive, and we look toward an exciting future where these become industry standards.

Intelligent regulation—A future state of automated AI assessment of chemicals

As we develop AI and begin to use it as part of daily life and process, we tend to stop addressing the technology as AI. For example, the AI behind making purchase recommendations in online stores or that which sits behind the networking of mobile (cell) phones is now largely considered GPT. Concerns for AI tend to fall as it becomes GPT and safe and trusted technology underpins progress with task-based AI. In our field of working to protect the environment and support societal need, there is significant opportunity to develop task-based AI. For example, it is the role of a regulatory authority to make a recommendation to a government (and therefore the people) as to the safety (including the environmental risk) of a pharmaceutical. AI offers the potential for iterative learning and development of intelligent environmental assessment beyond the human expert considerations used today. The current situation is that expert scientists and environmental risk assessors decide which tests are appropriate, commission them, and await the results. In contract laboratories, under highly complex reporting and data recording rules, the studies are conducted, and the data reported. A person then interprets the results, checks for anomalies, and reports the findings. These findings go back to the risk assessor to make the calculations, who then reports the risk as part of registration of the new pharmaceutical. This report goes to the regulatory authorities, who share with their staff experts who in turn review the data, reports, calculations, and conclusions. Based on their expertise, further tests may be required, data recalculated, and conclusions develop. At some point, there is a decision of whether it represents a risk or not. At many points along this pipeline, many individuals make decisions and interpretations based on their personal knowledge and understanding. We suggest that a future option might be to use a task-based AI to drive this process. If it were an open system that transparently showed how and why it made decisions, both the regulator and industry expert could speed up this process and make better science-based decisions if an AI could take the information and build on its existing understanding.

Such an innovative approach could be extended to the regulation of all chemicals. Currently, the process of assessment is complex and slow with the registration and processing often taking years for each chemical. Indeed, the regulators are working to find faster ways for chemical risk assessment.[74] This remains regional rather than global, so the process takes time to repeat and runs significant risk of conflicting conclusions. Given that there are at least 350,000 different chemicals used and sold around the world,[42] and only a small fraction of those are currently reliably assessed, an AI approach to intelligent assessment might better protect the environment. A globally centralized list and regionally appropriate assessments could help identify those classes and individual chemicals of concern. Setting up such a future state will require significant change and represents a huge challenge to the status quo. But given the complex challenges we face in terms of the impacts of climate change and its inherent exacerbation of pollution, this needs to happen.

Although we would advocate a rapid transition, we realistically expect this to develop over significant time. Simple software solutions are currently under development to store and collate information. Intelligent systems that can extract data from reports and make calculations through more complex decision trees are also available. We have partnered a company (Simomics, United Kingdom) to help develop a decision-based software to do just this.[75] They have developed a software that is transparent to show how decisions are made in an explainable way, referring to existing case studies and precedents, pulling data from available resources and generating logical arguments. This is an early prototype of what we hope could be more widely adopted. We anticipate that as this tool develops and gains wider datasets, it will be the precursor to a more sophisticated system that ultimately uses a task-driven AI that better protects the environment.

Key points for future development

Points to consider that are currently limiting the application of innovative techniques to better predict the risk of pharmaceuticals to the environment:

- The limited diversity of molecular properties of small molecule pharmaceuticals that have environmental data available makes poorly represented characteristics a significant limitation.
- Training data are sparse, often poorly curated, and require quality checking before use.
- There is a skills shortage and knowledge gap of individuals that understand the environmental risks and innovative ML and AI methodologies.
- There appears to be little agreement on validation methods for ML techniques; in the environmental field, we need to see the OECD QSAR toolbox updated to help this field progress and provide a baseline for regulatory adoption.

References

1. Gunnarsson L, et al. Pharmacology beyond the patient—the environmental risks of human drugs. *Environ Int* 2019;**129**:320–32.

2. aus der Beek T, et al. Pharmaceuticals in the environment—global occurrences and perspectives. *Environ Toxicol Chem* 2016;**35**:823–35.

3. Miller TH, Bury NR, Owen SF, MacRae JI, Barron LP. A review of the pharmaceutical exposome in aquatic fauna. *Environ Pollut* 2018;**239**:129–46.

4. Boxall ABA, et al. Pharmaceuticals and personal care products in the environment: what are the big questions? *Environ Health Perspect* 2012;**120**:1221–9.

5. Johnson AC, Sumpter JP. Putting pharmaceuticals into the wider context of challenges to fish populations in rivers. *Philos Trans R Soc B Biol Sci* 2014;**369**, 20130581.

6. Miller TH, et al. Biomonitoring of pesticides, pharmaceuticals and illicit drugs in a freshwater invertebrate to estimate toxic or effect pressure. *Environ Int* 2019;**129**:595–606.

7. Larsson DGJ. Pollution from drug manufacturing: review and perspectives. *Philos Trans R Soc B Biol Sci* 2014;**369**, 20130571.

8. Munro K, et al. Evaluation of combined sewer overflow impacts on short-term pharmaceutical and illicit drug occurrence in a heavily urbanised tidal river catchment (London, UK). *Sci Total Environ* 2019;**657**:1099–111.

9. Larsson DGJ, de Pedro C, Paxeus N. Effluent from drug manufactures contains extremely high levels of pharmaceuticals. *J Hazard Mater* 2007;**148**:751–5.

10. Johnson AC, Jin X, Nakada N, Sumpter JP. Learning from the past and considering the future of chemicals in the environment. *Science* 2020;**367**:384–7.

11. Murray-Smith RJ, Coombe VT, Grönlund MH, Waern F, Baird JA. Managing emissions of active pharmaceutical ingredients from manufacturing facilities: an environmental quality standard approach. *Integr Environ Assess Manag* 2012;**8**:320–30.

12. Bengtsson-Palme J, Larsson DGJ. Concentrations of antibiotics predicted to select for resistant bacteria: proposed limits for environmental regulation. *Environ Int* 2016;**86**:140–9.

13. Tell J, et al. Science-based targets for antibiotics in receiving waters from pharmaceutical manufacturing operations. *Integr Environ Assess Manag* 2019;**15**:312–9.

14. Oaks JL, et al. Diclofenac residues as the cause of vulture population decline in Pakistan. *Nature* 2004;**427**:630–3.

15. Galligan TH, et al. Partial recovery of critically endangered Gyps vulture populations in Nepal. *Bird Conserv Int* 2020;**30**:87–102.

16. Winter MJ, et al. Using data from drug discovery and development to aid the aquatic environmental risk assessment of human pharmaceuticals: concepts, considerations, and challenges. *Integr Environ Assess Manag* 2010;**6**:38–51.

17. Holm G, et al. Implementing ecopharmacovigilance in practice: challenges and potential opportunities. *Drug Saf* 2013;**36**:533–46.

18. Corcoran J, Winter MJ, Tyler CR. Pharmaceuticals in the aquatic environment: a critical review of the evidence for health effects in fish. *Crit Rev Toxicol* 2010;**40**:287–304.

19. Wheeler JR, Segner H, Weltje L, Hutchinson TH. Interpretation of sexual secondary characteristics (SSCs) in regulatory testing for endocrine activity in fish. *Chemosphere* 2020;**240**, 124943.

20. Kidd KA, et al. Collapse of a fish population after exposure to a synthetic estrogen. *Proc Natl Acad Sci USA* 2007;**104**:8897–901.

21. Blanchfield PJ, et al. Recovery of a wild fish population from whole-lake additions of a synthetic estrogen. *Environ Sci Technol* 2015;**49**:3136–44.

22. Thorpe KL, et al. Relative potencies and combination effects of steroidal estrogens in fish. *Environ Sci Technol* 2003;**37**:1142–9.

23. Hamilton PB, et al. Population-level consequences for wild fish exposed to sublethal concentrations of chemicals—a critical review. *Fish Fish* 2016;**17**:545–66.

24. Thrupp TJ, et al. The consequences of exposure to mixtures of chemicals: something from 'nothing' and 'a lot from a little' when fish are exposed to steroid hormones. *Sci Total Environ* 2018;**619–620**:1482–92.

25. Matthiessen P, Wheeler JR, Weltje L. A review of the evidence for endocrine disrupting effects of current-use chemicals on wildlife populations. *Crit Rev Toxicol* 2018;**48**:195–216.

26. Godfray HCJ, et al. A restatement of the natural science evidence base on the effects of endocrine disrupting chemicals on wildlife. *Proc R Soc B Biol Sci* 2019;**286**, 20182416.

27. Verbruggen B, et al. ECOdrug: a database connecting drugs and conservation of their targets across species. *Nucleic Acids Res* 2018;**46**:D930–6.

28. Miller TH, et al. Machine learning for environmental toxicology: a call for integration and innovation. *Environ Sci Technol* 2018;**52**:12952–5.

29. European Medicines Agency. *Guideline on the environmental risk assessment of medicinal products for human use.* European Medicines Agency; 2018. Available at: https://www.ema.europa.eu/en/environmental-risk-assessment-medicinal-products-human-use.

30. US Department of Health and Human Services. Environmental assessment of human drug and biologics applications. Guidance for industry. CMC-6, Rev. 1. U.S. Department of Health and Human Services Food and Drug Administration, Center for Drug Evaluation and Research (CDER), Center for Biologics Evaluation and Research (CBER); 1998. https://www.fda.gov/media/70809/download.

31. EMA. *Questions and answers on 'Guideline on the environmental risk assessment of medicinal products for human use'. EMA/CHMP/SWP/44609/2010*; 2015.

32. *Environmental assessment: questions and answers regarding drugs with estrogenic, androgenic, or thyroid activity guidance for industry environmental assessment: questions and answers regarding drugs with estrogenic, androgenic, or thyroid activity G. contains nonbinding recommendations.* U.S. Department of Health and Human Services Food and Drug Administration Center for Drug Evaluation and Research (CDER); 2016. https://www.fda.gov/media/91941/download.

33. OECD. *Pharmaceutical residues in freshwater. OECD studies on water*; 2019. https://doi.org/10.1787/c936f42d-en.

34. EU 2010/63. Directive 2010/63/EU of the European Parliament and of the Council of 22 September 2010 on the protection of animals used for scientific purposes Text with EEA relevance. http://data.europa.eu/eli/dir/2010/63/oj.

35. Miller TH, et al. Assessing the reliability of uptake and elimination kinetics modelling approaches for estimating bioconcentration factors in the freshwater invertebrate, *Gammarus pulex. Sci Total Environ* 2016;**547**:396–404.

36. Miller TH, Bury NR, Owen SF, Barron LP. Uptake, biotransformation and elimination of selected pharmaceuticals in a freshwater invertebrate measured using liquid chromatography tandem mass spectrometry. *Chemosphere* 2017;**183**:389–400.

37. Miller TH, et al. Prediction of bioconcentration factors in fish and invertebrates using machine learning. *Sci Total Environ* 2019;**648**:80–9.

38. Miller TH. *Measurement and modelling of pharmaceutical bioconcentration in an aquatic invertebrate, Gammarus pulex. (Predictive ecotoxicology: artificial neural networks for the prediction of xenobiotic bioconcentration in invertebrates).* London, UK: Kings College London; 2017.

39. EPA. Ecological Effects Test Guidelines OPPTS 850.1500 Fish life cycle toxicity. In: *United States Environmental Protection Agency*; 1996. EPA712-C-96-122 https://www.epa.gov/sites/production/files/2015-07/documents/850-1500.pdf.

40. Rivetti C, et al. Vision of a near future: bridging the human health–environment divide. Toward an integrated strategy to understand mechanisms across species for chemical safety assessment. *Toxicol Vitr* 2020;**62**, 104692.

41. Burns EE, Carter LJ, Snape J, Thomas-Oates J, Boxall ABA. Application of prioritization approaches to optimize environmental monitoring and testing of pharmaceuticals. *J Toxicol Environ Health B Crit Rev* 2018;**21**:115–41.

42. Wang Z, Walker GW, Muir DCG, Nagatani-Yoshida K. Toward a global understanding of chemical pollution: a first comprehensive analysis of national and regional chemical inventories. *Environ Sci Technol* 2020;**54**:2575–84.

43. Brooke DN, Crookes MJ, Merckel DAS. Methods for predicting the rate constant for uptake of organic chemicals from water by fish. *Environ Toxicol Chem* 2012;**31**:2465–71.

44. Moe SJ, et al. Development of a hybrid Bayesian network model for predicting acute fish toxicity using multiple lines of evidence. *Environ Model Softw* 2020;**126**, 104655.

45. Ebbrell DJ, Cronin MTD, Ellison CM, Firman JW, Madden JC. Development of baseline quantitative structure-activity relationships (QSARs) for the effects of active pharmaceutical ingredients (APIs) to aquatic species. In: *Ecotoxicological QSARs. Methods in pharmacology and toxicology*; 2020. p. 331–56. https://doi.org/10.1007/978-1-0716-0150-1_15.

46. Moermond CTA, Kase R, Korkaric M, Ågerstrand M. CRED: criteria for reporting and evaluating ecotoxicity data. *Environ Toxicol Chem* 2016;**35**:1297–309.

47. Lipinski CA. Avoiding investment in doomer drugs, is poor solubility an industry wide problem? *Curr Drug Discov* 2001;**46**(1–3):17–9.

48. Su C, Tong J, Zhu Y, Cui P, Wang F. Network embedding in biomedical data science. *Brief Bioinform* 2018. https://doi.org/10.1093/bib/bby117.

49. Stott LC, Schnell S, Hogstrand C, Owen SF, Bury NR. A primary fish gill cell culture model to assess pharmaceutical uptake and efflux: evidence for passive and facilitated transport. *Aquat Toxicol* 2015;**159**:127–37.

50. Huggett DB, Cook JC, Ericson JF, Williams RT. A theoretical model for utilizing mammalian pharmacology and safety data to prioritize potential impacts of human pharmaceuticals to fish. *Hum Ecol Risk Assess* 2003;**9**:1789–99.

51. Owen SF, et al. Uptake of propranolol, a cardiovascular pharmaceutical, from water into fish plasma and its effects on growth and organ biometry. *Aquat Toxicol* 2009;**93**:217–24.

52. Rand-Weaver M, et al. The read-across hypothesis and environmental risk assessment of pharmaceuticals. *Environ Sci Technol* 2013;**47**:11384–95.

53. Owen SF, et al. The value of repeating studies and multiple controls: replicated 28-day growth studies of rainbow trout exposed to clofibric acid. *Environ Toxicol Chem* 2010;**29**:2831–9.

54. Patel A, et al. Testing the "read-across hypothesis" by investigating the effects of ibuprofen on fish. *Chemosphere* 2016;**163**:592–600.

55. Bickley LK, et al. Bioavailability and kidney responses to diclofenac in the fathead minnow (*Pimephales promelas*). *Environ Sci Technol* 2017;**51**:1764–74.

56. Oldenkamp R, et al. A high-resolution spatial model to predict exposure to pharmaceuticals in European surface waters: EPiE. *Environ Sci Technol* 2018;**52**:12494–503.

57. Margiotta-Casaluci L, et al. Quantitative cross-species extrapolation between humans and fish: the case of the anti-depressant fluoxetine. *PLoS One* 2014;**9**:e110467.

58. Margiotta-Casaluci L, et al. Internal exposure dynamics drive the adverse outcome pathways of synthetic glucocorticoids in fish. *Nat Sci Rep* 2016;**6**:21978.

59. Weil M, Falkenhain AM, Scheurer M, Ryan JJ, Coors A. Uptake and effects of the beta-adrenergic agonist salbutamol in fish: supporting evidence for the fish plasma model. *Environ Toxicol Chem* 2019;**38**:2509–19.

60. Huerta B, et al. Presence of pharmaceuticals in fish collected from urban rivers in the U.S. EPA 2008–2009 National Rivers and Streams Assessment. *Sci Total Environ* 2018;**634**:542–9.

61. Brooks BW. Urbanization, environment and pharmaceuticals: advancing comparative physiology, pharmacology and toxicology. *Conserv Physiol* 2018;**6**. cox079.

62. Brown AR, et al. Climate change and pollution speed declines in zebrafish populations. *Proc Natl Acad Sci USA* 2015;**112**:E12.

63. Wood CM. Internal spatial and temporal CO_2 dynamics: fasting, feeding, drinking, and the alkaline tide. In: Grosell M, Munday PL, Farrell AP, Brauner CJ, editors. *Carbon dioxide. Fish physiology*, vol 37. London: Elsevier; 2019. p. 245–86 [chapter 7].

64. Zhu Y, Snape J, Jones K, Sweetman A. Spatially explicit large-scale environmental risk assessment of pharmaceuticals in surface water in China. *Environ Sci Technol* 2019;**53**:2559–69.

65. WHO medicines, vaccines and pharmaceutical annual report 2018 – promoting access to safe, effective, quality and affordable essential medical products for all. WHO/MVP/EMP/2019.03; 2018. Available at: https://www.who.int/medicines/publications/annual-reports/mvp_annual-report2018/en/.

66. WaterAid. *Water, toilets and hygiene facts and statistics*. Available at: https://www.wateraid.org/uk/the-crisis/facts-and-statistics. [Accessed 11 March 2020].

67. Sharma BM, et al. Health and ecological risk assessment of emerging contaminants (pharmaceuticals, personal care products, and artificial sweeteners) in surface and groundwater (drinking water) in the Ganges River Basin. *India Sci Total Environ* 2019;**646**:1459–67.

68. Bagnis S, et al. Characterization of the Nairobi River catchment impact zone and occurrence of pharmaceuticals: implications for an impact zone inclusive environmental risk assessment. *Sci Total Environ* 2020;**703**, 134925.

69. Barron L, et al. Predicting sorption of pharmaceuticals and personal care products onto soil and digested sludge using artificial neural networks. *Analyst* 2009;**134**:621–808.

70. Berthod L. *Mechanistic approach to predicting the sorption characteristics of pharmaceuticals—Portsmouth Research Portal*. Portsmouth, UK: University of Portsmouth; 2015.

71. Berthod L, et al. Quantitative structure-property relationships for predicting sorption of pharmaceuticals to sewage sludge during waste water treatment processes. *Sci Total Environ* 2017;**579**:1512–20.

72. Umweltbundesamt. *Database—pharmaceuticals in the environment*; 2014. Available at: https://www.umweltbundesamt.de/en/database-pharmaceuticals-in-the-environment-0. [Accessed 11 March 2020].

73. Speight VL, Mounce SR, Boxall JB. Identification of the causes of drinking water discolouration from machine learning analysis of historical datasets. *Environ Sci Water Res Technol* 2019;**5**:747–55.

74. Kavlock RJ, et al. Accelerating the pace of chemical risk assessment. *Chem Res Toxicol* 2018;**31**:287–90.

75. Timmis J, et al. Virtual fish ecotoxicology laboratory. *Ind Pharm* 2016;**50**(1):8–10.

Index

Note: Page numbers followed by *f* indicate figures and *t* indicate tables.

237

Printed in the United States
by Baker & Taylor Publisher Services